Tri Tan

2 Apr 2004

Free with Course.

CW00495150

Analysis of Survey Data

WILEY SERIES IN SURVEY METHODOLOGY

Established in part by WALTER A. SHEWHART AND SAMUEL S. WILKS

Editors: *Robert M. Groves, Graham Kalton, J. N. K. Rao, Norbert Schwarz, Christopher Skinner*

A complete list of the titles in this series appears at the end of this volume.

Analysis of Survey Data

Edited by

R. L. CHAMBERS and **C. J. SKINNER**
University of Southampton, UK

WILEY

Copyright © 2003 John Wiley & Sons, Ltd, The Atrium, Southern Gate,
Chichester, West Sussex PO19 8SQ, England

Telephone (+44) 1243 779777

Email (for orders and customer service enquiries): cs-books@wiley.co.uk
Visit our Home Page www.wileyeurope.com or www.wiley.com

Reprinted March 2004

Other Wiley Editorial Offices

John Wiley & Sons Inc., 111 River Street, Hoboken, NJ 07030, USA

Jossey-Bass, 989 Market Street, San Francisco, CA 94103-1741, USA

Wiley-VCH Verlag GmbH, Boschstr. 12, D-69469 Weinheim, Germany

John Wiley & Sons Australia Ltd, 33 Park Road, Milton, Queensland 4064, Australia

John Wiley & Sons (Asia) Pte Ltd, 2 Clementi Loop #02-01, Jin Xing Distripark,
Singapore 129809

John Wiley & Sons (Canada) Ltd, 22 Worcester Road, Etobicoke, Ontario M9W 1L1

Wiley also publishes its books in a variety of electronic formats. Some content that appears in
print may not be available in electronic books.

Library of Congress Cataloguing-in-Publication Data

Analysis of survey data / edited by R. L. Chambers and C. J. Skinner.
    p. cm. – (Wiley series in survey methodology)
    Includes bibliographical references and indexes.
    ISBN 0-471-89987-9 (acid-free paper)
    1. Mathematical statistics–Methodology. I. Chambers, R. L. (Ray L.) II. Skinner, C. J.
III. Series.

    QA276.A485 2003
    001.4′22–dc21   2002033132

British Library Cataloguing in Publication Data

A catalogue record for this book is available from the British Library

ISBN 0 471 89987 9

Typeset in 10/12pt Times by Kolam Information Services Pvt. Ltd, Pondicherry, India
Printed and bound in Great Britain by Biddles Ltd, King's Lynn, Norfolk
This book is printed on acid-free paper responsibly manufactured from sustainable forestry
in which at least two trees are planted for each one used for paper production.

To T. M. F. Smith

Contents

Preface

The book is dedicated to T. M. F. (Fred) Smith, and marks his 'official' retirement from the University of Southampton in 1999. Fred's deep influence on the ideas presented in this book is witnessed by the many references to his work. His publications up to 2002 are listed at the back of the book. Fred's most important early contributions to survey sampling were made from the late 1960s into the 1970s, in collaboration with Alastair Scott, a colleague of Fred, when he was a lecturer at the London School of Economics between 1960 and 1968. Their joint papers explored the foundations and advanced understanding of the role of models in inference from sample survey data, a key element of survey analysis. Fred's review of the foundations of survey sampling in Smith (1976), read to the Royal Statistical Society, was a landmark paper.

Fred moved to a lectureship position in the Department of Mathematics at the University of Southampton in 1968, was promoted to Professor in 1976 and has stayed there until his recent retirement. The 1970s saw the arrival of Tim Holt in the University's Department of Social Statistics and the beginning of a new collaboration. Fred and Tim's paper on poststratification (Holt and Smith, 1979) is particularly widely cited for its discussion of the role of conditional inference in survey sampling. Fred and Tim were awarded two grants for research on the analysis of survey data between 1977 and 1985, and the grants supported visits to Southampton by a number of authors in this book, including Alastair Scott, Jon Rao, Wayne Fuller and Danny Pfeffermann. The research undertaken was disseminated at a conference in Southampton in 1986 and via the book edited by Skinner, Holt and Smith (1989).

Fred has clearly influenced the development of survey statistics through his own publications, listed here, and by facilitating other collaborations, such as the work of Alastair Scott and Jon Rao on tests with survey data, started on their visit to Southampton in the 1970s. From our University of Southampton perspective, however, Fred's support of colleagues, visitors and students has been equally important. He has always shown tremendous warmth and encouragement towards his research students and to other colleagues and visitors undertaking research in survey sampling. He has also always promoted interactions and cooperation, whether in the early 1990s through regular informal discussion sessions on survey sampling in his room or, more recently, with the increase in numbers interested, through regular participation in Friday lunchtime workshops on sample survey methods. Fred is well known as an excellent and inspiring teacher at both undergraduate and postgraduate levels and his own research seminars and lectures have always been eagerly attended, not only for their subtle insights, but also for their self-deprecating humour. We look forward to many more years of his involvement.

Fred's positive approach and his interested support of others ranges far beyond his interaction with colleagues at Southampton. He has been a strong supporter of his graduate students and through conferences and meetings has interacted widely with others. Fred has a totally open approach and while he loves to argue and debate a point he is never defensive and always open to persuasion if the case can be made. His positive commitment and openness was reflected in his term as President of the Royal Statistical Society – which he carried out with great distinction.

This book originates from a conference on 'Analysis of Survey Data' held in honour of Fred in Southampton in August 1999. All the chapters, with the exception of the introductions, were presented as papers at that conference.

Both the conference and the book were conceived of as a follow-up to Skinner, Holt and Smith (1989) (referred to henceforth as 'SHS'). That book addressed a number of statistical issues arising in the application of methods of statistical analysis to sample survey data. This book considers a somewhat wider set of statistical issues and updates the discussion, in the light of more recent research in this field. The relation between these two books is described further in Chapter 1 (see Section 1.4).

The book is aimed at a statistical audience interested in methods of analysing sample survey data. The development builds upon two statistical traditions, first the tradition of modelling methods, such as regression modelling, used in all areas of statistics to analyse data and, second, the tradition of survey sampling, used for sampling design and estimation in surveys. It is assumed that readers will already have some familiarity with both these traditions. An understanding of regression modelling methods, to the level of Weisberg (1985), is assumed in many chapters. Familiarity with other modelling methods would be helpful for other chapters, for example categorical data analysis (Agresti, 1990) for Part B, generalized linear models (McCullagh and Nelder, 1989) in Parts B and C, survival analysis (Lawless, 2002; Cox and Oakes, 1984) for Part D. As to survey sampling, it is assumed that readers will be familiar with standard sampling designs and related estimation methods, as described in Särndal, Swensson and Wretman (1992), for example. Some awareness of sources of non-sampling error, such as nonresponse and measurement error (Lessler and Kalsbeek, 1992), will also be relevant in places, for example in Part E.

As in SHS, the aim is to discuss and develop the statistical principles and theory underlying methods, rather than to provide a step-by-step guide on how to apply methods. Nevertheless, we hope the book will have uses for researchers only interested in analysing survey data in practice.

Finally, we should like to acknowledge support in the preparation of this book. First, we thank the Economic and Social Research Council for support for the conference in 1999. Second, our thanks are due to Anne Owens, Jane

Schofield, Kathy Hooper and Debbie Edwards for support in the organization of the conference and handling manuscripts. Finally, we are very grateful to the chapter authors for responding to our requests and putting up with the delay between the conference and the delivery of the final manuscript to Wiley.

Ray Chambers and Chris Skinner
Southampton, July 2002

Contributors

D. R. Bellhouse
*Department of Statistical and
Actuarial Sciences
University of Western Ontario
London
Ontario N6A 5B7
Canada*

David A. Binder
*Methodology Branch
Statistics Canada
120 Parkdale Avenue
Ottawa
Ontario K1A 0T6
Canada*

R. L. Chambers
*Department of Social Statistics
University of Southampton
Southampton
SO17 1BJ
UK*

A. H. Dorfman
*Office of Survey Methods Research
Bureau of Labor Statistics
2 Massachusetts Ave NE
Washington, DC 20212-0001
USA*

Wayne A. Fuller
*Statistical Laboratory and Department
of Statistics
Iowa State University
Ames
IA 50011
USA*

C. M. Goia
*Department of Statistical and
Actuarial Sciences
University of Western Ontario
London
Ontario N6A 5B7
Canada*

D. J. Holmes
*Department of Social Statistics
University of Southampton
Southampton
SO17 1BJ
UK*

D. Holt
*Department of Social Statistics
University of Southampton
Southampton
SO17 1BJ
UK*

J. F. Lawless
*Department of Statistics and Actuarial
Science
University of Waterloo
200 University Avenue West
Waterloo
Ontario N2L 3G1
Canada*

Roderick J. Little
*Department of Biostatistics
University of Michigan School of
Public Health
1003 M4045 SPH II Washington
Heights
Ann Arbor
MI 48109–2029
USA*

Fabrizia Mealli
*Dipartamento di Statistica
Università di Firenze
Viale Morgagni
50134 Florence
Italy*

Danny Pfeffermann
*Department of Statistics
Hebrew University
Jerusalem
Israel*
and
*Department of Social Statistics
University of Southampton
Southampton
SO17 1BJ
UK*

Stephen Pudney
*Department of Economics
University of Leicester
Leicester
LE1 7RH
UK*

J. N. K. Rao
*School of Mathematics and Statistics
Carleton University
Ottawa
Ontario K1S 5B6
Canada*

Georgia R. Roberts
*Statistics Canada
120 Parkdale Avenue
Ottawa
Ontario K1A 0T6
Canada*

Richard Royall
*Department of Biostatistics
School of Hygiene and Public Health
Johns Hopkins University
615 N. Wolfe Street
Baltimore
MD 21205
USA*

Alastair Scott
*Department of Statistics
University of Auckland
Private Bag 92019
Auckland
New Zealand*

C. J. Skinner
*Department of Social Statistics
University of Southampton
Southampton
SO17 1BJ
UK*

J. E. Stafford
*Department of Public Health Sciences
McMurrich Building
University of Toronto
12 Queen's Park Crescent West
Toronto
Ontario M5S 1A8
Canada*

D. G. Steel
*School of Mathematics and Applied
Statistics
University of Wollongong
NSW 2522
Australia*

M. Yu. Sverchkov
*Burean of Labor Statistics
2 Massachusetts Ave NE
Washington, DC 20212-0001
USA*

D. R. Thomas
*School of Business
Carleton University
Ottawa
Ontario K1S 5B6
Canada*

M. Tranmer
*Centre for Census and Survey
Research
Faculty of Social Sciences and Law
University of Manchester
Manchester
M13 9PL
UK*

Chris Wild
*Department of Statistics
University of Auckland
Private Bag 92019
Auckland
New Zealand*

CHAPTER 1

Introduction

R. L. Chambers and C. J. Skinner

1.1. THE ANALYSIS OF SURVEY DATA

Many statistical methods are now used to analyse sample survey data. In particular, a wide range of generalisations of regression analysis, such as generalised linear modelling, event history analysis and multilevel modelling, are frequently applied to survey microdata. These methods are usually formulated in a statistical framework that is not specific to surveys and indeed these methods are often used to analyse other kinds of data. The aim of this book is to consider how statistical methods may be formulated and used appropriately for the analysis of sample survey data. We focus on issues of statistical inference which arise specifically with surveys.

The primary survey-related issues addressed are those related to sampling. The selection of samples in surveys rarely involves just simple random sampling. Instead, more complex sampling schemes are usually employed, involving, for example, stratification and multistage sampling. Moreover, these complex sampling schemes usually reflect complex underlying population structures, for example the geographical hierarchy underlying a national multistage sampling scheme. These features of surveys need to be handled appropriately when applied statistical methods. In the standard formulations of many statistical methods, it is assumed that the sample data are generated directly from the population model of interest, with no consideration of the sampling scheme. It may be reasonable for the analysis to ignore the sampling scheme in this way, but it may not. Moreover, even if the sampling scheme is ignorable, the stochastic assumptions involved in the standard formulation of the method may not adequately reflect the complex population structures underlying the sampling. For example, standard methods may assume that observations for different individuals are independent, whereas it may be more realistic to allow for correlated observations within clusters. Survey data arising from complex sampling schemes or reflecting associated underlying complex population structures are referred to as *complex survey data*.

While the analysis of complex survey data constitutes the primary focus of this book, other methodological issues in surveys also receive some attention.

Analysis of Survey Data Edited by R. L. Chambers and C. J. Skinner
© 2003 John Wiley & Sons, Ltd

In particular, there will be some discussion of nonresponse and measurement error, two aspects of surveys which may have important impacts on estimation. *Analytic* uses of surveys may be contrasted with *descriptive* uses. The latter relate to the estimation of summary measures for the population, such as means, proportions and rates. This book is primarily concerned with analytic uses which relate to inference about the parameters of models used for analysis, for example regression coefficients. For descriptive uses, the targets of inference are taken to be finite population characteristics. Inference about these parameters could in principle be carried out with certainty given a 'perfect' census of the population. In contrast, for analytic uses the targets of inference are usually taken to be parameters of models, which are hypothesised to have generated the values in the surveyed population. Even under a perfect census, it would not be possible to make inference about these parameters with certainty. Inference for descriptive purposes in complex surveys has been the subject of many books in survey sampling (e.g. Cochran, 1977; Särndal, Swensson and Wretman, 1992; Valliant, Dorfman and Royall, 2000). Several of the chapters in this book will build on that literature when addressing issues of inference for analytic purposes.

The survey sampling literature, relating to the descriptive uses of surveys, provides one key reference source for this book. The other key source consists of the standard (non-survey) statistical literature on the various methods of analysis, for example regression analysis or categorical data analysis. This literature sets out what we refer to as *standard procedures* of analysis. These procedures will usually be the ones implemented in the most commonly used general purpose statistical software. For example, in regression analysis, ordinary least squares methods are the standard procedures used to make inference about regression coefficients. For categorical data analysis, maximum likelihood estimation under the assumption of multinomial sampling will often be the standard procedure. These standard methods will typically ignore the complex nature of the survey. The impact of ignoring features of surveys will be considered and ways of building on standard procedures to develop appropriate methods for survey data will be investigated.

After setting out some statistical foundations of survey analysis in Sections 1.2 and 1.3, we outline the contents of the book and its relation to Skinner, Holt and Smith (1989) (referred to henceforth as SHS) in Sections 1.4 and 1.5.

1.2. FRAMEWORK, TERMINOLOGY AND SPECIFICATION OF PARAMETERS

In this section we set out some of the basic framework and terminology and consider the definition of the parameters of interest.

A *finite population*, U, consists of a set of N units, labelled $1, \ldots, N$. We write $U = \{1, \ldots, N\}$. Mostly, it is assumed that U is fixed, but more generally, for example in the context of a longitudinal survey, we may wish to allow the population to change over time. A *sample*, s, is a subset of U. The *survey variables*, denoted by the $1 \times J$ vector Y, are variables which are measured in the survey and which are of interest in the analysis. It is supposed that an aim of the survey is to record the value of Y for each unit in the sample. Many chapters assume that this aim is realised. In practice, it will usually not be possible to record Y without error for all sample units, either because of non-response or because of measurement error, and approaches to dealing with these problems are also discussed. The values which Y takes in the finite population are denoted $y_1, \ldots, y_N$. The process whereby these values are transformed into the data available to the analyst will be called the *observation process*. It will include both the sampling mechanism as well as the nonresponse and measurement processes. Some more complex temporal features of the observation process arising in longitudinal surveys are discussed in Chapter 15 by Lawless.

For the descriptive uses of surveys, the definition of parameters is generally straightforward. They consist of specified functions of $y_1, \ldots, y_N$, for example the vector of finite population means of the survey variables, and are referred to as *finite population parameters*.

In analytic surveys, parameters are usually defined with respect to a specified model. This model will usually be a *superpopulation model*, that is a model for the stochastic process which is assumed to have generated the finite population values $y_1, \ldots, y_N$. Often this model will be parametric or semi-parametric, that is fully or partially characterised by a finite number of parameters, denoted by the vector θ, which defines the parameter vector of interest. Sometimes the model will be non-parametric (see Chapters 10 and 11) and the target of inference may be a regression function or a density function.

In practice, it will often be unreasonable to assume that a specified parametric or semi-parametric model holds exactly. It may therefore be desirable to define the parameter vector in such a way that it is equal to θ if the model holds, but remains interpretable under (mild) misspecification of the model. Some approaches to defining the parameters in this way are discussed in Chapter 3 by Binder and Roberts and in Chapter 8 by Scott and Wild. In particular, one approach is to define a *census parameter*, θ_U, which is a finite population parameter and is 'close' to θ according to some metric. This provides a link with the descriptive uses of surveys. There remains the issue of how to define the metric and Chapters 3 and 8 consider somewhat different approaches.

Let us now consider the nature of possible superpopulation models and their relation to the sampling mechanism. Writing y_U as the $N \times J$ matrix with rows $y_1, \ldots, y_N$ and $f(.)$ as a generic probability density function or probability mass function, a basic superpopulation model might be expressed as $f(y_U; \theta)$. Here it is supposed that y_U is the realisation of a random matrix, Y_U, the distribution of which is governed by the parameter vector θ.

It is natural also to express the sampling mechanism probabilistically, especially if the sample is obtained by a conventional probability sampling design. It is convenient to represent the sample by a random vector with the same number of rows as Y_U. To do this, we define the *sample inclusion indicator*, i_t, for $t = 1, \ldots, N$, by

$$i_t = 1 \text{ if } t \in s, \ i_t = 0 \text{ otherwise.}$$

Assume sampling unknown — assume parametric dependence — how?

$$Y_{ku} \sim bin(Pk_a, N_{ka}) = f(y_u; \theta) \quad f(i_u) = ? \; why?$$

The N values $i_1, \ldots, i_N$ form the elements of the $N \times 1$ vector i_U. Since i_U determines the set s and the set s determines i_U, the sample may be represented alternatively by s or i_U. We denote the sampling mechanism by $f(i_U)$, with i_U being a realisation of the random vector I_U. Thus, $f(i_U)$ specifies the probability of obtaining each of the 2^N possible samples from the population. Under the assumption of a known probability sampling design, $f(i_U)$ is known for all possible values of i_U and thus no parameter is included in this specification. When the sampling mechanism is unknown, some parametric dependence of $f(i_U)$ might be desirable.

We thus have expressions, $f(y_U; \theta)$ and $f(i_U)$, for the distributions of the population values Y_U and the sample units I_U. If we are to proceed to use the sample data to make inference about θ it is necessary to be able to represent (y_U, i_U) as the joint outcome of a single process, that is the joint realisation of the random matrix (Y_U, I_U). How can we express the joint distribution of (Y_U, I_U) in terms of the distributions $f(y_U; \theta)$ and $f(i_U)$, which we have considered so far? Is it reasonable to assume independence and write $f(y_U; \theta)f(i_U)$ as the joint distribution? To answer these questions, we need to think more carefully about the sampling mechanism.

At the simplest level, we may ask whether the sampling mechanism depends directly on the realised value y_U of Y_U. One situation where this occurs is in case–control studies, discussed by Scott and Wild in Chapter 8. Here the outcome y_t is binary, indicating whether unit t is a case or a control, and the cases and controls define separate strata which are sampled independently. The way in which the population is sampled thus depends directly on y_U. In this case, it is natural to indicate this dependence by writing the sampling mechanism as $f(i_U|Y_U = y_U)$. We may then write the joint distribution of Y_U and I_U as $f(i_U|Y_U = y_U)f(y_U; \theta)$, where it is necessary not only to specify the model $f(y_U; \theta)$ for Y_U but also to 'model' what the sampling design $f(i_U|Y_U)$ would be under alternative outcomes Y_U than the observed one y_U. Sampling schemes which depend directly on y_U in this way are called *informative sampling schemes*. Sampling schemes, for which we may write the joint distribution of Y_U and I_U as $f(y_U; \theta)f(i_U)$, are called *noninformative*. An alternative but related definition of informative sampling will be used in Section 11.2.3 and in Chapter 12. Sampling is said there to be informative with respect to Y if the 'sample distribution' of Y differs from the population distribution of Y, where the idea of 'sample distribution' is introduced in Section 2.3.

Schemes where sampling is directly dependent upon a survey variable of interest are relatively rare. It is very common, however, for sampling to depend upon some other characteristics of the population, such as strata. These characteristics are used by the sample designer and we refer to them as *design variables*. The vector of values of the design variables for unit t is denoted z_t and the matrix of all population values $z_1, \ldots, z_N$ is denoted z_U. Just as the matrix y_U is viewed as a realisation of the random matrix Y_U, we may view z_U as a realisation of a random matrix Z_U. To emphasise the dependence of the sampling design on z_U, we may write the sampling mechanism as $f(i_U|Z_U = z_U)$. If we are to hold Z_U fixed at its actual value z_U when specifying the sampling mechanism $f(i_U|Z_U = z_U)$, then we must also hold it fixed when we specify the joint distribution of I_U and Y_U. We write the distribution of Y_U with Z_U fixed at z_U as $f(y_U|Z_U = z_U; \phi)$ and interpret it as the conditional distribution of Y_U given $Z_U = z_U$. The distribution is indexed by the parameter vector ϕ, which may differ from θ, since this conditional distribution may differ from the original distribution $f(y_U; \theta)$. Provided there is no additional direct dependence of sampling on y_U, it will usually be reasonable to express the joint distribution of Y_U and I_U (with z_U held fixed) as $f(I_U|Z_U = z_U)$ $f(Y_U|Z_U = z_U; \phi)$, that is to assume that Y_U and I_U are conditionally independent given $Z_U = z_U$. In this case, sampling is said to be *noninformative conditional on z_U*.

We see that the need to 'condition' on z_U when specifying the model for Y_U has implications for the definition of the target parameter. Conditioning on z_U may often be reasonable. Consider, for illustration, a sample survey of individuals in Great Britain, where sampling in England and Wales is independent of sampling in Scotland, that is these two parts of Great Britain are separate strata. Ignoring other features of the sample selection, we may thus conceive of z_t as a binary variable identifying these two strata, Suppose that we wish to conduct a regression analysis with some variables in this survey. The requirement that our model should condition on z_U in this context means essentially that we must include z_t as a potential covariate (perhaps with interaction terms) in our regression model. For many socio-economic outcome variables it may well be scientifically sensible to include such a covariate, if the distribution of the outcome variable varies between these regions.

In other circumstances it may be less reasonable to condition on z_U when defining the distribution of Y_U of interest. The design variables are chosen to assist in the selection of the sample and their nature may reflect administrative convenience more than scientific relevance to possible data analyses. For example, in Great Britain postal geography is often used for sample selection in surveys of private households involving face-to-face interviews. The design variables defining the postal geography may have little direct relevance to possible scientific analyses of the survey data. The need to condition on the design variables used for sampling involves changing the model for Y_U from $f(y_U; \theta)$ to $f(y_U|Z_U = z_U; \phi)$ and changing the parameter vector from θ to ϕ. This implies that the method of sampling is actually driving the specification of the target parameter, which seems inappropriate as a general approach. It seems generally more desirable to define the target parameter first, in the light of the scientific questions of interest, before considering what bearing the sampling scheme may have in making inferences about the target parameter using the survey data.

We thus have two possible broad approaches, which SHS refer to as *disaggregated* and *aggregated* analyses. A disaggregated analysis conditions on the values of the design variables in the finite population with ϕ the target parameter. In many social surveys these design variables define population subgroups, such as strata and clusters, and the disaggregated analysis essentially disaggregates the analysis by these subgroups, specifying models

which allow for different patterns within and between subgroups. Part C of SHS
provides illustrations.

In an aggregated analysis the target parameters θ are defined in a way that is
unrelated to the design variables. For example, one might be interested in a
factor analysis of a set of attitude variables in the population. For analytic
inference in an aggregated analysis it is necessary to conceive of z_U as a
realisation of a random matrix Z_U with distribution $f(z_U;\psi)$ indexed by a
further parameter vector ψ and, at least conceptually, to model the sampling
mechanism $f(I_U|Z_U)$ for different values of Z_U than the realised value z_U.
Provided the sampling is again noninformative conditional on z_U, the joint
distribution of I_U, Y_U and Z_U is given by $f(i_U|z_U)f(y_U|z_U;\phi)f(z_U;\psi)$. The
target parameter θ characterises the marginal distribution of Y_U:

$$f(y_U;\theta) = \int f(y_U|z_U;\phi)f(z_U;\psi)\mathrm{d}z_U.$$

Aggregated analysis may therefore alternatively be referred to as *marginal
modelling* and the distinction between aggregated and disaggregated analysis is
analogous, to a limited extent, to the distinction between population-averaged
and subject-specific analysis, widely used in biostatistics (Diggle *et al.*, 2002,
Ch. 7) when clusters of repeated measurements are made on subjects. In this
analogy, the design variables z_t consist of indicator variables or random effects
for these clusters.

1.3.   STATISTICAL INFERENCE

In the previous section we discussed the different kinds of parameters of
interest. We now consider alternative approaches to inference about these
parameters, referring to inference about finite population parameters as *descriptive inference* and inference about model parameters, our main focus, as *analytic
inference*.

Descriptive inference is the traditional concern of survey sampling and a
basic distinction is between *design-based* and *model-based* inference. Under
design-based inference the only source of random variation considered is that
induced in the vector i_U by the sampling mechanism, assumed to be a known
probability sampling design. The matrix of finite population values y_U is
treated as fixed, avoiding the need to specify a model which generates y_U.
A frequentist approach to inference is adopted. The aim is to find a point
estimator $\hat{\theta}$ which is approximately unbiased for θ and has 'good efficiency',
both bias and efficiency being defined with respect to the distribution of $\hat{\theta}$
induced by the sampling mechanism. Point estimators are often formed using
survey weights, which may incorporate auxiliary population information perhaps based upon the design variables z_t, but are usually not dependent upon the
values of the survey variables y_t (e.g. Deville and Särndal, 1992). Large-sample
arguments are often then used to justify a normal approximation $\hat{\theta} \sim N(\theta, \Sigma)$.
An estimator $\hat{\Sigma}$ is then sought for Σ, which enables interval estimation and

testing of hypotheses about θ to be conducted. In the simplest approach, $\hat{\Sigma}$ is sought such that inference statements about θ, based upon the assumptions that $\hat{\theta} \sim N(\theta, \Sigma)$ and Σ is known, remain valid, to a reasonable approximation, if Σ is replaced by $\hat{\Sigma}$. The design-based approach is the traditional one in many sampling textbooks such as Cochran (1977). Models may be used in this approach to motivate the choice of estimators, in which case the approach may be called *model-assisted* (Särndal, Swensson and Wretman, 1992).

The application of the design-based approach to analytic inference is less straightforward, since the parameters are necessarily defined with respect to a model and hence it is not possible to avoid the use of models entirely. A common approach is via the notion of a census parameter, referred to in the previous section. This involves specifying a finite population parameter θ_U corresponding to the model parameter θ and then considering design-based (descriptive) inference about θ_U. These issues are discussed further in Chapter 3 by Binder and Roberts and in Chapter 2. As noted in the previous section, a critical issue is the specification of the census parameter.

In classical model-based inference the only source of random variation considered is that from the model which generates y_U. The sample, represented by the vector i_U, is held fixed, even if it has been generated by a probability sampling design. Model-based inference may, like design-based inference, follow a frequentist approach. For example, a point estimator of a given parameter θ might be obtained by maximum likelihood, tests about θ might be based upon a likelihood ratio approach and interval estimation about θ might be based upon a normal approximation $\hat{\theta} \sim N(\theta, \Sigma)$, justified by large-sample arguments. More direct likelihood or Bayesian approaches might also be adopted, as discussed in the chapters of Part A.

Classical model-based inference thus ignores the probability distribution induced by the sampling design. Such an approach may be justified, under certain conditions, by considering a statistical framework in which stochastic variation arises from both the generation of y_U and the sampling mechanism. Conditions for the *ignorability* of the sampling design may then be obtained by demonstrating, for example, that the likelihood is free of the design or that the sample outcome i_U is an ancillary statistic, depending on the approach to statistical inference (Rubin, 1976; Little and Rubin, 1989). These issues are discussed by Chambers in Chapter 2 from a likelihood perspective in a general framework, which allows for stochastic variation not only from the sampling mechanism and a model, but also from nonresponse.

From a likelihood-based perspective, it is necessary first to consider the nature of the *data*. Consider a framework for analytic inference, in the notation introduced above, where I_U, Y_U and Z_U have a distribution given by

$$f(i_U | y_U, z_U) f(y_U | z_U; \phi) f(z_U; \psi).$$

Suppose that the rows of y_U corresponding to sampled units are collected into the matrix y_{obs} and that the remaining rows form the matrix y_{mis}. Supposing that i_U, y_{obs} and z_U are observed and that y_{mis} is unobserved, the data consist of (i_U, y_{obs}, z_U) and the likelihood for (ϕ, ψ) is given by

$$L(\phi, \psi) \propto \int f(i_U|y_U, z_U) f(y_U|z_U; \phi) f(z_U; \psi) dy_{mis}.$$

If sampling is noninformative given z_U so that $f(i_U|y_U, z_U) = f(i_U|z_U)$ and if this design $f(i_U|z_U)$ is known and free of (ϕ, ψ) then the term $f(i_U|y_U, z_U)$ may be dropped from the likelihood above since it is only a constant with respect to (ϕ, ψ). Hence, under these conditions sampling is *ignorable* for likelihood-based inference about (ϕ, ψ), that is we may proceed to make likelihood-based inference treating i_U as fixed. Classical model-based inference is thus justified in this case.

Chambers refers in Chapter 2 to information about the sample, i_U, the sample observations, y_{obs}, as well as the population values of the design variables, z_U, as the 'full information' case. The likelihood based upon this information is called the 'full information likelihood'. In practice, all the population values, z_U, will often not be available to the survey data analyst, as discussed further by Chambers, Dorfman and Sverchkov in Chapter 11 and by Chambers, Dorfman and Wang (1998). The only unit-level information available to the analyst about the sampling design might consist of the inclusion probabilities π_i of each sample unit and further information, for example identifiers of strata and primary sampling units, which enables suitable standard errors to be computed for descriptive inference. In these circumstances, a more limited set of inference options will be available. Even this amount of unit-level information about the design may not be available to some users and SHS describe some simpler methods of adjustment, such as the use of approximate design effects.

1.4. RELATION TO SKINNER, HOLT AND SMITH (1989)

As indicated in the Preface, this work is conceived of as a follow-up to Skinner, Holt and Smith (1989) (referred to as SHS). This book updates and extends SHS, but does not replace it. The discussion is intended to be at a similar level to SHS, focusing on statistical principles and general theory, rather than on the details of how to implement specific methods in practice. Some chapters, most notably Chapter 7 by Rao and Thomas, update SHS by including discussion of developments since 1989 of methods covered in SHS. Many chapters extend the topic coverage of SHS. For example, there was almost no reference to longitudinal surveys in SHS, whereas the whole of Part D of this book is devoted to this topic. There is also little overlap between SHS and many of the other chapters. In particular, SHS focused only on the issue of complex survey data, referred to in Section 1.1, whereas this book makes some reference to additional methodological issues in the analysis of survey data, such as nonresponse and measurement error.

There remains, however, much in SHS not covered here. In particular, only limited coverage is given to investigating the effects of complex designs and population structure on standard procedures or to simple adjustments to standard procedures to compensate for these effects, two of the major objectives of SHS. One reason we focus less on standard procedures is that appropriate methods for the analysis of complex survey data have increasingly become available in standard software since 1989. The need to discuss the properties of inappropriate standard methods may thus have declined, although we still consider it useful that measures such as misspecification effects (*meffs*), introduced in SHS (p. 24) to assess properties of standard procedures, have found their way into software, such as Stata (Stata Corporation, 2001). More importantly, we have not attempted to produce a second edition of SHS, that is we have chosen not to duplicate the discussion in SHS. Thus this book is intended to complement SHS, not to supersede it. References will be made to SHS where appropriate, but we attempt to make this book self-contained, especially via the introductory chapters to each part, and it is not essential for the reader to have access to SHS to read this book.

Given its somewhat different topic coverage, the structure of this book differs from that of SHS. That book was organised according to two particular features. First, a distinction was drawn between aggregated and disaggregated analysis (see Section 1.2). This feature is no longer used for organising this book, partly because methods of disaggregated analysis receive only limited attention here, but this distinction will be referred to in places in individual chapters. Second, there was a separation in SHS between the discussion of first point estimation and bias and second standard errors and tests. Again, this feature is no longer used for organising the chapters in this book. The structure of this book is set out and discussed in the next section.

1.5. OUTLINE OF THIS BOOK

Basic issues regarding the statistical framework and the approach to inference continue to be fundamental to discussions of appropriate methods of survey analysis. These issues are therefore discussed first, in the four chapters of Part A. These chapters adopt a similar finite population framework, contrasting analytic inference about a model parameter with descriptive inference about a finite population parameter. The main difference between the chapters concerns the approach to statistical inference. The different approaches are introduced first in Section 1.3 of this chapter and then by Chambers in Chapter 2, as an introduction to Part A. A basic distinction is between design-based and model-based inference (see Section 1.3). Binder and Roberts compare the two approaches in Chapter 3. The other chapters focus on model-based approaches. Little discusses the Bayesian approach in Chapter 4. Royall discusses an approach to descriptive inference based upon the likelihood in Chapter 5. Chambers focuses on a likelihood-based approach to analytic inference in Chapter 2, as well as discussing the approaches in Chapters 3–5.

Following Part A, the remainder of the book is broadly organised according to the type of survey data. Parts B and C are primarily concerned with the analysis of cross-sectional survey data, with a focus on the analysis of relationships between variables, and with the separation between the two parts corresponding to the usual distinction between discrete and continuous response variables. Rao and Thomas provide an overview of methods for discrete response data in Chapter 7, including both the analysis of multiway tables and regression models for microdata. Scott and Wild deal with the special case of logistic regression analysis of survey data from case–control studies in Chapter 8. Regression models for continuous responses are discussed in Part C, especially in Chapter 11 by Chambers, Dorfman and Sverchkov and in Chapter 12 by Pfeffermann and Sverchkov. These include methods of non-parametric regression, and non-parametric graphical methods for displaying continuous data are also covered in Chapter 10 by Bellhouse, Goia and Stafford.

The extension to the analysis of longitudinal survey data is considered in Part D. Skinner and Holmes discuss the use of random effects models in Chapter 14 for data on continuous variables recorded at repeated waves of a panel survey. Lawless and Mealli and Pudney discuss the use of methods of event history analysis in Chapters 15 and 16.

Finally, Part E is concerned with data structures which are more complex than the largely 'rectangular' data structures considered in previous chapters. Little discusses the treatment of missing data from nonresponse in Chapter 18. Fuller considers the nested data structures arising from multiphase sampling in Chapter 19. Steel, Tranmer and Holt consider analyses which combine survey microdata and geographically aggregated data in Chapter 20.

PART A

Approaches to Inference

CHAPTER 2

Introduction to Part A

R. L. Chambers

2.1. INTRODUCTION

Recent developments have served to blur the distinction between the design-based and model-based approaches to survey sampling inference. As noted in the previous chapter, it is now recognised that the distribution induced by the sampling process has a role to play in model-based inference, particularly where this distribution depends directly on the distribution of the population values of interest, or where there is insufficient information to justify the assumption that the sampling design can be ignored in inference. Similarly, the increased use of model-assisted methods (Särndal, Swensson and Wretman, 1992) has meant that the role of population models is now well established in design-based inference.

The philosophical division between the two approaches has not disappeared completely, however. This becomes clear when one reads the following three chapters that make up Part A of this book. The first, by Binder and Roberts, argues for the use of design-based methods in analytic inference, while the second, by Little, presents the Bayesian approach to model-based analytic and descriptive inference. The third chapter, by Royall, focuses on application of the likelihood principle in model-based descriptive inference.

The purpose of this chapter is to provide a theoretical background for these chapters, and to comment on the arguments put forward in them. In particular we first summarise current approaches to survey sampling inference by describing three basic, but essentially different, approaches to analytic inference from sample survey data. All three make use of likelihood ideas within a frequentist inferential framework (unlike the development by Royall in Chapter 5), but define or approximate the likelihood in different ways. The first approach, described in the next section, develops the estimating equation and associated variance estimator for what might be called a *full information* maximum likelihood approach. That is, the likelihood is defined by the probability of observing all the relevant data available to the survey analyst. The second, described in Section 2.3, develops the *maximum sample likelihood* estimator.

Analysis of Survey Data Edited by R. L. Chambers and C. J. Skinner
© 2003 John Wiley & Sons, Ltd

This estimator maximises the likelihood defined by the sample data only, excluding population information. The third approach, described in Section 2.4, is based on the *maximum pseudo-likelihood* estimator (SHS, section 3.4.4), where the unobservable population-level likelihood of interest is estimated using methods for descriptive inference. In Section 2.5 we then explore the link between the pseudo-likelihood approach and the total variation concept that underlies the approach to analytic inference advocated by Binder and Roberts in Chapter 3. In contrast, in Chapter 4 Little advocates a traditional Bayesian approach to sample survey inference, and we briefly set out his arguments in Section 2.6. Finally, in Section 2.7 we summarise the arguments put forward by Royall for extending likelihood inference to finite population inference.

2.2. FULL INFORMATION LIKELIHOOD

The development below is based on Breckling *et al.* (1994). We start by reiterating the important point made in Section 1.3, i.e. application of the likelihood idea to sample survey data first requires one to identify what these data *are*. Following the notation introduced in Section 1.2, we let Y denote the vector of survey variables of interest, with matrix of population values y_U, and let s denote the set of 'labels' identifying the sampled population units, with a subscript of *obs* denoting the subsample of these units that respond. Each unit in the population, if selected into the sample, can be a respondent or a non-respondent on any particular component of Y. We model this behaviour by a multivariate zero–one response variable R of the same dimension as Y. The population values of R are denoted by the matrix r_U. Similarly, we define a sample inclusion indicator I that takes the value one if a unit is selected into the sample and is zero otherwise. The vector containing the population values of I is denoted by i_U. Also, following convention, we do not distinguish between a random variable and its realisation unless this is necessary for making a point. In such cases we use upper case to identify the random variable, and lower case to identify its realisation.

If we assume the values in y_U can be measured without error, then the survey data are the array of respondents' values y_{obs} in y_U, the matrix r_s corresponding to the values of R associated with the sampled units, the vector i_U and the matrix z_U of population values of a multivariate design variable Z. It is assumed that the sample design is at least partially based on the values in z_U.

We note in passing that this situation is an ideal one, corresponding to the data that would be available to the survey analyst responsible for both selecting the sample and analysing the data eventually obtained from the responding sample units. In many cases, however, survey data analysts are *secondary analysts*, in the sense that they are not responsible for the survey design and so, for example, do not have access to the values in z_U for both non-responding sample units as well as non-sampled units. In some cases they may not even have access to values in z_U for the sampled units. Chambers, Dorfman and Wang (1998) explore likelihood-based inference in this limited data situation.

In what follows we use f_U to denote a density defined with respect to a stochastic model for a population of interest. We assume that the target of inference is the parameter θ characterising the *marginal* population density $f_U(y_U;\theta)$ of the population values of Y and our aim is to develop a maximum likelihood estimation procedure for θ. In general θ will be a vector. In order to do so we assume that the random variables Y, R, I and Z generating y_U, r_U, i_U and z_U are (conceptually) observable over the entire population, representing the density of their *joint* distribution over the population by $f_U(y_U, r_U, i_U, z_U; \gamma)$, where γ is the vector of parameters characterising this joint distribution. It follows that θ either is a component of γ or can be obtained by a one-to-one transformation of components of γ. In either case if we can calculate the maximum likelihood estimate (MLE) for γ we can then calculate the MLE for θ.

This estimation problem can be made simpler by using the following equivalent factorisations of $f_U(y_U, r_U, i_U, z_U; \gamma)$ to define γ:

$$f_U(y_U, r_U, i_U, z_U) = f_U(r_U | y_U, i_U, z_U)f_U(i_U | y_U, z_U)f_U(y_U | z_U)f_U(z_U) \qquad (2.1)$$

$$= f_U(r_U | y_U, i_U, z_U)f_U(i_U | y_U, z_U)f_U(z_U | y_U)f_U(y_U) \qquad (2.2)$$

$$= f_U(i_U | y_U, r_U, z_U)f_U(r_U | y_U, z_U)f_U(y_U | z_U)f_U(z_U). \qquad (2.3)$$

The choice of which factorisation to use depends on our knowledge and assumptions about the sampling method and the population generating process. Two common simplifying assumptions (see Section 1.4) are

Noninformative sampling given z_U: $Y_U \perp I_U | Z_U = z_U.$

Noninformative nonresponse given z_U: $Y_U \perp R_U | Z_U = z_U.$

Here upper case denotes a random variable, $\perp$ denotes independence and $|$ denotes conditioning. Under noninformative sampling $f_U(i_U | y_U, z_U) = f_U(i_U | z_U)$ in (2.1) and (2.2) and so i_U is *ancillary* as far as inference about θ is concerned. Similarly, under noninformative nonresponse $f_U(r_U | y_U, i_U, z_U) = f_U(r_U | i_U, z_U)$, in which case r_U is ancillary for inference about θ. Under both noninformative sampling and noninformative nonresponse both r_U and i_U are ancillary and γ is defined by the joint population distribution of just y_U and z_U. When nonresponse becomes informative (but sampling remains noninformative) we see from (2.3) that our survey data distribution is now the joint distribution of y_U, r_U and z_U, and so γ parameterises this distribution. We reverse the roles of r_U and i_U in (2.3) when sampling is informative and nonresponse is not. Finally, when both sampling and nonresponse are informative we have no choice but to model the full joint distribution of y_U, r_U, i_U and z_U in order to define γ.

The likelihood function for γ is then the function $L_s(\gamma) = f_U(y_{obs}, r_s, i_U, z_U; \gamma)$. The MLE of γ is the value $\hat{\gamma}$ that maximises this function. In order to calculate this MLE and to construct an associated confidence interval for the value of γ, we note two basic quantities: the score function for γ, i.e. the first derivative with respect to γ of $\log(L_s(\gamma))$, and the information function for γ, defined as the negative of the first derivative with respect to γ of this score function. The MLE of γ is the value where the score function is zero, while the inverse of the value of the information function evaluated at this MLE (often referred to as the *observed information*) is an estimate of its large-sample variance–covariance matrix.

Let g be a real-valued differentiable function of a vector-valued argument x, with $\partial_x g$ denoting the vector of first-order partial derivatives of $g(x)$ with respect to the components of x, and $\partial_{xx} g$ denoting the matrix of second-order partial derivatives of $g(x)$ with respect to the components of x. Then the score function $sc_s(\gamma)$ for γ generated by the survey data is the conditional expectation, given these data, of the score for γ generated by the population data. That is,

$$sc_s(\gamma) = E_U[\partial_\gamma \log f_U(Y_U, R_U, I_U, Z_U; \gamma) | y_{obs}, r_s, i_U, z_U]. \tag{2.4}$$

Similarly, the information function $infos_s(\gamma)$ for γ generated by the survey data is the conditional expectation, given these data, of the information for γ generated by the population data *minus* the corresponding conditional variance of the population score. That is,

$$\begin{aligned} info_s(\gamma) =& E_U[-\partial_{\gamma\gamma} \log f_U(Y_U, R_U, I_U, Z_U; \gamma) | y_{obs}, r_s, i_U, z_U] \\ &- var_U[\partial_\gamma \log f_U(Y_U, R_U, I_U, Z_U; \gamma) | y_{obs}, r_s, i_U, z_U]. \end{aligned} \tag{2.5}$$

We illustrate the application of these results using two simple examples. Our notation will be such that population quantities will subscripted by U, while their sample and non-sample equivalents will be subscripted by s and $U-s$ respectively and E_s and var_s will be used to denote expectation and variance conditional on the survey data.

Example 1

Consider the situation where Z is univariate and Y is bivariate, with components Y and X, and where Y, X and Z are independently and identically distributed as $N(\mu, \Sigma)$ over the population of interest. We assume $\mu' = (\mu_Y, \mu_X, \mu_Z)$ is unknown but

$$\Sigma = \begin{bmatrix} \sigma_{YY} & \sigma_{YX} & \sigma_{YZ} \\ \sigma_{XY} & \sigma_{XX} & \sigma_{XZ} \\ \sigma_{ZY} & \sigma_{ZY} & \sigma_{ZZ} \end{bmatrix}$$

is known. Suppose further that there is full response and sampling is noninformative, so we can ignore i_U and r_U when defining the population likelihood. That is, in this case the survey data consist of the sample values of Y and X and the population values of Z, and $\gamma = \mu$. The population score for μ is easily seen to be

$$sc_U(\mu) = \Sigma^{-1} \sum_{t=1}^{N} \begin{bmatrix} y_t - \mu_Y \\ x_t - \mu_X \\ z_t - \mu_Z \end{bmatrix}$$

where y_t is the value of Y for population unit t, with x_t and z_t defined similarly, and the summation is over the N units making up the population of interest. Applying (2.4), the score for μ defined by the survey data is

$$sc_s(\mu) = \Sigma^{-1} \sum_{t=1}^{N} E\left\{ \begin{bmatrix} y_t - \mu_Y \\ x_t - \mu_X \\ z_t - \mu_Z \end{bmatrix} \,\middle|\, y_s, x_s, z \right\}$$

$$= \Sigma^{-1}\left[n\begin{pmatrix} \bar{y}_s - \mu_Y \\ \bar{x}_s - \mu_X \\ \bar{z}_s - \mu_Z \end{pmatrix} + \frac{(N-n)}{\sigma_{ZZ}} \begin{pmatrix} \sigma_{YZ} \\ \sigma_{XZ} \\ \sigma_{ZZ} \end{pmatrix} (\bar{z}_{U-s} - \mu_Z) \right]$$

using well-known properties of the normal distribution. The MLE for μ is obtained by setting this score to zero and solving for μ:

$$\hat{\mu} = \begin{pmatrix} \hat{\mu}_Y \\ \hat{\mu}_X \\ \hat{\mu}_Z \end{pmatrix} = \begin{pmatrix} \bar{y}_s + \sigma_{YZ}\sigma_{ZZ}^{-1}(\bar{z}_U - \bar{z}_s) \\ \bar{x}_s + \sigma_{XZ}\sigma_{ZZ}^{-1}(\bar{z}_U - \bar{z}_s) \\ \bar{z}_U \end{pmatrix}. \qquad (2.6)$$

Turning now to the corresponding information for μ, we note that the population information for this parameter is $info_U(\mu) = N\Sigma^{-1}$. Applying (2.5), the information for μ defined by the survey data is

$$info_s(\mu) = E_s(info_U(\mu)) - var_s(sc_U(\mu)) = N\Sigma^{-1}\left[\Sigma - \left(1 - \frac{n}{N}\right)C\right]\Sigma^{-1} \qquad (2.7)$$

where

$$C = \begin{bmatrix} var_U\left(\begin{pmatrix} Y \\ X \end{pmatrix}\middle|Z\right) & 0 \\ & 0 \\ 0 \quad 0 & 0 \end{bmatrix} = \begin{bmatrix} \sigma_{YY} - \sigma_{YZ}\sigma_{ZZ}^{-1}\sigma_{ZY} & \sigma_{YX} - \sigma_{YZ}\sigma_{ZZ}^{-1}\sigma_{ZX} & 0 \\ \sigma_{XY} - \sigma_{XZ}\sigma_{ZZ}^{-1}\sigma_{ZY} & \sigma_{XX} - \sigma_{XZ}\sigma_{ZZ}^{-1}\sigma_{ZX} & 0 \\ 0 & 0 & 0 \end{bmatrix}.$$

Now suppose that the target of inference is the regression of Y on X in the population. This regression is defined by the parameters α, β and $\sigma^2_{Y|X}$ of the homogeneous linear model defined by $E(Y|X) = \alpha + \beta X$ and $var(Y|X) = \sigma^2_{Y|X}$. Since

$$\left.\begin{aligned} \mu_Y &= \alpha + \beta\mu_X \\ \sigma_{YY} &= \sigma^2_{Y|X} + \beta^2\sigma_{XX} \\ \sigma_{YX} &= \beta\sigma_{XX} \end{aligned}\right\} \Rightarrow \left\{\begin{aligned} \beta &= \sigma_{XX}^{-1}\sigma_{YX} \\ \alpha &= \mu_Y - \beta\mu_X \\ \sigma^2_{Y|X} &= \sigma_{YY} - \beta^2\sigma_{XX} \end{aligned}\right. \qquad (2.8)$$

and since Σ is assumed known, the only parameter that needs to be estimated is the intercept α. Using the invariance of the maximum likelihood approach we see that the MLE for this parameter is

$$\hat{\alpha} = \hat{\mu}_Y - \frac{\sigma_{YX}}{\sigma_{XX}}\hat{\mu}_X = \left(\bar{y}_s - \frac{\sigma_{YX}}{\sigma_{XX}}\bar{x}_s\right) + \left(\sigma_{YZ} - \frac{\sigma_{YX}}{\sigma_{XX}}\sigma_{XZ}\right)\left(\frac{\bar{z}_U - \bar{z}_s}{\sigma_{ZZ}}\right).$$

The first term in brackets on the right hand side above is the *face value* MLE for α. That is, it is the estimator we would use if we ignored the information in Z. From (2.7), the asymptotic variance of $\hat{\alpha}$ above is then

$$var_s(\hat{\alpha}) = N^{-1}\left(1 - \frac{\sigma_{YX}}{\sigma_{XX}}\, 0\right)\Sigma\left[\Sigma - \left(1 - \frac{n}{N}\right)C\right]^{-1}\Sigma\left(1 - \frac{\sigma_{YX}}{\sigma_{XX}}\, 0\right)'.$$

What about if Σ is not known? The direct approach would be then to use (2.4) and (2.5) simultaneously to estimate μ and Σ. However, since the notation required for this is rather complex, we adopt an indirect approach. To start, from (2.8) we can write

$$\beta = [E_U(var_U(X|Z)) + var_U(E_U(X|Z))]^{-1}[E_U(cov_U(X, Y|Z))$$
$$+ cov_U(E_U(X|Z), E_U(Y|Z))]$$
$$\alpha = E_U(E_U(Y|Z)) - E_U(E_U(X|Z))\beta$$
$$\sigma_{Y|X}^2 = [E_U(var_U(Y|Z)) + var_U(E_U(Y|Z))] - \beta^2[E_U(var_U(X|Z))$$
$$+ var_U(E(X|Z))].$$

We also observe that, under the joint normality of Y, X and Z,

$$E_U\left(\begin{pmatrix} Y \\ X \end{pmatrix}\Big|Z\right) = \begin{pmatrix} \lambda_{0Y} + \lambda_{1Y}Z \\ \lambda_{0X} + \lambda_{1X}Z \end{pmatrix} \text{ and } var\left(\begin{pmatrix} Y \\ X \end{pmatrix}\Big|Z\right) = \begin{bmatrix} \chi_{YY} & \chi_{XY} \\ \chi_{XY} & \chi_{XX} \end{bmatrix}. \quad (2.9)$$

Combining (2.4) and (2.5) with the factorisation $f_U(y_U, x_U, z_U) = f_U(y_U, x_U|z_U)f_U(z_U)$, one can see that the MLEs of the parameters in (2.9) are just their face value MLEs based on the sample data vectors y_s, x_s and z_s,

$$\hat{\lambda}_{1Y} = S_{sZZ}^{-1}S_{sZY} \qquad\qquad \hat{\lambda}_{1X} = S_{sZZ}^{-1}S_{sZX}$$
$$\hat{\lambda}_{0Y} = \bar{y}_s - \hat{\lambda}_{1Y}\bar{z}_s \qquad\qquad \hat{\lambda}_{0X} = \bar{y}_s - \hat{\lambda}_{1X}\bar{z}_s$$
$$\hat{\chi}_{YY} = S_{sYY} - S_{sYZ}S_{sZZ}^{-1}S_{sZY} \quad \hat{\chi}_{XX} = S_{sXX} - S_{sXZ}S_{sZZ}^{-1}S_{sZX}$$
$$\hat{\chi}_{XY} = S_{sXY} - S_{sXZ}S_{sZZ}^{-1}S_{sZY}$$

where S_{sAB} denotes the sample covariance of the variables A and B. Also, it is intuitively obvious (and can be rigorously shown) that the MLEs of μ_z and σ_{ZZ} are the mean $\bar{z}_U$ and variance S_{UZZ} of the values in the known population vector z_U. Consequently, invoking the invariance properties of the MLE once more, we obtain MLEs for α, β and $\sigma_{Y|X}^2$ by substituting the MLEs for the parameters of (2.9) as well as the MLEs for the marginal distribution of Z into the identities preceding (2.9). This leads to the estimators

$$\hat{\beta} = \frac{S_{sXY} + S_{sXZ}S_{sZZ}^{-1}\{S_{UZZ} - S_{sZZ}\}S_{sZZ}^{-1}S_{sZY}}{S_{sXX} + S_{sXZ}S_{sZZ}^{-1}\{S_{UZZ} - S_{sZZ}\}S_{sZZ}^{-1}S_{sZX}}$$
$$\hat{\alpha} = (\bar{y}_s + S_{sZZ}^{-1}S_{sZY}(\bar{z}_U - \bar{z}_s)) - (\bar{x}_s + S_{sZZ}^{-1}S_{sZX}(\bar{z}_U - \bar{z}_s))\hat{\beta}$$
$$\hat{\sigma}_{Y|X}^2 = [S_{sYY} + S_{sYZ}S_{sZZ}^{-1}(S_{UZZ} - S_{sZZ})S_{sZZ}^{-1}S_{sZY}]$$
$$- \hat{\beta}^2[S_{sXX} + S_{sXZ}S_{sZZ}^{-1}(S_{UZZ} - S_{sZZ})S_{sZZ}^{-1}S_{sZX}]$$

These estimators are often referred to as the *Pearson-adjusted* MLEs for α, β and $\sigma_{Y|X}^2$, and have a long history in the statistical literature. See the discussion in SHS, section 6.4.

Example 2

This example is not meant to be practical, but it is useful for illustrating the impact of informative sampling on maximum likelihood estimation. We assume that the population distribution of the survey variable Y is one-parameter exponential, with density

$$f_U(y; \theta) = \theta \exp - \theta y.$$

The aim is to estimate the population mean $\mu = E(Y) = \theta^{-1}$ of this variable. We further assume that the sample itself has been selected in such a way that the sample inclusion probability of a population unit is proportional to its value of Y. The survey data then consist of the vectors y_s and π_s, where the latter consists of the sample inclusion probabilities $\pi_t = n y_t / N \bar{y}_U$ of the sampled population units.

Given π_s, it is clear that the value $\bar{y}_U$ of the finite population mean of Y is *deducible* from sample data. Consequently, these survey data are actually the sample values of Y, the sample values of π *and* the value of $\bar{y}_U$. Again, applying (2.4), we see that the score for θ is then

$$sc_s(\theta) = \sum_{t=1}^{N} E(\theta^{-1} - y_t | \text{sample data}) = N(\theta^{-1} - E_s(\bar{y}_U)) = N(\mu - \bar{y}_U)$$

so the MLE for μ is just $\bar{y}_U$. Since this score has zero conditional variability given the sample data, the information for θ is the same as the population information for θ, $info_U(\theta) = N/\theta^2$, so the estimated variance of $\hat{\theta}$ is $N^{-1}\hat{\theta}^2$. Finally, since $\mu = 1/\theta$, the estimated variance of $\hat{\mu}$ is $N^{-1}\hat{\theta}^{-2} = N^{-1}\bar{y}_U^2$.

An alternative informative sampling scheme is where the sample is selected using cut-off sampling, so that $\pi_t = I(y_t > K)$, for known K. In this case

$$E_s(\bar{y}) = \frac{1}{N}(n\bar{y}_s + (N-n)E_U(Y|Y \le K)) = \frac{1}{N}\left(n\bar{y}_s + (N-n)\left[\frac{1}{\theta} - \frac{Ke^{-\theta K}}{1 - e^{-\theta K}}\right]\right)$$

so the score for θ becomes

$$sc_s(\theta) = N\left(\frac{1}{\theta} - E_s(\bar{y})\right) = n\left(\frac{1}{\theta} - \bar{y}_s\right) + (N-n)\left(\frac{Ke^{-\theta K}}{1 - e^{-\theta K}}\right).$$

There is no closed form expression for the MLE $\hat{\theta}$ in this case, but it is relatively easy to calculate its value using numerical approximation. The corresponding estimating equation for the MLE for μ is

$$\hat{\mu} = \bar{y}_s - \left(\frac{N-n}{n}\right)\left(\frac{Ke^{-\hat{\theta}K}}{1 - e^{-\hat{\theta}K}}\right).$$

It is easiest to obtain the information for θ by direct differentiation of its score. This leads to

$$info_s(\theta) = n\theta^{-2} + (N-n)K^2 e^{-\theta K}\left(1 - e^{-\theta K}\right)^{-2}.$$

The estimated variance of $\hat{\mu}$ is then $\hat{\theta}^{-4}info_s^{-1}(\hat{\theta})$.

2.3. SAMPLE LIKELIHOOD

Motivated by inferential methods used with size-biased sampling, Krieger and Pfeffermann (1992, 1997) introduced an alternative model-based approach to analytic likelihood-based inference from sample survey data. Their basic idea is to model the distribution of the sample data by modelling the impact of the sampling method on the population distribution of interest. Once this *sample distribution* is defined, standard likelihood-based methods can be used to obtain an estimate of the parameter of interest. See Chapters 11 and 12 in Part C of this book for applications based on this approach.

In order to describe the fundamental idea behind the sample distribution, we assume full response, supposing the population values of the survey variable Y and the covariate Z correspond to N independent and identically distributed realisations of a bivariate random variable with density $f_U(y|z; \beta) f_U(z; \phi)$. Here β and ϕ are unknown parameters. A key assumption is that sample inclusion for any particular population unit is independent of that of any other unit and is determined by the outcome of a zero–one random variable I, whose distribution depends only on the unit's values of Y and Z. It follows that the sample values y_s and z_s of Y and Z are realisations of random variables that, conditional on the outcome i_U of the sampling procedure, are independent and identically distributed with density parameterised by β and ϕ:

$$f_s(y, z; \beta, \phi) = f_U(y, z | I = 1) = \frac{Pr(I = 1 | Y = y, Z = z) f_U(y|z; \beta) f_U(z; \phi)}{Pr(I = 1; \beta, \phi)}$$

This leads to a *sample likelihood* for β and ϕ,

$$L_s(\beta, \phi) = \prod_{t \in s} \frac{Pr(I_t = 1 | Y_t = y_t, Z_t = z_t) f_U(y_t|z_t; \beta) f_U(z_t; \phi)}{Pr(I_t = 1; \beta, \phi)}$$

that can be maximised with respect to β and ϕ to obtain *maximum sample likelihood estimates* of these parameters. Maximum sample likelihood estimators of other parameters (e.g. the parameter θ characterising the marginal population distribution of Y) are defined using the invariance properties of the maximum likelihood approach.

Example 2 (continued)

We return to the one-parameter exponential population model of Example 2, but now assume that the sample was selected using a probability proportional to Z sampling method, where Z is a positive-valued auxiliary variable correlated with Y. In this case $Pr(I_t = 1 | Z_t = z_t) \propto z_t$, so $Pr(I_t = 1) \propto E(Z) = \lambda$, say. We further assume that the conditional population distribution of Y given Z can be modelled, and let $f_U(y|z; \beta)$ denote the resulting conditional population density of Y given Z. The population marginal density of Z is denoted $f_U(z; \phi)$, so $\lambda = \lambda(\phi)$. The joint population density of Y and Z is then $f_U(y|z; \beta) f_U(z; \phi)$, and the logarithm of the sample likelihood for β and ϕ defined by this approach is

$$\ln\left(L_s(\beta,\phi)\right) = \ln\left(\prod_{t\in s} \frac{z_t f_U(y_t|Z_t = z_t;\beta)f_U(z_t;\phi)}{\lambda(\phi)}\right)$$

$$= \sum_s \ln\left(z_t\right) + \sum_s \ln\left(f_U(y_t|Z_t = z_t;\beta)\right)$$

$$+ \sum_s \ln\left(f_U(z_t;\phi)\right) - n\ln\lambda(\phi).$$

It is easy to see that the value $\hat{\beta}_s$ maximising the second term in this sample likelihood is the *face value* MLE of this parameter (i.e. the estimator that would result if we ignored the method of sample selection and just treated the sample values of Y as independent draws from the conditional population distribution of Y given Z). However, it is also easy to see that the value $\hat{\phi}_s$ maximising the third and fourth terms is *not* the face value estimator of ϕ. In fact, it is defined by the estimating equation

$$\sum_s \frac{\partial_\phi f_U(z_t;\phi)}{f_U(z_t;\phi)} - n\frac{\partial_\phi \lambda(\phi)}{\lambda(\phi)} = 0.$$

Recollect that our aim here is estimation of the marginal population expectation μ of Y. The maximum sample likelihood estimate of this quantity is then

$$\hat{\mu}_s = \int\int y f_U(y|z;\hat{\beta}_s) f_U(z;\hat{\phi}_s)\mathrm{d}y\mathrm{d}z.$$

This can be calculated via numerical integration.

Now suppose $Y = Z$. In this case $Pr(I_t = 1|Y_t = y_t) \propto y_t$, so $Pr(I_t = 1) \propto E(Y_t) = 1/\theta$, and the logarithm of the sample likelihood becomes

$$\ln\left(L_s(\theta)\right) = \ln\left(\prod_{t\in s} \frac{Pr(I_t = 1|Y_t = y_t)f_U(y_t;\theta)}{Pr(I_t = 1;\theta)}\right) \propto \ln\left(\prod_{t\in s} y_t \theta^2 \mathrm{e}^{-\theta y_t}\right)$$

$$= \sum_s \ln\left(y_t\right) + 2n\ln\theta - \theta n\bar{y}_s.$$

The value of θ maximising this expression is $\hat{\theta}_s = 2/\bar{y}_s$ so the maximum sample likelihood estimator of μ is $\hat{\mu}_s = \bar{y}_s/2$. This can be compared with the full information MLE, which is the *known* population mean of Y.

Finally we consider cut-off sampling, where $Pr(I_t = 1) = Pr(Y_t > K) = \mathrm{e}^{-\theta K}$. Here

$$\ln\left(L_s(\theta)\right) = \ln\left(\prod_{t\in s} \theta\mathrm{e}^{-\theta(y_t - K)}\right) = n\ln\theta - n\theta(\bar{y}_s - K).$$

It is easy to see that this function is maximised when $\hat{\theta}_s^{-1} = \hat{\mu}_s = \bar{y}_s - K$. Again, this is not the full information MLE, but it *is* unbiased for μ since

$$E_U(\hat{\mu}_s) = E_U(E_s(\hat{\mu}_s)) = E_U(E(\bar{y}_s|y_t > K; t\in s) - K) = \mu.$$

2.4. PSEUDO-LIKELIHOOD

This approach is now widely used, forming as it does the basis for the methods implemented in a number of software packages for the analysis of complex survey data. The basic idea had its origin in Kish and Frankel (1974), with Binder (1983) and Godambe and Thompson (1986) making major contributions. SHS (section 3.4.4) provides an overview of the method.

Essentially, pseudo-likelihood is a descriptive inference approach to likelihood-based analytic inference. Let $f_U(y_U; \theta)$ denote a statistical model for the probability density of the matrix y_U corresponding to the N population values of the survey variables of interest. Here θ is an unknown parameter and the aim is to estimate its value from the sample data. Now suppose that y_U is observed. The MLE for θ would then be defined as the solution to an estimating equation of the form $sc_U(\theta) = 0$, where $sc_U(\theta)$ is the score function for θ defined by y_U. However, for any value of θ, the value of $sc_U(\theta)$ is also a finite population parameter that can be estimated using standard methods. In particular, let $\hat{s}_U(\theta)$ be such an estimator of $sc_U(\theta)$. Then the maximum pseudo-likelihood estimator of θ is the solution to the estimating equation $\hat{s}_U(\theta) = 0$. Note that this estimator is not unique, depending on the method used to estimate $sc_U(\theta)$.

Example 2 (continued)

Continuing with our example, we see that the population score function in this case is

$$sc_U(\theta) = \sum_{t=1}^{N} (\theta^{-1} - y_t)$$

which is the population total of the variable $\theta^{-1} - Y$. We can estimate this total using the design-unbiased Horvitz–Thompson estimator

$$\hat{s}_{HT}(\theta) = \sum_s \pi_t^{-1}(\theta^{-1} - y_t).$$

Here π_t is the sample inclusion probability of unit t. Setting this estimator equal to zero and solving for θ, and hence (by inversion) μ, we obtain the Horvitz–Thompson maximum pseudo-likelihood estimator of μ. This is

$$\hat{\mu}_{HT} = \sum_s \pi_t^{-1} y_t \Big/ \sum_s \pi_t^{-1}$$

which is the Hajek estimator of the population mean of Y. Under probability proportional to Z sampling this estimator reduces to

$$\hat{\mu}_{HT} = \left(\sum_s z_t^{-1} \right)^{-1} \sum_s y_t z_t^{-1}$$

while for the case of size-biased sampling ($Y = Z$) it reduces to the harmonic mean of the sample Y-values

$$\hat{\mu}_{HT} = n\left(\sum_s y_t^{-1}\right)^{-1}.$$

This last expression can be compared with the full information maximum likelihood estimator (the known population mean of Y) and the maximum sample likelihood estimator (half the sample mean of Y) for this case. As an aside we note that where cut-off sampling is used, so population units with Y greater than a known constant K are sampled with probability one with the remaining units having zero probability of sample inclusion, no design-unbiased estimator of $sc_U(\theta)$ can be defined and so no design-based pseudo-likelihood estimator exists.

Inference under pseudo-likelihood can be design based or model based. Thus, variance estimation is usually carried out using a combination of a Taylor series linearisation argument and an appropriate method (design based or model based) for estimating the variance of $\hat{s}_U(\theta)$ (see Binder, 1983; SHS, section 3.4.4). We write

$$0 = \hat{s}_U(\hat{\theta}) \cong \hat{s}_U(\theta_N) + (\hat{\theta} - \theta_N)\left[\frac{\mathrm{d}\hat{s}_U}{\mathrm{d}\theta}\right]_{\theta=\theta_N}$$

where θ_N is defined by $s_U(\theta_N) = 0$. Let $var(\hat{\theta} - \theta_N)$ denote an appropriate variance for the estimation error $\hat{\theta} - \theta_N$. Then

$$var(\hat{\theta} - \theta_N) \cong \left[\frac{\mathrm{d}\hat{s}_U}{\mathrm{d}\theta}\bigg|_{\theta=\theta_N}\right]^{-1} var[\hat{s}_U(\theta_N) - s_U(\theta_N)]\left[\frac{\mathrm{d}\hat{s}_U}{\mathrm{d}\theta}\bigg|_{\theta=\theta_N}\right]^{-1}$$

which leads to a 'sandwich' variance estimator of the form

$$\hat{V}(\hat{\theta}) \cong \left\{\left[\frac{\mathrm{d}\hat{s}_U}{\mathrm{d}\theta}\right]^{-1} \hat{V}[\hat{s}_U(\theta) - s_U(\theta)]\left[\frac{\mathrm{d}\hat{s}_U}{\mathrm{d}\theta}\right]^{-1}\right\}_{\theta=\hat{\theta}_N} \qquad (2.8)$$

where $\hat{V}[\hat{s}_U(\theta) - s_U(\theta)]$ is a corresponding estimator of the variance of $\hat{s}_U(\theta) - s_U(\theta)$.

2.5. PSEUDO-LIKELIHOOD APPLIED TO ANALYTIC INFERENCE

From the development in the previous section, one can see that pseudo-likelihood is essentially a method for descriptive, rather than analytic, inference, since the target of inference is the census value of the parameter of interest (sometimes referred to as the *census parameter*). However, as Binder and Roberts show in Chapter 3, one can develop an analytic perspective that justifies the pseudo-likelihood approach. This perspective takes account of the *total variation* from both the population generating process and the sample selection process.

In particular, Binder and Roberts explore the link between analytic (i.e. model-based) and design-based inference for a class of *linear parameters* corresponding to the expected values of population sums (or means) under an appropriate model for the population. From a design-based perspective, design-unbiased or design-consistent estimators of these population sums should then be good estimators of these expectations for large-sample sizes, and so should have a role to play in analytic inference. Furthermore, since a solution to a population-level estimating equation can usually be approximated by such a sum, the class of pseudo-likelihood estimators can be represented in this way. Below we show how the total variation theory developed by Binder and Roberts applies to these estimators.

Following these authors, we assume that the sampling method is noninformative given Z. That is, conditional on the values z_U of a known population covariate Z, the population distributions of the variables of interest Y and the sample inclusion indicator I are independent. An immediate consequence is that the joint distribution of Y and I given z_U is the product of their two corresponding 'marginal' (i.e. conditional on z_U) distributions.

To simplify notation, conditioning on the values in i_U and z_U is denoted by a subscript ξ, and conditioning on the values in y_U and z_U by a subscript p. The situation where conditioning is only with respect to the values in z_U is denoted by a subscript ξp and we again do not distinguish between a random variable and its realisation. Under noninformative sampling, the ξp-expectation of a function $g(y_U, i_U)$ of both y_U and i_U (its *total expectation*) is

$$E_{\xi p}[g(y_U, i_U)] = E_U[E_U(g(y_U, i_U)|y_U, z_U)|z_U] = E_\xi[E_p(g(y_U, i_U))]$$

since the random variable inside the square brackets on the right hand side only depends on y_U and z_U, and so its expectation given z_U and its expectation conditional on i_U and z_U are the same. The corresponding total variance for $g(y_U, i_U)$ is

$$\begin{aligned} var_{\xi p}[g(y_U, i_U)] &= var_U[E_U(g(y_U, i_U)|y_U, z_U)|z_U] \\ &\quad + E_U[var_U(g(y_U, i_U)|y_U, z_U)|z_U] \\ &= var_\xi[E_p(g(y_U, i_U))] + E_\xi[var_p(g(y_U, i_U)). \end{aligned}$$

Now suppose the population values of Y are mutually independent given Z, with the conditional density of Y given Z over the population parameterised by β. The aim is to estimate the value of this parameter from the sample data. A Horvitz Thompson maximum pseudo-likelihood approach is used for this purpose, so β is estimated by $\hat{B}$, where

$$\sum_s \pi_t^{-1} sc_t(\hat{B}) = 0.$$

Here $sc_t(\beta)$ is the contribution of unit t to the population score function for β. Let B_N denote the population maximum likelihood estimator of β. Then the design-based consistency of $\hat{B}$ for B_N allows us to write

$$\hat{B} - B_N = \left[\sum_s \pi_t^{-1} ((\partial_\beta sc_t(\beta))_{\beta = B_N} \right]^{-1} \sum_s \pi_t^{-1} sc_t(B_N) + o(n^{-\frac{1}{2}})$$

$$= \sum_s \pi_t^{-1} U_t(B_N) + o(n^{-\frac{1}{2}})$$

where

$$U_t(B_N) = \left[\sum_{t=1}^N (\partial_\beta sc_t(\beta))_{\beta = B_N} \right]^{-1} sc_t(B_N)$$

so that the design expectation of $\hat{B}$ is

$$E_p(\hat{B}) = B_N + \sum_{t=1}^N U_t(B_N) + o(n^{-\frac{1}{2}}) = B_N + o(n^{-\frac{1}{2}}).$$

The total expectation of $\hat{B}$ is therefore

$$E_{\xi p}(\hat{B}) = E_\xi(B_N) + o(n^{-\frac{1}{2}}) = \beta + o(n^{-\frac{1}{2}})$$

when we apply the same Taylor series linearisation argument to show that the expected value of B_N under the assumed population model is β plus a term of order $N^{-\frac{1}{2}}$. The corresponding total variance of $\hat{B}$ is

$$var_{\xi p}(\hat{B}) = var_{\xi p} \left[B_N + \sum_s \pi_t^{-1} U_t(B_N) + o(n^{-\frac{1}{2}}) \right]$$

$$= \sum_{t=1}^N \Delta_{tt} var_\xi(U_t(B_N)) + o(n^{-1})$$

where $\Delta_{tu} = cov_p(I_t \pi_t^{-1}, I_u \pi_u^{-1})$ and we have used the fact that population units are mutually independent under ξ and that $E_\xi(U_t(B_N)) = 0$. Now let $\hat{V}_p(B_N)$ denote the large-sample limit of the estimator (2.8) when it estimates the design variance of $\hat{B}$. Then $\hat{V}_p(B_N)$ is approximately design unbiased for the design variance of the Horvitz–Thompson estimator of the population total of the $U_t(B_N)$,

$$E_p(\hat{V}_p(B_N)) = \sum_{t=1}^N \sum_{u=1}^N \Delta_{tu} U_t(B_N) U_u(B_N) + o(n^{-1})$$

and so is unbiased for the total variance of $\hat{B}$ to the same degree of approximation,

$$E_{\xi p}(\hat{V}_p(B_N)) = \sum_{t=1}^N \Delta_{tt} var_\xi(U_t(B_N)) + o(n^{-1}).$$

Furthermore $\hat{V}_p(B_N)$ is also approximately model-unbiased for the prediction variance of $\hat{B}$ under ξ. To show this, put $\tau_{tu} = cov_\xi(U_t(B_N), U_u(B_N))$. Then

$$var_\xi(\hat{B} - B_N) = \sum_{t=1}^{N}\sum_{u=1}^{N} I_t I_u \pi_t^{-1} \pi_u^{-1} \tau_{tu}$$

$$= \sum_{t=1}^{N}\sum_{t=1}^{N} E_p(I_t I_u \pi_t^{-1} \pi_u^{-1})\tau_{tu} + o(n^{-1})$$

$$= \sum_{t=1}^{N}\sum_{u=1}^{N} (\Delta_{tu} + E_p(I_t \pi_t^{-1})E_p(I_u \pi_u^{-1}))\tau_{tu} + o(n^{-1})$$

$$= \sum_{t=1}^{N} \Delta_{tt}\tau_{tt} + o(n^{-1})$$

and so the standard design-based estimator (2.8) of the design variance of the pseudo-likelihood estimator $\hat{B}$ is also an estimator of its model variance.

So far we have assumed noninformative sampling and a correctly specified model. What if either (or both) of these assumptions are wrong? It is often said that design-based inference remains valid in such a situation because it does not depend on model assumptions. Binder and Roberts justify this claim in Chapter 3, provided one accepts that the expected value $E_\xi(B_N)$ of the finite population parameter B_N under the true model remains the target of inference under such mis-specification. It is interesting to speculate on practical conditions under which this would be the case.

2.6.   BAYESIAN INFERENCE FOR SAMPLE SURVEYS

So far, the discussion in this chapter has been based on frequentist arguments. However, the Bayesian method has a strong history in survey sampling theory (Ericson, 1969; Ghosh and Meeden, 1997) and offers an integrated solution to both analytic and descriptive survey sampling problems, since no distinction is made between population quantities (e.g. population sums) and model parameters. In both cases, the inference problem is treated as a prediction problem. The Bayesian approach therefore has considerable theoretical appeal. Unfortunately, its practical application has been somewhat limited to date by the need to specify appropriate priors for unknown parameters, and by the lack of closed form expressions for estimators when one deviates substantially from normal distribution population models. Use of improper *noninformative* priors is a standard way of getting around the first problem, while modern, computationally intensive techniques like Markov chain Monte Carlo methods now allow the fitting of extremely sophisticated non-normal models to data. Consequently, it is to be expected that Bayesian methods will play an increasingly significant role in survey data analysis.

In Chapter 4 Little gives an insight into the power of the Bayesian approach when applied to sample survey data. Here we see again the need to model the joint population distribution of the values y_U of the survey variables and the sample inclusion indicator i_U when analysing complex survey data, with analytic inference about the parameters (θ, ω) of the joint population distribution of these values then based on their joint posterior distribution. This posterior distribution is defined as the product of the joint prior distribution of these parameters given the values z_U of the design variable Z times their likelihood, obtained by integrating the joint population density of y_U and i_U over the unobserved non-sample values in y_U. Descriptive inference about a characteristic Q of the finite population values of Y (e.g. their sum) is then based on the posterior density of Q. This is the expected value, relative to the posterior distribution for θ and ω, of the conditional density of Q given the population values z_U and i_U and the sample values y_s.

Following Rubin (1976), Little defines the selection process to be ignorable when the marginal posterior distribution of the parameter θ characterising the population distribution of y_U obtained from this joint posterior reduces to the usual posterior for θ obtained from the product of the marginal prior and marginal likelihood for this parameter (i.e. ignoring the outcome of the sample selection process). Clearly, a selection process corresponding to simple random sampling does not depend on y_U or on any unknown parameters and is therefore ignorable. In contrast, stratified sampling is only ignorable when the population model for y_U conditions on the stratum indicators. Similarly, a two-stage sampling procedure is ignorable provided the population model conditions on cluster information, which in this case corresponds to cluster indicators, and allows for within-cluster correlation.

2.7. APPLICATION OF THE LIKELIHOOD PRINCIPLE IN DESCRIPTIVE INFERENCE

Section 2.2 above outlined the maximum likelihood method for analytic inference from complex survey data. In Chapter 5, Royall discusses the related problem of applying the likelihood method to descriptive inference. In particular, he develops the form of the likelihood function that is appropriate when the aim is to use the likelihood principle (rather than frequentist arguments) to measure the evidence in the sample data for a particular value for a finite population characteristic.

To illustrate, suppose there is a single survey variable Y and the descriptive parameter of interest is its finite population total T and a sample of size n is taken from this population and values of Y observed. Let y_s denote the vector of these sample values. Using an argument based on the fact that the ratio of values taken by the likelihood function for one value of a parameter compared with another is the factor by which the prior probability ratio for these values of the parameter is changed by the observed sample data to yield the corresponding posterior probability ratio, Royall defines the value of likelihood function for T at an arbitrary value q of this parameter as proportional to the value of the conditional density of y_s given $T = q$. That is, using $L_U(q)$ to denote this likelihood function,

$$L_U(q) = f_U(y_s | T = q) = \frac{f_U(q | y_s) f_U(y_s)}{f_U(q)}.$$

Clearly, this definition is easily extended to give the likelihood for any well-defined function of the population values of Y. In general $L_U(q)$ will depend on the values of nuisance parameters associated with the various densities above. That is, $L_U(q) = L_U(q; \theta)$ where θ is unknown. Here Royall suggests calculation of a profile likelihood, defined by

$$L_U^{profile}(q) = \max_\theta L_U(q; \theta).$$

Under the likelihood-based approach described by Royall in Chapter 5, the concept of a confidence interval is irrelevant. Instead, one can define regions around the value where the likelihood function is maximised that correspond to alternative parameter values whose associated likelihood values are not too different from the maximum. Conversely, values outside this region are then viewed as rather unlikely candidates for being the actual parameter value of interest. For example, Royall suggests that a value for T whose likelihood ratio relative to the maximum likelihood value $\hat{T}_{MLE}$ is less than 1/8 should be considered as the value for which the strength of the evidence in favour of $\hat{T}_{MLE}$ being the correct value is moderately strong. A value whose likelihood ratio relative to the MLE is less than 1/32 is viewed as one where the strength of the evidence in favour of the MLE being the true value is strong.

Incorporation of sample design information (denoted by the known population matrix z_U) into this approach is conceptually straightforward. One just replaces the various densities defining the likelihood function L_U by conditional densities, where the conditioning is with respect to z_U. Also, although not expressly considered by Royall, the extension of this approach to the case of informative sampling and/or informative nonresponse would require the nature of the informative sampling method and the nonresponse mechanism also to be taken account of explicitly in the modelling process. In general, the conditional density of T given y_s and the marginal density of y_s in the expression for $L_U(q)$ above would be replaced by the conditional density of T given the actual survey data, i.e. y_{obs}, r_s, i_U and z_U, and the joint marginal density of these data.

CHAPTER 3

Design-based and Model-based Methods for Estimating Model Parameters

David A. Binder and Georgia R. Roberts

3.1. CHOICE OF METHODS

One of the first questions an analyst asks when fitting a model to data that have been collected from a complex survey is whether and how to account for the survey design in the analysis. In fact, there are two questions that the analyst should address. The first is whether and how to use the sampling weights for the point estimates of the unknown parameters; the second is how to estimate the variance of the estimators required for hypothesis testing and for deriving confidence intervals. (We are assuming that the sample size is sufficiently large that the sampling distribution of the parameter estimators is approximately normal.)

There are a number of schools of thought on these questions. The pure model-based approach would demand that if the model being fitted is true, then one should use an optimal model-based estimator. Normally this would result in ignoring the sample design, unless the sample design is an inherent part of the model, such as for a stratified design where the model allows for different parameter values in different strata. The Bayesian approach discussed in Little (Chapter 4) is an example of this model-based perspective.

As an example of the model-based approach, suppose that under the model it is assumed that the sample observations, $y_1, \ldots, y_n$, are random variables which, given $x_1, \ldots, x_n$, satisfy

$$y_t = x_t'\beta + \varepsilon_t, \quad \text{for } t = 1, \ldots, n, \tag{3.1}$$

Analysis of Survey Data Edited by R. L. Chambers and C. J. Skinner
© 2003 John Wiley & Sons, Ltd

where ε_t has mean 0, variance σ^2, and is uncorrelated with $\varepsilon_{t'}$ for $t \neq t'$. Standard statistical theory would imply that the ordinary least squares estimator for β is the best linear unbiased estimator. In particular, standard theory yields the estimator

$$\hat{\beta} = \left(X_s' X_s \right)^{-1} X_s' y_s, \tag{3.2}$$

with variance

$$var(\hat{\beta}) = \sigma^2 (X_s' X_s)^{-1}, \tag{3.3}$$

where X_s and y_s are based on the sample observations.

Example 1.   Estimating the mean

We consider the simplest case of model (3.1) where x_t is a scalar equal to one for all t. In this case, β, a scalar, is the expected value of the random variables $y_t (t = 1, \ldots, n)$ and $\hat{\beta}$ is $\bar{y}_s$, the unweighted mean of the observed sample y-values. Here we ignore any sample design information used for obtaining the n units in our sample. Yet, from the design-based perspective, we know that the unweighted mean can be biased for estimating the finite population mean, $\bar{y}_U$. Is this contradictory? The answer lies both in what we are assuming to be the parameter of interest and in what we are assuming to be the randomisation mechanism.

In the model-based approach, where we are interested in making inferences about β, we assume that we have a conceptual infinite superpopulation of y-values, each with scalar mean, β, and variance σ^2. The observations, $y_1, \ldots, y_n$, are assumed to be independent realisations from this superpopulation. The model-based sample error is $\hat{\beta} - \beta$, which has mean 0 and variance σ^2/n. The sample design is assumed to be ignorable as discussed in Rubin (1976), Scott (1977a) and Sugden and Smith (1984).

In the design-based approach, on the other hand, where the parameter of interest is $\bar{y}_U$, the finite population mean, we assume that the n observations are a probability sample from the finite population, $y_1, \ldots, y_N$. There is no reference to a superpopulation. The randomisation mechanism is dictated by the chosen sampling design, which may include unequal probabilities of selection, clustering, stratification, and so on. We denote by I_t the 0–1 random variable indicating whether or not the tth unit is in the sample. In the design-based approach, all inferences are made with respect to the properties of the random variables, I_t $(t = 1, \ldots, N)$, since the quantities y_t $(t = 1, \ldots, N)$, are assumed to be fixed, rather than random, as would be assumed in the model-based setting. From the design-based perspective, the finite population mean is regarded as the descriptive parameter to be estimated. Considerations, such as asymptotic design-unbiasedness, would lead to including the sampling weights in the estimate of $\bar{y}_U$. Much of the traditional sampling theory adopts this approach, since it is the finite population mean (or total) that is of primary interest to the survey statistician.

At this point, it might appear that the design-based and model-based approaches are irreconcilable. We will show this not to be true. In Section 3.2 we introduce linear parameters and their estimators in the design-based and model-based contexts. We review the biases of these estimators under both the design-based and model-based randomisation mechanisms.

In Section 3.3 we introduce design-based and total variances and we compare the results under the design-based and model-based approaches. The total variance is based on considering the variation due to both the model and the survey design. This comparison leads to some general results on the similarities and differences of the two approaches. We show that the design-based approach often leads to variances that are close to the total variance, even though a model is not used in the design-based approach. These results are similar to those in Molina, Smith and Sugden (2001).

The basic results for linear parameters and linear estimators are extended to more complex cases in Section 3.4. The traditional model-based approach is explored in Section 3.5, where it is assumed that the realisation of the randomisation mechanism used to select the units in the sample may be ignored for making model-based inferences. We also introduce an estimating function framework that leads to a closer relationship between the pure model-based and the pure design-based approaches. In Section 3.6 we study the implications of taking the 'wrong' approach and we discuss why the design-based approach may be more robust for model-based inferences, at least for large samples and small sampling fractions. We summarise our conclusions in Section 3.7.

3.2. DESIGN-BASED AND MODEL-BASED LINEAR ESTIMATORS

We consider the case where there is a superpopulation model which generates the finite population values, $y_1, \ldots, y_N$. We suppose that, given the realised finite population, a probability sample is selected from it using some, possibly complex, sampling scheme. This sampling scheme may contain stratification, clustering, unequal probabilities of selection, and so on. Our sample can thus be viewed as a second phase of sampling from the original superpopulation.

In our discussions about asymptotics for finite populations, we are assuming that we have a sequence of finite populations of sizes increasing to infinity and that for each finite population, we take a sample such that the sample size also increases to infinity. In the case of multi-stage (or multi-phase) sampling, it is the number of first-stage (or first-phase) units that we are assuming to be increasing to infinity. We do not discuss the regularity conditions for the asymptotic normality of the sampling distributions; instead we simply assume that for large samples, the normal distribution is a reasonable approximation. For more details on asymptotics from finite populations, see, for example, Sen (1988). In this and subsequent sections, we make frequent use of the o-notation to denote various types of convergence to zero. First of all, we use $a_n = b_n + o(n^{-q})$ to mean $n^q |a_n - b_n| \to 0$ almost surely as $n \to \infty$. Next, to distinguish among various types of convergence in probability under different randomisation mechanisms, we use the somewhat unconventional notation $a_n = b_n + o_p(n^{-q})$, $a_n = b_n + o_{\xi p}(n^{-q})$, and $a_n = b_n + o_\xi(n^{-q})$ for convergence in probability under the p-, ξp- and ξ-randomisations, respectively. (We define these randomisation mechanisms in subsequent sections.)

We first focus on linear estimators. As we will see in Section 3.4, many of the results for more complex estimators are based on the asymptotic distributions of linear estimators. To avoid complexities of notation, we consider the simplest case of univariate parameters of interest.

3.2.1. Parameters of interest

Under the model, we define μ_t as

$$\mu_t \equiv E_\xi(y_t), \qquad (3.4)$$

where we use the subscript ξ to denote expectation under the model. (To denote the expectation under the design we will use the subscript p.) The model parameter that we consider is

$$\beta \equiv \sum I_t \mu_t / n, \qquad (3.5)$$

which is the expectation of the sample mean, $\bar{y}_s = \sum I_t y_t / n$, when the sample design is ignorable. Normally, modellers are interested in β only when the μ_t are all equal; however, to allow more generality for complex situations, we are not restricting ourselves to that simple case here. Defining $\bar{\mu} \equiv \sum \mu_t / N$, we assume that β has the property that

$$\beta = \bar{\mu} + o(1); \qquad (3.6)$$

that is, that β converges to $\bar{\mu}$. This important assumption is reasonable when the sample design is ignorable. We discuss the estimation of β for the case of nonignorable sample designs in Section 3.6.

The design-based parameter of interest that is analogous to β is the finite population mean,

$$b \equiv \bar{y}_U = \sum y_t / N. \qquad (3.7)$$

This is commonly considered as the finite population parameter of interest in the design-based framework.

3.2.2. Linear estimators

In general, a model-based linear estimator of β can be expressed as

$$\hat{\beta} = \sum I_t c_t y_t, \qquad (3.8)$$

where the c_t are determined so that $\hat{\beta}$ has good model-based properties. In particular, we assume that the c_t are chosen so that

$$\hat{\beta} = \beta + o_\xi(1), \qquad (3.9)$$

which, when combined with (3.6), implies that

$$\hat{\beta} = \bar{\mu} + o_\xi(1).\tag{3.10}$$

On the other hand, the usual design-based linear estimator of b has the form

$$\hat{b} = \bar{y}_d \equiv \sum I_t d_t y_t,\tag{3.11}$$

where the d_t are chosen so that $\hat{b}$ has good design-based properties. In particular, we assume that

$$E_p(I_t d_t) = 1/N + o(N^{-1}),\tag{3.12}$$

so that

$$\hat{b} = b + o_p(1).\tag{3.13}$$

3.2.3. Properties of $\hat{\beta}$ and $\hat{b}$

Let us now consider $\hat{b}$ and $\hat{\beta}$ from the perspective of estimating b. It follows from (3.13) that

$$E_p(\hat{b}) = b + o(1) = \bar{y}_U + o(1).\tag{3.14}$$

Also,

$$E_p(\hat{\beta}) = \sum E_p(I_t c_t) y_t.\tag{3.15}$$

We see that $\hat{\beta}$ is not necessarily asymptotically design unbiased for b; the condition for this is that

$$E_p(\hat{\beta}) = b + o(1).\tag{3.16}$$

We now show that this condition holds when the model is true.
   Since, by the strong law of large numbers (under the model),

$$b = \bar{y}_U = E_\xi(\bar{y}_U) + o(1) = \bar{\mu} + o(1),\tag{3.17}$$

it follows from (3.10) and (3.17) that

$$E_\xi(\hat{\beta} - b) = o(1).\tag{3.18}$$

Again, by the strong law of large numbers (under the model) and (3.18), we have

$$\begin{aligned}
E_p(\hat{\beta}) - b &= E_\xi[E_p(\hat{\beta}) - b] + o(1) \\
&= E_p[E_\xi(\hat{\beta} - b)] + o(1) \\
&= o(1),
\end{aligned}\tag{3.19}$$

which means that condition (3.16) holds when the model is true.

We conclude, therefore, that when the data have been generated according to the assumed model, not only does $\hat{\beta}$ have good model-based properties for estimating β, but also $\hat{\beta}$ is approximately design unbiased for b.

Using similar arguments, it may be shown that $E_\xi(\hat{b}) \quad b = o(1)$, so that $\hat{b}$ is approximately model unbiased for the finite population quantity b. Additionally, we can see that the model expectations and the design expectations of both $\hat{\beta}$ and $\hat{b}$ all converge to the same quantity, $\bar{\mu}$.

Example 2

Suppose that under the model all the μ_t are equal to μ, and the y_t have equal variance, σ^2, and are uncorrelated. The best linear unbiased estimator of β is $\hat{\beta} = \bar{y}_s$, the unweighted sample mean. If our sample design is a stratified random sample from two strata with n_1 units selected from the first stratum (stratum size N_1) and n_2 units selected from the second stratum (stratum size N_2), then the usual design-based estimator for $b = \bar{y}_U$ is

$$\hat{b} = (N_1 \bar{y}_{1s} + N_2 \bar{y}_{2s})/N, \tag{3.20}$$

where $\bar{y}_{1s}$ and $\bar{y}_{2s}$ are the respective stratum sample means. Now,

$$E_p(\hat{\beta}) = (n_1 \bar{y}_{1U} + n_2 \bar{y}_{2U})/n, \tag{3.21}$$

where $\bar{y}_{1U}$ and $\bar{y}_{2U}$ are the respective stratum means. In general, for disproportionate sampling between strata, the design-based mean of $\hat{\beta}$ is not equal to

$$E_p(\hat{b}) = (N_1 \bar{y}_{1U} + N_2 \bar{y}_{2U})/N. \tag{3.22}$$

We see here that the model-based estimator may be design inconsistent for the finite population parameter of interest. However, when the model holds, $E_\xi(\bar{y}_{1U}) = E_\xi(\bar{y}_{2U}) = \mu$, so that $E_\xi[E_p(\hat{\beta})] = \mu$ and, by the strong law of large numbers, $E_p(\hat{\beta}) = \mu + o(1)$. As well, under the model, $E_\xi(b) = \mu$, so that by the strong law of large numbers, $b = \mu + o(1)$. Thus, $E_p(\hat{\beta}) = b + o(1)$, implying that when the model holds, we do achieve asymptotic design unbiasedness for $\hat{\beta}$.

Many modellers do not consider the design-based expectations to be relevant for their analysis; these modellers are more interested in the 'pure' model-based expectations which ignore the randomisation due to the sample design. This situation will be considered in Section 3.5.

3.3. DESIGN-BASED AND TOTAL VARIANCES OF LINEAR ESTIMATORS

3.3.1. Design-based and total variance of $\hat{b}$

The observed units could be considered to be selected through two phases of sampling – the first phase yielding the finite population generated by the superpopulation model, and the second phase yielding the sample selected from the finite population. Analysts may then be interested in three types of randomisation mechanisms. The pure design-based approach conditions on the outcome of the first phase, so that the finite population values are considered fixed constants. The pure model-based approach conditions on the sample design, so that the I_t are considered fixed; this approach will be discussed further in Section 3.5. An alternative to these approaches is to consider both phases of sampling to be random. This case has been examined in the literature; see, for example, Hartley and Sielken (1975) and Molina, Smith and Sugden (2001). We define the *total variance* to be the variance over the two phases of sampling.

We now turn to examine the variance of $\hat{b}$. The pure design-based variance is given by

$$var_p(\hat{b}) = var_p\left(\sum I_t d_t y_t\right) = \sum\sum \Delta_{tt'} y_t y_{t'}, \tag{3.23}$$

where $\Delta_{tt'}$ denotes the design-based covariance of $I_t d_t$ and $I_{t'} d_{t'}$.

Under a wide set of models, it can be assumed that $E_p(\hat{b} - \bar{y}_U) = O_\xi(N^{-\frac{1}{2}})$, where O_ξ refers to the probability limit under the model. This assumption is certainly true when the model is based on independent and identically distributed observations with finite variance, but it is also true under weaker conditions. Therefore, assuming the sampling fraction, n/N, to be $o(1)$, and denoting the total variance of $\hat{b}$ by $var_{\xi p}(\hat{b})$, we have

$$
\begin{aligned}
var_{\xi p}(\hat{b}) &= var_\xi E_p(\hat{b}) + E_\xi var_p(\hat{b}) \\
&= var_\xi(\bar{y}_U) + E_\xi\left(\sum\sum \Delta_{tt'} y_t y_{t'}\right) + o(n^{-1}) \\
&= \bar{\sigma}_{\bullet\bullet} + \sum\sum \Delta_{tt'}(\sigma_{tt'} + \mu_t \mu_{t'}) + o(n^{-1})
\end{aligned}
\tag{3.24}
$$

where

$$\sigma_{tt'} \equiv cov_\xi(y_t, y_{t'}) \tag{3.25}$$

and

$$\bar{\sigma}_{\bullet\bullet} \equiv \sum\sum \sigma_{tt'}/N^2 = var_\xi(\bar{y}_U). \tag{3.26}$$

If we assume that $\bar{\sigma}_{\bullet\bullet}$ is $O(N^{-1})$ then

$$var_{\xi p}(\hat{b}) = \sum\sum \Delta_{tt'}(\sigma_{tt'} + \mu_t \mu_{t'}) + o(n^{-1}). \tag{3.27}$$

Note that assuming that $\bar{\sigma}_{\bullet\bullet}$ is $O(N^{-1})$ is weaker than the commonly used assumption that the random variables associated with the finite population units are independent and identically distributed realisations with finite variance from the superpopulation.

When $\bar{\sigma}_{\bullet\bullet}$ is $O(N^{-1})$, we see that $var_{\xi p}(\hat{b})$ and $E_\xi var_p(\hat{b})$ are both asymptotically equal to $\sum\sum \Delta_{tt'}(\sigma_{tt'} + \mu_t \mu_{t'})$. Thus, a design-based estimator of $var_p(\hat{b})$, given by $v_p(\hat{b})$, would be suitable for estimating $var_{\xi p}(\hat{b})$, provided $v_p(\hat{b})$ is asymptotically model unbiased for $E_\xi var_p(\hat{b})$. This is the case when $v_p(\hat{b})$ converges to $var_p(\hat{b}) = \sum\sum \Delta_{tt'} y_t y_{t'}$. For a more complete discussion of design-based variance estimation, see Wolter (1985).

3.3.2. Design-based mean squared error of $\hat{\beta}$ and its model expectation

Design-based survey statisticians do not normally use $\hat{\beta}$ to estimate b, since $\hat{\beta}$ is often design-inconsistent for b. However, design-based survey statisticians should not discard $\hat{\beta}$ without first considering its other properties such as its mean squared error. Rather than examining the mean squared error in general, we continue with the situation introduced in Example 2.

Example 3

In Example 2, we had

$$\hat{\beta} = \bar{y}_s = (n_1 \bar{y}_{1s} + n_2 \bar{y}_{2s})/n, \tag{3.28}$$

the unweighted sample mean. Our sample design was a stratified random sample from two strata with n_1 units selected from the first stratum (stratum size N_1) and n_2 units selected from the second stratum (stratum size N_2). Since $E_p(\hat{\beta}) = (n_1 \bar{y}_{1U} + n_2 \bar{y}_{2U})/n$, the design-based bias of $\hat{\beta}$ in estimating b is

$$E_p(\hat{\beta}) - b = \alpha(\bar{y}_{1U} - \bar{y}_{2U}), \tag{3.29}$$

where

$$\alpha = \frac{n_1}{n} - \frac{N_1}{N}. \tag{3.30}$$

The design-based variance of $\hat{\beta}$, ignoring, for simplicity, the finite population correction factors (by assuming the stratum sample sizes to be small relative to the stratum sizes), is

$$var_p(\hat{\beta}) = (n_1 v_{1U} + n_2 v_{2U})/n^2, \tag{3.31}$$

where v_{1U} and v_{2U} are the respective stratum population variances. The design-based mean squared error of $\hat{\beta}$ would therefore be smaller than the design-based variance of $\hat{b}$ when

$$\alpha^2(\bar{y}_{1U} - \bar{y}_{2U})^2 + \frac{n_1 v_{1U} + n_2 v_{2U}}{n^2} < \frac{N_1^2 v_{1U}}{N^2 n_1} + \frac{N_2^2 v_{2U}}{N^2 n_2}. \tag{3.32}$$

Condition (3.32) is generally not easy to verify; however, we may consider the model expectation of the design-based mean squared error of $\hat{\beta}$ and of the design-based variance of $\hat{b}$. Under the assumed model, we have

$$\begin{aligned} E_\xi mse_p(\hat{\beta}) &= E_\xi \left[\alpha^2(\bar{y}_{1U} - \bar{y}_{2U})^2 + \frac{n_1 v_{1U} + n_2 v_{2U}}{n^2} \right] \\ &= \alpha^2 \sigma^2 \left(\frac{1}{N_1} + \frac{1}{N_2} \right) + \frac{\sigma^2}{n}, \end{aligned} \tag{3.33}$$

which is approximately equal to σ^2/n when N_1 and N_2 are large compared with n. On the other hand, the model expectation of the design-based variance is given by

$$E_\xi var_p(\hat{b}) = E_\xi \left[\frac{N_1^2 v_{1U}}{N^2 n_1} + \frac{N_2^2 v_{2U}}{N^2 n_2} \right]$$

$$= \sigma^2 \left(\frac{N_1^2}{N^2 n_1} + \frac{N_2^2}{N^2 n_2} \right) \geq \frac{\sigma^2}{n}. \tag{3.34}$$

Thus, when the model is true and when the sampling fractions in the two strata are unequal, the design-based mean squared error of $\hat{\beta}$ is expected to be smaller than the design-based variance of $\hat{b}$.

Example 3 illustrates that when the model holds, $\hat{\beta}$ may be better than $\hat{b}$ as an estimator of b, from the point of view of having the smaller design-based mean squared error. However, as pointed out by Hansen, Madow and Tepping (1983), model-based methods can lead to misleading results even when the data seem to support the validity of the model.

3.4. MORE COMPLEX ESTIMATORS

3.4.1. Taylor linearisation of non-linear statistics

In Sections 3.2 and 3.3 we restricted our discussion to the case of linear estimators of the parameter of interest. However, many methods of data analysis deal with more complex statistics. In this section we show that the properties of many of these more complex quantities are asymptotically equivalent to the properties of linear estimators, so that the discussions in Sections 3.2 and 3.3 can be extended to the more complex situations. We take a Taylor linearisation approach here.

3.4.2. Ratio estimation

To introduce the basic notions, we start with the example of the estimation of a ratio. The quantities y_t and x_t are assumed to have model expectations μ_Y and μ_X, respectively. We first consider the properties of the model-based estimator, $\bar{y}_s/\bar{x}_s$, of the ratio parameter, $\theta = \mu_Y/\mu_X$. Assuming the validity of a first-order Taylor expansion, such as when the y-values and x-values have finite model variances, we have

$$\frac{\bar{y}_s}{\bar{x}_s} - \theta = \frac{1}{\mu_X} (\bar{y}_s - \theta \bar{x}_s) + o_\xi(n^{-1/2}). \tag{3.35}$$

We now introduce the *pseudo-variables*, $u_t(\mu_X, \mu_Y)$, given by

$$u_t(\mu_X, \mu_Y) \equiv \frac{1}{\mu_X} \left(y_t - \frac{\mu_Y}{\mu_X} x_t \right) = \frac{1}{\mu_X} (y_t - \theta x_t), \quad \text{for } t = 1, \ldots, N. \tag{3.36}$$

(While the u_t are defined for all N units in the finite population, only those for the n units in the sample are required in the model-based framework.) The sample mean of the u_t has, asymptotically, the same model expectation and variance as the left hand side of expression (3.35) since

$$\frac{\bar{y}_s}{\bar{x}_s} - \theta = \frac{1}{n}\sum I_t u_t(\mu_X, \mu_Y) + o_\xi(n^{-\frac{1}{2}}), \tag{3.37}$$

so that the model-based properties of $\bar{y}_s/\bar{x}_s$ are asymptotically equivalent to linear estimators which are discussed in Sections 3.2 and 3.3. The design-based expectation of $\bar{y}_s/\bar{x}_s$ is

$$E_p(\bar{y}_s/\bar{x}_s) = \theta + \frac{1}{n}\sum E_p(I_t)u_t(\mu_X, \mu_Y) + o_\xi(n^{-\frac{1}{2}})$$

$$= \theta + \frac{1}{n}\sum \pi_t u_t(\mu_X, \mu_Y) + o_\xi(n^{-\frac{1}{2}}), \tag{3.38}$$

where π_t is the probability that the tth unit is in the sample. Since $E_\xi[u_t(\mu_X, \mu_Y)] = 0$ for $t = 1,\ldots,n$, then $E_p(\bar{y}_s/\bar{x}_s)$, in turn, has model expectation

$$E_{\xi p}(\bar{y}_s/\bar{x}_s) = \theta + o(n^{-\frac{1}{2}}). \tag{3.39}$$

Therefore, the ξp-expectation of $\bar{y}_s/\bar{x}_s$ converges to the parameter of interest, θ, when the model holds.

Similarly, $\bar{y}_d/\bar{x}_d = \sum I_t d_t y_t / \sum I_t d_t x_t$, a design-based estimator of the ratio of the finite population means, may be linearised as

$$\frac{\bar{y}_d}{\bar{x}_d} - \frac{\bar{y}_U}{\bar{x}_U} = \sum I_t d_t u_t(\bar{x}_U, \bar{y}_U) + o_p(n^{-\frac{1}{2}}), \tag{3.40}$$

where $u_t(\bar{x}_U, \bar{y}_U)$ is defined analogously to $u_t(\mu_X, \mu_Y)$ in (3.36) for $t = 1,\ldots,N$. Since $\sum u_t(\bar{x}_U, \bar{y}_U) = 0$, the design-based expectation and variance of $\bar{y}_d/\bar{x}_d$ are

$$E_p(\bar{y}_d/\bar{x}_d) = \bar{y}_U/\bar{x}_U + \sum E_p(I_t d_t)u_t(\bar{x}_U, \bar{y}_U) + o(n^{-\frac{1}{2}})$$

$$= \bar{y}_U/\bar{x}_U + \frac{1}{N}\sum u_t(\bar{x}_U, \bar{y}_U) + o(n^{-\frac{1}{2}})$$

$$= \bar{y}_U/\bar{x}_U + o(n^{-\frac{1}{2}}), \tag{3.41}$$

and

$$var_p(\bar{y}_d/\bar{x}_d) = var_p\left[\sum I_t d_t u_t(\bar{x}_U, \bar{y}_U)\right] + o(n^{-1})$$

$$= \sum\sum \Delta_{tt'}u_t(\bar{x}_U, \bar{y}_U)u_{t'}(\bar{x}_U, \bar{y}_U) + o(n^{-1}). \tag{3.42}$$

We can now derive the total variance of $\bar{y}_d/\bar{x}_d$. Assuming that $var_\xi E_p(\bar{y}_d/\bar{x}_d) = O(N^{-1})$ and that n/N is negligible, we have

$$var_{\xi p}(\bar{y}_d/\bar{x}_d) = \sum\sum \Delta_{tt'}E_\xi[u_t(\bar{x}_U, \bar{y}_U)u_{t'}(\bar{x}_U, \bar{y}_U)] + o(n^{-1}). \tag{3.43}$$

Therefore, the design-based variance of $\bar{y}_d/\bar{x}_d$ is asymptotically model unbiased for its total variance.

3.4.3. Non-linear statistics – explicitly defined statistics

The same linearisation method can be used for a wide class of non-linear statistics. For example, suppose that we have a simple random sample of size n and that our estimator can be expressed as a differentiable function, g, of sample means, $\bar{y}_s = (\bar{y}_{1s}, \ldots, \bar{y}_{ms})'$. Letting μ_j be the model expectation of y_{tj}, where y_{tj} is the value of the jth variable for the tth unit, we have

$$g(\bar{y}_s) - g(\mu) = \sum_{j=1}^{m} \frac{\partial g}{\partial \mu_j} (\bar{y}_{js} - \mu_j) + o_\xi(n^{-\frac{1}{2}}), \tag{3.44}$$

where $\mu = (\mu_1, \ldots, \mu_m)'$. We define pseudo-variables, $u_t(\mu)$ for $t = 1, \ldots, N$ as

$$u_t(\mu) = \sum_{j=1}^{m} \frac{\partial g}{\partial \mu_j} (y_{tj} - \mu_j), \tag{3.45}$$

so that

$$g(\bar{y}_s) - g(\mu) = \sum I_t u_t(\mu)/n + o_\xi(n^{-\frac{1}{2}}). \tag{3.46}$$

This implies that inferences about $g(\bar{y}_s) - g(\mu)$ are asymptotically equivalent to inferences on $\bar{u}_s(\mu) \equiv \sum I_t u_t(\mu)/n$. For example,

$$E_p[g(\bar{y}_s)] = g(\mu) + \sum \pi_t u_t(\mu)/n + o_\xi(n^{-\frac{1}{2}}) \tag{3.47}$$

and

$$E_{\xi p}[g_s(\bar{y}_s)] = g(\mu) + o(n^{-\frac{1}{2}}) \tag{3.48}$$

since $E_\xi[u_t(\mu)] = 0$.

The design-based analogues of $\bar{y}_s$ and μ are $\bar{y}_d = (\bar{y}_{1d}, \ldots, \bar{y}_{md})'$ and $\bar{y}_U = (\bar{y}_{1U}, \ldots, \bar{y}_{mU})'$ respectively, where we denote the finite population means by $\bar{y}_{jU}$, for $j = 1, \ldots, m$, and we let $\bar{y}_{jd} \equiv \sum I_t d_t y_{tj}$ be the design-based estimator of $\bar{y}_{jU}$. We have

$$g(\bar{y}_d) - g(\bar{y}_U) = \sum I_t d_t u_t(\bar{y}_U) + o_p(n^{-\frac{1}{2}}). \tag{3.49}$$

Taking the design-based mean, we have

$$E_p[g(\bar{y}_d)] = g(\bar{y}_U) + \sum u_t(\bar{y}_U)/N + o(n^{-\frac{1}{2}})$$

$$= g(\bar{y}_U) + \sum_{j=1}^{m} \sum_{t=1}^{N} \frac{\partial g}{\partial \bar{y}_{jU}} (y_{tj} - \bar{y}_{jU})/N + o(n^{-\frac{1}{2}})$$

$$= g(\bar{y}_U) + o(n^{-\frac{1}{2}}) \tag{3.50}$$

and its model expectation is

$$E_{\xi p}[g(\bar{y}_d)] = E_\xi[g(\bar{y}_U)] + o(n^{-\frac{1}{2}})$$

$$= g(\mu) + o(n^{-\frac{1}{2}}). \tag{3.51}$$

For the design-based variance, we have

$$var_p[g(\bar{y}_d)] = var_p\left[\sum I_t d_t u_t(\bar{y}_U)\right] + o(n^{-1})$$

$$= \sum\sum \Delta_{tt'} u_t(\bar{y}_U) u_{t'}(\bar{y}_U) + o(n^{-1}). \qquad (3.52)$$

For the case where n/N is negligible, and assuming that $var_\xi E_p[g(\bar{y}_d)] = O(N^{-1})$, the total variance of $g(\bar{y}_d)$ is

$$var_{\xi p}[g(\bar{y}_d)] = \sum\sum \Delta_{tt'} u_t(\mu) u_{t'}(\mu) + o(n^{-1}) \qquad (3.53)$$

since, under the model, $\bar{y}_U = \mu + o_\xi(N^{-\frac{1}{2}})$. Therefore, the design-based variance of $g(\bar{y}_d)$ is asymptotically model unbiased for its total variance.

3.4.4.   Non-linear statistics – defined implicitly by score statistics

We can now extend our discussion to parameters that are defined implicitly through estimating equations. We suppose that, under the model, the vector-valued y_t are independent with density function given by $f_t(y_t; \beta)$. A model-based approach using maximum likelihood estimation, assuming ignorable sampling, yields $\hat{\beta}$ as the solution to the likelihood equations:

$$n^{-1}\sum I_t \frac{\partial \log f_t(y_t; \hat{\beta})}{\partial \hat{\beta}} = 0. \qquad (3.54)$$

Taking a first-order Taylor expansion of the left hand side of expression (3.54) around $\hat{\beta} = \beta$ and rearranging terms, we obtain

$$n^{-1}\sum I_t \frac{\partial^2 \log f_t(y_t; \beta)}{\partial\beta\partial\beta'}(\hat{\beta} - \beta) = -n^{-1}\sum I_t \frac{\partial \log f_t(y_t; \beta)}{\partial\beta} + o_\xi(n^{-\frac{1}{2}}). \quad (3.55)$$

Since, for most models, we have

$$n^{-1}\left[\sum I_t \frac{\partial^2 \log f_t(y_t; \beta)}{\partial\beta\partial\beta'}\right] = O_\xi(1), \qquad (3.56)$$

it follows that

$$\hat{\beta} - \beta = -\left[\frac{\sum I_t \partial^2 \log f_t(y_t; \beta)}{n\,\partial\beta\partial\beta'}\right]^{-1}\sum I_t \frac{\partial \log f_t(y_t; \beta)}{n\partial\beta} + o_\xi(n^{-\frac{1}{2}}). \qquad (3.57)$$

Therefore, by defining vector-valued pseudo-variables,

$$u_t(\beta) = -\left[\sum I_t \frac{\partial^2 \log f_t(y_t; \beta)}{n\,\partial\beta\partial\beta'}\right]^{-1}\frac{\partial \log f_t(y_t; \beta)}{\partial\beta}, \qquad (3.58)$$

we see that the sample mean of these pseudo-variables has the same asymptotic expectation and variance as $\hat{\beta} - \beta$. In particular, we have

$$\hat{\beta} - \beta = \sum I_t u_t(\beta)/n + o_\xi(n^{-\frac{1}{2}}). \qquad (3.59)$$

This implies that inferences about $\hat{\beta} - \beta$ are asymptotically equivalent to inferences on $\bar{u}_s(\beta) \equiv \sum I_t u_t(\beta)/n$. For example, we have

$$E_p(\hat{\beta} - \beta) = \sum \pi_t u_t(\beta)/n + o_\xi(n^{-\frac{1}{2}}) \tag{3.60}$$

and

$$E_{\xi p}(\hat{\beta} - \beta) = o(n^{-\frac{1}{2}}), \tag{3.61}$$

when we assume that $E_\xi[u_t(\beta)] = 0$, which is one of the usual regularity conditions for maximum likelihood estimation.

For the case of design-based inference, following the approach in Binder (1983), our finite population parameter of interest is b, the solution to the likelihood equations (3.54) when all the population units are observed. Our estimate, $\hat{b}$, of the finite population parameter of interest, b, is the solution to the design-based version of (3.54), given by

$$\sum I_t d_t \frac{\partial \log f_t(y_t; \hat{b})}{\partial \hat{b}} = 0. \tag{3.62}$$

Taking a Taylor expansion of the left hand side of (3.62) around $\hat{b} = b$, we obtain

$$\hat{b} - b = \sum I_t d_t u_t(b) + o_p(n^{-\frac{1}{2}}), \tag{3.63}$$

where

$$u_t(b) = -\left[\sum I_t d_t \frac{\partial^2 \log f_t(y_t; b)}{\partial b \partial b'}\right]^{-1} \frac{\partial \log f_t(y_t; b)}{\partial b}. \tag{3.64}$$

Since b is the maximum likelihood estimate of β, defined analogously to (3.54) for the finite population, we have the design expectation of $\hat{b}$, given by

$$E_p(\hat{b}) = b + \sum u_t(b)/N + o(n^{-\frac{1}{2}}) = b + o(n^{-\frac{1}{2}}), \tag{3.65}$$

which has model expectation given by

$$\begin{aligned} E_{\xi p}(\hat{b}) &= E_\xi(b) + o(n^{-\frac{1}{2}}) \\ &= \beta + o(n^{-\frac{1}{2}}). \end{aligned} \tag{3.66}$$

For the design-based variance, we have

$$\begin{aligned} var_p(\hat{b}) &= var_p\left[\sum I_t d_t u_t(b)\right] + o(n^{-1}) \\ &= \sum \sum \Delta_{tt'} u_t(b) u_{t'}(b)' + o(n^{-1}). \end{aligned} \tag{3.67}$$

We can now derive the total variance matrix of $\hat{b}$. For the case where n/N is negligible, we have

$$var_{\xi p}(\hat{b}) = \sum \sum \Delta_{tt'} E_\xi[u_t(\beta) u_{t'}(\beta)'] + o(n^{-1}), \tag{3.68}$$

since $b = \beta + o(N^{-\frac{1}{2}})$ and it has been assumed that $var_\xi E_p(\hat{b}) = O(N^{-1})$.

We see, therefore, that the discussion in Section 3.3 can often be extended to non-linear parameters and estimators. This implies that the ξp-expectations of both $\hat{b}$ and $\hat{\beta}$ are often equal to the model parameter of interest and that the design-based variance matrix of $\hat{b}$ is asymptotically model unbiased for its total variance matrix.

3.4.5. Total variance matrix of $\hat{b}$ for non-negligible sampling fractions

Much of the discussion with respect to the behaviour of the sampling variances and covariances has assumed that the sampling fractions are negligible. We conclude this section with an examination of the total variance matrix of $\hat{b}$ when the sampling fractions are not small.

We suppose that $\hat{b}$ has the property that, conditional on the realised finite population, the asymptotic design-based sampling distribution of $\sqrt{n}(\hat{b} - b)$ has mean 0 and has variance matrix $var_p[\sqrt{n}(\hat{b} - b)]$, which is $O(1)$. Now, conditional on the finite population, the asymptotic design-based sampling distribution of $\sqrt{n}(\hat{b} - \beta)$ has mean $E_p[\sqrt{n}(\hat{b} - \beta)] = \sqrt{n}(b - \beta)$ and variance matrix $var_p[\sqrt{n}(\hat{b} - b)]$. As a result, the asymptotic distribution of $\sqrt{n}(\hat{b} - \beta)$ has total variance matrix

$$var_{\xi p}\left[\sqrt{n}(\hat{b} - \beta)\right] = E_\xi var_p\left[\sqrt{n}(\hat{b} - b)\right] + (n/N)var_\xi\left[\sqrt{N}(b - \beta)\right]. \quad (3.69)$$

For most models, $var_\xi[\sqrt{N}(b - \beta)]$ is $O(1)$. We can thus see that when the sampling fraction, n/N, is small, the second term in Equation (3.69) may be ignored, and an asymptotically design-unbiased estimator of $var_p[\sqrt{n}(\hat{b} - b)]$ is asymptotically model unbiased for $var_{\xi p}[\sqrt{n}(\hat{b} - \beta)]$. However, when n/N is not negligible – for example, in situations where the sampling fractions are large in at least some strata – $var_\xi[\sqrt{N}(b - \beta)]$ must also be estimated in order to estimate the total variance matrix of $\hat{b}$. Model-based methods must be used to estimate this term since the design-based variance matrix of the finite population constant, b, is zero. Examples of correcting the estimates of total variance under more complex survey designs are given in Korn and Graubard (1998a).

3.5. CONDITIONAL MODEL-BASED PROPERTIES

3.5.1. Conditional model-based properties of $\hat{b}$

Many model-based analysts would not consider the ξp-expectation or the total variance as the relevant quantities of interest for inferences, since they assume that the sample design is ignorable; see Rubin (1976), Scott (1977a), Sugden and Smith (1984) and Pfeffermann (1993) for discussions on ignorable sample designs. Since the model-based properties of $\hat{\beta}$ are discussed at length in the standard literature, we focus here, instead, on the properties of $\hat{b}$ from the perspective of these model-based analysts. We thus consider the expectations and variances conditional on the sample design; that is, conditional on the random variables I_t $(t = 1, \ldots, N)$, and, therefore, implicitly conditional on auxiliary variables associated with the selected units.

3.5.2. Conditional model-based expectations

Considering the I_t to be fixed quantities, so that we are conditioning on the randomisation under the design, we have

$$E_\xi(\hat{b}) = E_\xi\left[\sum I_t d_t y_t\right] = \sum I_t d_t \mu_t. \tag{3.70}$$

For large samples, by the strong law of large numbers (under the design), $E_\xi(\hat{b})$ converges to its design expectation. Therefore,

$$\begin{aligned} E_\xi(\hat{b}) &= E_p[E_\xi(\hat{b})] + o(1) \\ &= E_\xi[E_p(\hat{b})] + o(1) \\ &= E_\xi[\bar{y}_U] + o(1) \\ &= \sum \mu_t/N + o(1). \end{aligned} \tag{3.71}$$

We see that if $\sum \mu_t/N = \sum I_t\mu_t/n + o(1)$, as is normally the case, then $\hat{b}$ is asymptotically conditionally model unbiased for β.

3.5.3. Conditional model-based variance for $\hat{b}$ and the use of estimating functions

The model-based variance for $\hat{b}$ is given by

$$var_\xi(\hat{b}) = var_\xi\left(\sum I_t d_t y_t\right) = \sum\sum I_t I_{t'} d_t d_{t'} \sigma_{tt'}. \tag{3.72}$$

If we make the further assumption that, for large samples, $var_\xi(\hat{b})$ converges to its design expectation, that is $var_\xi(\hat{b}) = E_p var_\xi(\hat{b}) + o(n^{-1})$, then we have

$$\begin{aligned} var_\xi(\hat{b}) &= \sum\sum E_p(I_t d_t I_{t'} d_{t'})\sigma_{tt'} + o(n^{-1}) \\ &= \sum\sum \Delta_{tt'}\sigma_{tt'} + o(n^{-1}). \end{aligned} \tag{3.73}$$

Now, comparing (3.27) and (3.73), we see that if $\sum\sum \mu_t\mu_{t'}\Delta_{tt'} = o(n^{-1})$, then the total variance of $\hat{b}$, given by $var_{\xi p}(\hat{b})$, and the model-based variance of $\hat{b}$, given by $var_\xi(\hat{b})$, are both asymptotically equal to $\sum\sum \Delta_{tt'}\sigma_{tt'}$. From (3.27) we have that $var_{\xi p}(\hat{b}) = E_\xi var_p(\hat{b}) + o(n^{-1})$. Therefore, if, under the model, the design-based variance of $\hat{b}$, given by $var_p(\hat{b})$, converges to its model expectation, $E_\xi var_p(\hat{b})$, then we also have that $var_p(\hat{b})$ is asymptotically equal to $\sum\sum \Delta_{tt'}\sigma_{tt'}$ when $\sum\sum \mu_t\mu_{t'}\Delta_{tt'} = o(n^{-1})$.

In general, the condition, $\sum\sum \mu_t\mu_{t'}\Delta_{tt'} = o(n^{-1})$, depends on both the model and the design. For example, if $\mu_t = \mu$ for all t and if we have a fixed sample size design, then we have the even stronger condition that $\sum\sum \mu_t\mu_{t'}\Delta_{tt'} = 0$.

An important case where this stronger condition holds regardless of the sample design is where all the μ_t are zero. We examine how to achieve this case by putting the estimation problem into an estimating function framework.

Instead of considering the population values, $y_1, \ldots, y_N$, we consider the quantities

$$u_t(\mu_t) \equiv y_t - \mu_t \ (t = 1, \ldots, N). \tag{3.74}$$

Under the model, the u_t have mean 0 and have the same variance matrix as the y_t given by

$$cov_\xi[u_t(\mu_t), u_{t'}(\mu_{t'})] = \sigma_{tt'}. \tag{3.75}$$

Note that we make u_t explicitly a function of μ_t since the u_t are not directly observable when the μ_t are unknown. We define

$$\bar{u}_d(\mu) \equiv \sum I_t d_t u_t(\mu_t), \tag{3.76}$$

where $\mu = (\mu_1, \ldots, \mu_N)'$. From our results in (3.27) and (3.73), we have that

$$var_{\xi p}[\bar{u}_d(\mu)] = var_\xi[\bar{u}_d(\mu)] + o(n^{-1}) = \sum\sum \Delta_{tt'}\sigma_{tt'} + o(n^{-1}), \tag{3.77}$$

because $E_\xi[u_t(\mu_t)] = 0 \ (t = 1, \ldots, N)$. Therefore, an asymptotically model-unbiased estimator of $var_{\xi p}[\bar{u}_d(\mu)]$ is asymptotically model unbiased for $var_\xi[\bar{u}_d(\mu)]$.

From (3.23), the design-based variance of $\bar{u}_d$ is given by

$$var_p[\bar{u}_d(\mu)] = var_p\left[\sum I_t d_t u_t(\mu_t)\right] = \sum\sum \Delta_{tt'} u_t(\mu_t) u_{t'}(\mu_{t'}), \tag{3.78}$$

which has model expectation $\sum\sum \Delta_{tt'}\sigma_{tt'}$. Assuming that a design-based estimator, $v_p[\bar{u}_d(\mu)]$, converges to $var_p[\bar{u}_d(\mu)]$, and that $var_p[\bar{u}_d(\mu)]$ converges to its model expectation, then the model expectation of $v_p[\bar{u}_d(\mu)]$ would also be asymptotically equal to $var_\xi[\bar{u}_d(\mu)]$.

Generally, there are many choices for obtaining design-based estimators for $var_p[\bar{u}_d(\mu)]$, as reviewed in Wolter (1985), for example. However, to use $v_p[\bar{u}_d(\mu)]$, the μ_t would usually need to be estimated. We denote by $v_p[\bar{u}_d(\hat{\mu})]$ the design-based estimator of the design-based variance of $\bar{u}_d(\mu)$ evaluated at $\hat{\mu}$. We choose $\hat{\mu}$ such that $v_p[\bar{u}_d(\hat{\mu})]$ is asymptotically design unbiased for $var_p[\bar{u}_d(\mu)]$, and is, therefore, asymptotically model unbiased for $var_p[\bar{u}_d(\mu)]$.

From the discussion above, we see that, in practice, $v_p[\bar{u}_d(\hat{\mu})]$ provides an estimate of the variance of $\bar{u}_d(\mu)$ for a wide class of inference problems, both in the design-based and in the model-based frameworks. Generally, even when the μ_t are not all zero, the design-based variance of $\bar{u}_d(\mu)$ is asymptotically equal to its model-based counterpart, providing the model is valid. However, this equality does not necessarily hold for design-based and model-based variances of $\hat{b}$. Fortunately, as we have seen in Section 3.4, for many situations where we are estimating model parameters, we have all the μ_t equal to zero after linearisation. In these situations, the design-based approach has appropriate model-based properties.

3.6.   PROPERTIES OF METHODS WHEN THE ASSUMED MODEL IS INVALID

We now examine the properties of design-based and model-based methods when the model we have assumed is violated. When the true model has been misspecified, there are two important elements to consider. These are whether the point estimate of β is consistent under the true model, or at least asymptotically model unbiased, and whether the estimated variance of this point estimate is asymptotically model unbiased under the true model.

3.6.1.   Critical model assumptions

We made some critical model assumptions in Sections 3.2 and 3.3. First, we assumed that the sample design was ignorable; as well, we assumed a certain structure for the model means and covariances of the observations. When the sampling is not ignorable, it is no longer appropriate to assume that the parameter of interest, $\beta \equiv \sum I_t \mu_t / n$, is the same as $E_\xi(\sum I_t y_t / n)$, since the conditional expected value of y_t, given $I_t = 1$, under the model, is not necessarily equal to $\mu_t = E_\xi(y_t)$. However, we still suppose, as we did in the ignorable case (3.6), that β converges to the model expectation of the finite population mean, $\bar{y}_U$; that is,

$$\beta = \sum \mu_t / N + o(1). \tag{3.79}$$

3.6.2.   Model-based properties of $\hat{b}$

We first consider the model-based properties of the estimator, $\hat{b}$, under the true model. We have already seen that $\hat{b}$ is asymptotically model unbiased for the true β; see (3.71) and (3.79). Also, we have shown that the design-based estimator of the design variance of $\hat{b}$ provides a reasonable estimator for the total variance of $\hat{b}$, when the sample is large and the sampling fraction is small. However, we mentioned in Section 3.5 that, since under certain conditions the design-based variance of $\bar{u}_d(\mu)$ is asymptotically equal to its model-based variance, and since the model-based variances of $\bar{u}_d(\mu)$ and $\hat{b}$ are asymptotically equal, it may be more appropriate to estimate the model variance of $\hat{b}$ using the estimating equation approach, described in Section 3.5.3, rather than by deriving the model variance explicitly. This would still be true when the assumed model is incorrect, provided the structure for the means of the y_t is correctly specified in the assumed model, even if the sample design is nonignorable. (For nonignorable sample designs we consider the means conditional on which units have been selected in the sample.) If, however, the structure for the means is incorrectly specified, then the design-based variance of $\bar{u}_d(\mu)$ is not, in general, asymptotically equal to its model variance, so that a design-based variance estimate of $\bar{u}_d(\mu)$ may not be a good estimate of the model variance of $\hat{b}$.

3.6.3. Model-based properties of $\hat{\beta}$

Since the model-based estimator, $\hat{\beta} = \sum I_t c_t y_t$, has been derived using model considerations, such as best linear unbiasedness, we now consider the situations where $\hat{b}$ would be preferred because some of the assumed properties of $\hat{\beta}$ and of its estimated variance are incorrect. In particular, we consider the following three cases in turn:

(i)   the sample design is nonignorable;
(ii) the structure for the model means given by the μ_t is incorrect; and
(iii) the structure for the model variances given by the $\sigma_{tt'}$ is incorrect.

First, we consider the case where the sample design is nonignorable. Under the true model, the expectation of $\hat{\beta}$, conditional on the units selected in the sample, is given by

$$E_\xi(\hat{\beta}|I_U) = \sum E_\xi(I_t c_t y_t|I_U) = \sum I_t c_t \mu_{t1}, \qquad (3.80)$$

where $I_U = (I_1, \ldots, I_N)'$, $\mu_{t1} = E_\xi[y_t|I_t = 1]$, provided that the c_t depend only on the units included in the sample and not on the y-values. When the sample design is nonignorable, $\mu_{t1} \neq \mu_t$ in general, so that $\hat{\beta}$ may be asymptotically biased under the model.

Example 4

Pfeffermann (1996) gives the following artificial example. Suppose that the finite population consists of zeros and ones where, under the model, $Pr(y_t = 1) = \mu$, with $0 < \mu < 1$. The parameter of interest, β, is equal to $E_\xi(\bar{y}_U) = \mu$. We now assume that for each unit where $y_t = 0$, the probability of inclusion in the sample is p_0, but for each unit where $y_t = 1$, the probability of inclusion in the sample is p_1. Therefore, using Bayes' theorem, we have

$$\mu_{t1} = Pr(y_t = 1|I_t = 1) = \frac{\mu p_1}{(1 - \mu)p_0 + \mu p_1}, \qquad (3.81)$$

which is not necessarily equal to μ. Under this nonignorable design, the model bias of the sample mean, $\bar{y}_s$, as an estimate of μ is

$$\mu_{t1} - \mu = \frac{\mu(1 - \mu)(p_1 - p_0)}{(1 - \mu)p_0 + \mu p_1}. \qquad (3.82)$$

We see, therefore, that the sample mean, $\bar{y}_s$, is model biased and model inconsistent for μ under the true model if $p_0 \neq p_1$. (A modeller who knows about the non-ignorability of the inclusion probabilities would use this model information and would not normally use $\bar{y}_s$ to estimate μ.)

We now discuss the second case, where the structure for the assumed model means given by the μ_t is incorrect. The condition for $\hat{\beta}$ to be asymptotically unbiased for β under the true model is

$$\sum I_t c_t \mu_t = \sum I_t \mu_t / n + o_p(1). \tag{3.83}$$

We now give an example where a bias would be introduced when the model means are incorrect.

Example 5

Suppose that there are two observations, y_1 and y_2, and that the modeller assumes that y_1 and y_2 have the same mean and have variance matrix

$$\frac{\sigma^2}{n} \begin{bmatrix} 1 & 0 \\ 0 & 4 \end{bmatrix}. \tag{3.84}$$

In this case, the best linear model-unbiased estimator of the common mean is given by $0.8y_1 + 0.2y_2$. If, however, the true means, $\mu_t = E_\xi(y_t)$, for $t = 1, 2$, are different, then the bias of this estimator in estimating $\beta = (\mu_1 + \mu_2)/2$ is $0.3(\mu_1 - \mu_2)$ and $\hat{\beta}$ is model inconsistent.

Some model-based methods to protect against biases when the model mean structure is incorrectly specified have been suggested in the literature; see, for example, the application of non-parametric calibration in a model-based setting to estimate finite population characteristics, given in Chambers, Dorfman and Wehrly (1993).

Finally, we consider the third case where the model variance structure is incorrectly specified. Here, the estimator for the variance of $\hat{\beta}$ under the model may be biased and inconsistent. The following is a simple example of this case.

Example 6

Suppose that, under the assumed model, the observations, $y_1, \ldots, y_n$, are independent and identically distributed Poisson random variables with mean μ. The usual estimate for μ is the sample mean, $\bar{y}_s$, which has variance μ/n under the Poisson model. Therefore, a commonly used model-based estimator of the variance of $\bar{y}_s$ is $\bar{y}_s/n$. Clearly, this can be biased if the Poisson model is incorrect and the true variance of the y_t is not μ.

The statistical literature contains many model-based examples of model-based methods for robust estimation of variances which address this issue: see, for example, McCullagh and Nelder (1983), Royall (1986) and Liang and Zeger (1986). Many of these methods are analogous to the design-based approach that we have been discussing.

3.6.4. Summary

In summary, when our model parameter of interest, β, is asymptotically equal to $\sum \mu_t / N$, we find that the model-based estimator, $\hat{\beta}$, may have poor properties when some of the model assumptions are violated. On the other hand, $\hat{b}$ is asymptotically model unbiased for β, and the design-based estimator of its design variance provides a reasonable estimator for the total variance of $\hat{b}$, when the sample is large and the sampling fraction is small, even when the model assumptions are incorrect. This provides some rationale for preferring $\hat{b}$ over $\hat{\beta}$, and for using a design-based estimator for the variance of $\hat{b}$. If the assumed model is true, then the design-based estimator, $\hat{b}$, may have larger model variance than the model-based estimator, $\hat{\beta}$. However, for large sample sizes, this loss of efficiency may not be serious, compared with the gain in robustness. As well, the design-based estimator of the variance of $\bar{u}_d(\mu)$ provides an asymptotically model-unbiased estimate of $var_\xi(\hat{b})$ provided that the conditional means of the y_t (conditional on having been selected into the sample) are correctly specified. This is the case even when the model variance structure is incorrectly specified.

3.7. CONCLUSION

The choice between model-based and design-based estimation should be based on whether or not the assumed model is true. If the model could have been misspecified, and if a model-based approach has been taken, one needs to consider carefully what is being estimated and whether or not the variance estimators are robust to misspecification. On the other hand, the corresponding design-based estimates tend to give valid inferences, even in some cases when the model is misspecified. However, the use of design-based methods tends to lead to a loss of efficiency when the assumed model is indeed true. For very large samples, this loss may not be too serious, given the extra robustness achieved from the design-based approach. When the sample size is not large or when the sampling fractions are non-negligible, some modification of a pure design-based approach would be necessary.

We have proposed the use of an estimating function framework to obtain variance estimates that can be justified under either a design-based or model-based framework. These variance estimates use design-based methods applied to estimating functions which are obtained through the model mean structure. We believe this approach would have wide appeal for many analysts.

CHAPTER 4

The Bayesian Approach to Sample Survey Inference

Roderick J. Little

4.1. INTRODUCTION

This chapter outlines the Bayesian approach to statistical inference for finite population surveys (Ericson, 1969, 1988; Basu, 1971; Scott, 1977b; Binder, 1982; Rubin, 1983, 1987; Ghosh and Meeden, 1997). Chapter 18 in this volume applies this perspective to the analysis of survey data subject to unit and item nonresponse. In the context of this conference honoring the retirement of Fred Smith, it is a pleasure to note his many important contributions to the model-based approach to survey inference over the years. His landmark paper with Scott was the first to describe random effects models for the analysis of clustered data (Scott and Smith, 1969). His paper with Holt (Holt and Smith, 1979) revealed the difficulties with the randomization-based approach to post-stratification, and his early assessment of the historical development of survey inference favored a model-based approach (Smith, 1976). The difficulties of formulating realistic models for human populations, and more recent work by Smith and his colleagues on effects of model misspecification (for example, see Holt, Smith and Winter, 1980; Pfeffermann and Holmes, 1985), have led Smith to a pragmatic approach that recognizes the strengths of the randomization approach for problems of descriptive inference while favoring models for analytic survey inference (Smith, 1994). My own view is that careful model specification, sensitive to the survey design, can address the concerns with model misspecification, and that Bayesian statistics provide a coherent and unified treatment of descriptive and analytic survey inference. Three main approaches to finite population inference can be distinguished (Little and Rubin, 1983):

1. Design-based inference, where probability statements are based on the distribution of sample selection, with the population quantities treated as fixed.

Analysis of Survey Data Edited by R. L. Chambers and C. J. Skinner
© 2003 John Wiley & Sons, Ltd

2. Frequentist superpopulation modeling, where the population values are assumed to be generated from a superpopulation model that conditions on fixed superpopulation parameters, and inferences are based on repeated sampling from this model. For analytic inference about superpopulation parameters, this is standard frequentist model-based parametric inference, with features of the complex sample design reflected in the model as appropriate. For descriptive inference, the emphasis is on prediction of nonsampled values and hence of finite population quantities.

3. Bayesian inference, where the population values are assigned a prior distribution, and the posterior distribution of finite population quantities is computed using Bayes' theorem. In practice, the prior is usually specified by formulating a parametric superpopulation model for the population values, and then assuming a prior distribution for the parameters, as detailed below. When the prior is relatively weak and dispersed compared with the likelihood, as when the sample size is large, this form of Bayesian inference is closely related to frequentist superpopulation modeling, except that a prior for the parameters is added to the model specification, and inferences are based on the posterior distribution rather than on repeated sampling from the superpopulation model.

Attractive features of the Bayesian approach including the following:

*   Bayesian methods provide a unified framework for addressing all the problems of survey inference, including inferences for descriptive or analytical estimands, small or large sample sizes, inference under planned, ignorable sample selection methods such as probability sampling, and problems where modeling assumptions play a more central role such as missing data or measurement error. Many survey statisticians favor design-based inference for some problems, such as large-sample descriptive inference for means and totals, and model-based inference for other problems, such as small-area estimation or nonresponse. The adoption of two potentially contradictory modes of inference is confusing, and impedes communication with substantive researchers such as economists who are accustomed to models and have difficulty reconciling model-based and randomization-based inferences. Consider, for example, the issue of weighting in regression models, and in particular the difficulties of explaining to substantive researchers the difference between weights based on the inverse of the sampling probability and weights for nonconstant variance (Konijn, 1962; Brewer and Mellor, 1973; Dumouchel and Duncan, 1983; Smith, 1988; Little, 1991; Pfeffermann, 1993).

*   As illustrated in Examples 1 and 2 below, many standard design-based inferences can be derived from the Bayesian perspective, using classical models with noninformative prior distributions. Hence tried and tested methods do not need to be abandoned. In general, Bayesian inferences under a carefully chosen model should have good frequentist properties,

and a method derived under the design-based paradigm does not become any less robust under a Bayesian etiology.

- Bayesian inference allows for prior information about a problem to be incorporated in the analysis in a simple and clear way, via the prior distribution. In surveys, noninformative priors are generally chosen for fixed parameters, reflecting absence of strong prior information; but multilevel models with random parameters also play an important role for problems such as small-area estimation. Informative priors are generally shunned in survey setting, but may play a useful role for some problems. For example, the treatment of outliers in repeated cross-sectional surveys is problematic, since information about the tails of distributions in a single survey is generally very limited. A Bayesian analysis can allow information about the tails of distributions to be pooled from prior surveys and incorporated into the analysis of the current survey via a proper prior distribution, with dispersion added to compensate for differences between current and prior surveys.

- With large samples and relatively dispersed priors, Bayesian inference yields similar estimates and estimates of precision to frequentist superpopulation modeling under the same model, although confidence intervals are replaced by posterior probability intervals and hence have a somewhat different interpretation. Bayes also provides unique solutions to small-sample problems under clearly specified assumptions where there are no exact frequentist answers, and the choice between approximate frequentist solutions is ambiguous. Examples include whether to condition on the margins in tests for independence in 2×2 tables (see, for example, Little, 1989), and in the survey context, inference from post-stratified survey data, where there is ambiguity in the frequentist approaches concerning whether or not to condition on the post-stratum counts (Holt and Smith, 1979; Little, 1993).

- Bayesian inference deals with nuisance parameters in a natural and appealing way.

- Bayesian inference satisfies the likelihood principle, unlike frequentist superpopulation modeling or design-based inference.

- Modern computation tools make Bayesian analysis much more practically feasible than in the past (for example, Tanner, 1996).

Perhaps the main barrier to the use of Bayesian methods in surveys is the perceived 'subjectivity' involved in the choice of a model and prior, and the problems stemming from misspecified models. It is true that Bayes lacks built-in 'quality control,' and may be disastrous if the prior or model is poorly specified. Frequentist superpopulation modeling avoids the specification of the prior, but is also vulnerable to misspecification of the model, which is often the more serious issue, particularly when samples are large. The design-based approach, which is based on the known probability distribution of sample selection and avoids overt specification of a model for the survey outcomes, is less vulnerable to gross errors of misspecification. However,

Bayesian models with noninformative priors can be formulated that lead to results equivalent to those of standard large-sample design-based inference, and these results retain their robustness properties regardless of their inferential origin. I believe that with carefully specified models that account for the sample design, Bayesian methods can enjoy good frequentist properties (Rubin, 1984). I also believe that the design-based approach lacks the flexibility to handle surveys with small samples and missing data. For such problems, the Bayesian approach under carefully chosen models yields principled approaches to imputation and smoothing based on clearly specified assumptions.

A common misconception of the Bayesian approach to survey inference is that it does not affirm the importance of random sampling, since the model rather than the selection mechanism is the basis for Bayesian probability statements. However, random selection plays an important role in the Bayesian approach in affording protection against the effects of model misspecification and biased selection. The role is illuminated by considering models that incorporate the inclusion indicators as well as the survey outcomes, as discussed in the next section. In Chapter 18 this formulation is extended to handle missing data, leading to Bayesian methods for unit and item nonresponse.

4.2.   MODELING THE SELECTION MECHANISM

I now discuss how a model for the selection mechanism is included within the Bayesian framework. This formulation illuminates the role of random sampling within the Bayesian inferential framework. The following account is based on Gelman *et al.* (1995). Following notation in Chapters 1 and 2, let

$y_U=(y_{tj})$, y_{tj} = value of survey variable j for unit t, $j = 1, \ldots, J; t \in U = \{1, \ldots, N\}$

$Q = Q(y_U)$ = finite population quantity

$i_U = (i_1, \ldots, i_N)$ = sample inclusion indicators; $i_t = 1$ if unit t included, 0 otherwise

$y_U=(y_{inc}, y_{exc})$, y_{inc} = included part of y_U, y_{exc} = excluded part of y_U

z_U = fully observed covariates, design variables.

We assume for simplicity that design variables are recorded and available for analysis; if not, then z_U denotes the subset of recorded variables. The expanded Bayesian approach specifies a model for both the survey data y_U and the sample inclusion indicators i_U. The model can be formulated as

$$p(y_U, i_U | z_U, \theta, \phi) = p(y_U | z_U, \theta) \times p(i_U | z_U, y_U, \phi),$$

where θ, ϕ are unknown parameters indexing the distributions of y_U and i_U given y_U, respectively. The likelihood of θ, ϕ based on the observed data (z_U, y_{inc}, i_U) is then

$$L(\theta, \phi | z_U, y_{inc}, i_U) \propto p(y_{inc}, i_U | z_U, \theta, \phi) = \int p(y_U, i_U | z_U, \theta, \phi) dy_{exc}.$$

In the Bayesian approach the parameters (θ, ϕ) are assigned a prior distribution $p(\theta, \phi | z_U)$. Analytical inference about the parameters is based on the posterior distribution:

$$p(\theta, \phi | z_U, y_{inc}, i_U) \propto p(\theta, \phi | z_U) L(\theta, \phi | z_U, y_{inc}, i_U).$$

Descriptive inference about $Q = Q(y_U)$ is based on its posterior distribution given the data, which is conveniently derived by first conditioning on, and then integrating over, the parameters:

$$p(Q(y_U) | z_U, y_{inc}, i_U) = \int p(Q(y_U) | z_U, y_{inc}, i_U, \theta, \phi) p(\theta, \phi | z_U, y_{inc}, i_U) d\theta d\phi.$$

The more usual likelihood does not include the inclusion indicators i_U as part of the model. Specifically, the likelihood *ignoring the selection process* is based on the model for y_U alone:

$$L(\theta | z_U, y_{inc}) \propto p(y_{inc} | z_U, \theta) = \int p(y_U | z_U, \theta) dy_{exc}.$$

The corresponding posterior distributions for θ and $Q(y_U)$ are

$$p(\theta | z_U, y_{inc}) \propto p(\theta | z_U) L(\theta | z_U, y_{inc})$$

$$p(Q(y_U) | z_U, y_{inc}) = \int p(Q(y_U) | z_U, y_{inc}, \theta) p(\theta | z_U, y_{inc}) d\theta.$$

When the full posterior reduces to this simpler posterior, the selection mechanism is called *ignorable* for Bayesian inference. Applying Rubin's (1976) theory, two general and simple conditions for ignoring the selection mechanism are:

Selection at random (SAR): $p(i_U | z_U, y_U, \phi) = p(i_U | z_U, y_{inc}, \phi)$ for all y_{exc}.

Bayesian distinctness: $p(\theta, \phi | z_U) = p(\theta | z_U) p(\phi | z_U)$.

It is easy to show that these conditions together imply that

$$p(\theta | z_U, y_{inc}) = p(\theta | z_U, y_{inc}, i_U) \text{ and } p(Q(y_U) | z_U, y_{inc}) = p(Q(y_U) | z_U, y_{inc}, i_U),$$

so the model for the selection mechanism does not affect inferences about θ and $Q(y_U)$.

Probability sample designs are generally both ignorable and known, in the sense that

$$p(i_U | z_U, y_U, \phi) = p(i_U | z_U, y_{inc}),$$

where z_U represents known sample design information, such as clustering or stratification information. In fact, many sample designs depend only on z_U and not on y_{inc}; an exception is double sampling, where a sample of units is obtained, and inclusion into a second phase of questions is restricted to a subsample with a design that depends on characteristics measured in the first phase.

Example 1. Simple random sampling

The distribution of the simple random sampling selection mechanism is

$$p(i_U|z_U, y_U, \phi) = \begin{cases} \binom{N}{n}^{-1}, & \text{if } \Sigma_{t=1}^{N} i_t = n; \\ 0, & \text{otherwise.} \end{cases}$$

This mechanism does not depend on y_U or unknown parameters, and hence is ignorable. A basic model for a single continuous survey outcome with simple random sampling is

$$[y_t|\theta, \sigma^2] \sim_{ind} N(\theta, \sigma^2), \tag{4.1}$$

where $N(a, b)$ denotes the normal distribution with mean a, variance b. For simplicity assume initially that σ^2 is known and assign an improper uniform prior for the mean:

$$p(\theta|z_U) \propto const. \tag{4.2}$$

Standard Bayesian calculations (Gelman *et al.*, 1995) then yield the posterior distribution of θ to be $N(\bar{y}_s, \sigma^2/n)$, where $\bar{y}_s$ is the sample mean. To derive the posterior distribution of the population mean $\bar{y}_U$, note that the posterior distribution of $\bar{y}_U$ given the data and θ is normal with mean $(n\bar{y}_s + (N-n)\theta)/N$ and variance $(N-n)\sigma^2/N^2$. Integrating over θ, the posterior distribution of $\bar{y}_U$ given the data is normal with mean

$$(n\bar{y}_s + (N-n)E(\theta|\bar{y}_s))/N = (n\bar{y}_s + (N-n)\bar{y}_s)/N = \bar{y}_s$$

and variance

$$E[Var(\bar{y}_U|\bar{y}_s, \theta)] + Var[E(\bar{y}_U|\bar{y}_s, \theta)] = (N-n)\sigma^2/N^2$$
$$+(1-n/N)^2\sigma^2/n = (1-n/N)\sigma^2/n.$$

Hence a 95% posterior probability interval for $\bar{y}_U$ is $\bar{y}_s \pm 1.96\sqrt{\sigma^2(1-n/N)/n}$, which is identical to the 95% confidence interval from design-based theory for a simple random sample. This correspondence between Bayes' and design-based results also applies asymptotically to nonparametric multinomial models with Dirichlet priors (Scott, 1977b; Binder, 1982).

With σ^2 unknown, the standard design-based approach estimates σ^2 by the sample variance s^2, and assumes large samples. The Bayesian approach yields small sample t corrections, under normality assumptions. In particular, if the variance is assigned Jeffreys' prior $p(\sigma^2) \propto 1/\sigma^2$, the posterior distribution of $\bar{y}_U$ is Student's t with mean $\bar{y}_s$, scale $\sqrt{s^2(1-n/N)/n}$ and degrees of freedom $n-1$. The resulting 95% posterior probability interval is $\bar{y} \pm t_{n-1, 0.975}$ $\sqrt{s^2(1-n/N)/n}$, where $t_{n-1, 0.975}$ is the 97.5th percentile of the t distribution with $n - 1$ degrees of freedom. Note that the Bayesian approach automatically incorporates the finite population correction $(1 - n/N)$ in the inference.

Example 2. Stratified random sampling

More generally, under stratified random sampling the population is divided into H strata and n_h units are selected from the population of N_h units in stratum h. Define z_U as a set of stratum indicators, with components

$$z_t = \begin{cases} 1, & \text{if unit } t \text{ is in stratum } h; \\ 0, & \text{otherwise.} \end{cases}$$

This selection mechanism is ignorable *providing* the model for y_t conditions on the stratum variables z_t. A simple model that does this is

$$[y_t|z_t = h, \{\theta_h, \sigma_h^2\}] \sim_{ind} N(\theta_h, \sigma_h^2), \tag{4.3}$$

where $N(a, b)$ denotes the normal distribution with mean a, variance b. For simplicity assume σ_h^2 is known and the flat prior

$$p(\theta_h|z_U) \propto const.$$

on the stratum means. Bayesian calculations similar to the first example lead to

$$[\bar{y}_U|z_U, data, \{\sigma_h^2\}] \sim N(\bar{y}_{st}, \sigma_{st}^2)$$

where

$$\bar{y}_{st} = \sum_{h=1}^{H} P_h \bar{y}_{sh}, \quad P_h = N_h/N, \bar{y}_{sh} = \text{sample mean in stratum } h,$$

$$\sigma_{st}^2 = \sum_{h=1}^{H} P_h^2(1 - f_h)\sigma_h^2/n_h, \quad f_h = n_h/N_h.$$

These Bayesian results lead to Bayes' probability intervals that are equivalent to standard confidence intervals from design-based inference for a stratified random sample. In particular, the posterior mean weights cases by the inverse of their inclusion probabilities, as in the Horvitz–Thompson estimator (Horvitz and Thompson, 1952):

$$\bar{y}_{st} = N^{-1} \sum_{h=1}^{H} N_h \bar{y}_{sh} = N^{-1} \sum_{h=1}^{H} \sum_{t:x_t=h} y_t i_t/\pi_h,$$

$$\pi_h = n_h/N_h = \text{selection probability in stratum } h.$$

The posterior variance equals the design-based variance of the stratified mean:

$$Var(\bar{y}_U|z_U, data) = \sum_{h=1}^{H} P_h^2(1 - n_h/N_h)\sigma_h^2/n_h.$$

Binder (1982) demonstrates a similar correspondence asymptotically for a stratified nonparametric model with Dirichlet priors. With unknown variances, the posterior distribution of $\bar{y}_U$ for this model with a uniform prior on $\log(\sigma_h^2)$ is a mixture of t distributions, thus propagating the uncertainty from estimating the stratum variances.

Suppose that we assumed the model (4.1) and (4.2) of Example 1, with no stratum effects. With a flat prior on the mean, the posterior mean of $\bar{y}_U$ is then the unweighted mean

$$E(\bar{y}_U|z_U, data, \sigma^2) = \bar{y}_s \equiv \sum_{h=1}^{H} p_h \bar{y}_{sh}, \quad p_h = n_h/n,$$

which is potentially very biased for $\bar{y}_U$ if the selection rates $\pi_h = n_h/N_h$ vary across the strata. The problem is that inferences from this model are nonrobust to violations of the assumption of no stratum effects, and stratum effects are to be expected in most settings. Robustness considerations lead to the model (4.3) that allows for stratum effects.

From the design-based perspective, an estimator is *design consistent* if (irrespective of the truth of the model) it converges to the true population quantity as the sample size increases, holding design features constant. For stratified sampling, the posterior mean based on the stratified normal model (4.1) converges to $\bar{y}_U$, and hence is design consistent. For the normal model that ignores stratum effects, the posterior mean converges to $\bar{y}_\pi = \sum_{h=1}^{H} \pi_h N_h \bar{y}_{Uh} / \sum_{h=1}^{H} \pi_h N_h$, and hence is not design consistent unless $\pi_h = const$. I generally favor models that yield design-consistent estimates, to limit effects of model misspecification (Little, 1983a,b).

An extreme form of stratification is to sample systematically with interval B from a list ordered on some stratifying variable Z. The sample can be viewed as selecting one unit at random from implicit strata that are blocks of size B defined by selection interval. This design is ignorable for Bayesian inference provided the model conditions on z_U. The standard stratified sampling model (4.3) does not provide estimates of variance since only one unit is selected per stratum, and inference requires a more structured model for the stratum effects. Modeling assumptions are also needed to derive sampling variances under the design-based approach in this situation; generally the approach taken is to derive variances from the differences of outcomes of pairs of units selected from adjacent strata, ignoring differences between adjacent strata.

Example 3. Two-stage sampling

Suppose the population is divided into C clusters, based for example on geographical areas. A simple form of two-stage sampling first selects a simple random sample of c clusters, and then selects a simple random sample of n_c of the N_c units in each sampled cluster c. The inclusion mechanism is ignorable conditional on cluster information, but the model needs to account for within-cluster correlation in the population. A normal model that does this is

y_{ct} = outcome for unit t in cluster $c, t = 1, \ldots, N_c; c = 1, \ldots, C,$

$$[y_{ct}|\theta_c, \sigma^2] \underset{ind}{\sim} N(\theta_c, \sigma^2), \qquad\qquad (4.4)$$

$$[\theta_c|\mu, \phi] \underset{ind}{\sim} N(\mu, \phi).$$

Unlike (4.3), the cluster means cannot be assigned a flat prior, $p(\theta_c) = const.$, because only a subset of the clusters is sampled; the uniform prior does not allow information from sampled clusters to predict means for nonsampled clusters. Maximum likelihood inference for models such as (4.4) can be implemented in SAS Proc Mixed (SAS, 1992).

Example 4. Quota Sampling

Market research firms often use quota sampling, where strata are formed based on characteristics of interest, and researchers select individuals purposively until a fixed number n_h of respondents are obtained in each stratum h. This scheme assures that distribution of the stratifying variable in the sample matches that in the population. However, the lack of control of selection of units within strata allows the selection mechanism to depend on unknown factors that might be associated with the survey outcomes, and hence be nonignorable. We might analyze the data assuming ignorable selection, but the possibility of unknown selection biases compromises the validity of the inference. Random sampling avoids this problem, and hence greatly enhances the plausibility of the inference, frequentist or Bayesian.

This concludes this brief overview of the Bayesian approach to survey inference. In Chapter 18 the Bayesian framework is extended to handle problems of survey nonresponse.

CHAPTER 5

Interpreting a Sample as Evidence about a Finite Population

Richard Royall

5.1. INTRODUCTION

Observations, when interpreted under a probability model, constitute statistical evidence. A key responsibility of the discipline of statistics is to provide science with theory and methods for objective evaluation of statistical evidence per se, i.e., for representing observations as evidence and measuring the strength of that evidence. Birnbaum (1962) named this problem area 'informative inference,' and contrasted it to other important problem areas of statistics, such as those concerned with how beliefs should change in the face of new evidence, or with how statistical evidence can be used most efficiently, along with other types of information, in decision-making procedures. In this chapter we briefly sketch the basic general theory and methods for evidential interpretation of statistical data and then apply them to problems where the data consist of observations on a sample from a finite population.

The fundamental rule for interpreting statistical evidence is what Hacking (1965) named the law of likelihood. It states that

If hypothesis A implies that the probability that a random variable X takes the value x is p_A, while hypothesis B implies that the probability is p_B, then the observation $X = x$ is evidence supporting A over B if $p_A > p_B$, and the likelihood ratio, p_A/p_B, measures the strength of that evidence.

This says simply that if an event is more probable under A than B, then the occurrence of that event is evidence supporting A over B – the hypothesis that did the better job of predicting the event is better supported by its occurrence. It further states that the *degree* to which occurrence of the event supports A over

Analysis of Survey Data Edited by R. L. Chambers and C. J. Skinner

B (the strength of the evidence) is quantified by the ratio of the two probabilities, i.e., the likelihood ratio.

When uncertainty about the hypotheses, before $X = x$ is observed, is measured by probabilities $Pr(A)$ and $Pr(B)$, the law of likelihood can be derived from elementary probability theory. In that case the quantity p_B is the conditional probability that $X = x$, given that A is true, $Pr(X = x|A)$, and P_B is $Pr(X = x|B)$. The definition of conditional probability implies that

$$\frac{Pr(A|X = x)}{Pr(B|X = x)} = \frac{p_A Pr(A)}{p_B Pr(B)}.$$

This formula shows that the effect of the statistical evidence (the observation $X = x$) is to change the probability ratio from $Pr(A)/Pr(B)$ to $Pr(A|X = x)/Pr(B|X = x)$. The likelihood ratio p_A/p_B is the exact factor by which the probability ratio is changed. If the likelihood ratio equals 5, it means that the observation $X = x$ constitutes evidence just strong enough to cause a five-fold increase in the probability ratio. Note that the strength of the evidence is independent of the magnitudes of the probabilities, $Pr(A)$ and $Pr(B)$, and of their ratio. The same argument applies when p_A and p_B are not probabilities, but probability densities at the point x.

The likelihood ratio is a precise and objective numerical measure of the strength of statistical evidence. Practical use of this measure requires that we learn to relate it to intuitive verbal descriptions such as 'weak', 'fairly strong', 'very strong', etc. For this purpose the values of 8 and 32 have been suggested as benchmarks – observations with a likelihood ratio of 8 (or 1/8) constitute 'moderately strong' evidence, and observations with a likelihood of ratio of 32 (or 1/32) are 'strong' evidence (Royall, 1997). These benchmark values are similar to others that have been proposed (Jeffreys, 1961; Edwards, 1972; Kass and Raftery, 1995). They are suggested by consideration of a simple experiment with two urns, one containing only white balls, and the other containing half white balls and half black balls. Suppose a ball is drawn from one of these urns and is seen to be white. While this observation surely represents evidence supporting the hypothesis that the urn is the all-white one (vs. the alternative that it is the half white one), it is clear that the evidence is 'weak'. The likelihood ratio is 2. If the ball is replaced and a second draw is made, also white, the two observations together represent somewhat stronger evidence supporting the 'all-white' urn hypothesis – the likelihood ratio is 4. A likelihood ratio of 8, which we suggest describing as 'moderately strong' evidence, has the same strength as three consecutive white balls. Observation of five white balls is 'strong' evidence, and gives a likelihood ratio of 32.

A key concept of evidential statistics is that of misleading evidence. Observations with a likelihood ratio of $p_A/p_B = 40$ (40 times as probable under A as under B) constitute strong evidence supporting A over B. Such observations can occur when B is true, and when that happens they constitute strong misleading evidence. No error has been made – the evidence has been properly interpreted. The evidence itself is misleading. *Statistical evidence, properly interpreted, can be misleading.* But the nature of statistical evidence is such that we cannot observe strong misleading evidence very often. There is a universal bound for the probability of misleading evidence: if A implies that X has destiny (or mass) function f_A, while B implies f_B, then for any $k > 1$, $p_B(f_A(X)/f_B(X) \geq k) \leq 1/k$. Thus when B is true, the probability of observing data giving a likelihood ratio of 40 or more in favour of A can be no greater than 0.025. Royall (2000) discusses this universal bound as well as much smaller ones that apply within many important parametric models.

A critical distinction in evidence theory and methods is that between the strength of the evidence represented by a given body of observations, which is measured by likelihood ratios, and the probabilities that a particular procedure for making observations (sampling plan, stopping rule) will produce observations that constitute weak or misleading evidence. The essential flaw in standard frequentist theory and methods such as hypothesis testing and confidence intervals, when they are used for evidential interpretation of data, is the failure to make such a distinction (Hacking, 1965, Ch. 7). Lacking an explicit concept of evidence like that embodied in the likelihood, standard statistical theory tries to use probabilities (Type I and Type II error probabilities, confidence coefficients, etc.) in both roles: (i) to describe the uncertainty in a statistical procedure before observations have been made; and (ii) to interpret the evidence represented by a given body of observations. It is the use of probabilities in the second role, which is incompatible with the law of likelihood, that produces the paradoxes pervading contemporary statistics (Royall, 1997, Ch. 5).

With respect to a model that consists of a collection of probability distributions indexed by a parameter θ, the statistical evidence in observations $X = x$ supporting any value, θ_1, vis-à-vis any other, θ_2, is measured by the likelihood ratio, $f(x; \theta_1)/f(x; \theta_2)$. Thus the likelihood function, $L(\theta) \propto f(x; \theta)$, is the mathematical representation of the statistical evidence under this model – if two instances of statistical evidence generate the same likelihood function, they represent evidence of the same strength with respect to all possible pairs (θ_1, θ_2), so they are equivalent as evidence about θ. This important implication of the law of likelihood is known as the likelihood principle. It in turn has immediate implications for statistical *methods* in the problem area of information inference. As Birnbaum (1962) expressed it, 'One basic consequence is that reports of experimental results in scientific journals should in principle be descriptions of likelihood functions.' Note that the likelihood function is, by definition, the function whose ratios measure the strength of the evidence in the observations.

When the elements of a finite population are modelled as realisations of random variables, and a sample is drawn from that population, the observations presumably represent evidence about both the probability model and the population. This chapter considers the definition, construction, and use of likelihood functions for representing and interpreting the observations as evidence about (i) parameters in the probability model and (ii) characteristics of the actual population (such as the population mean or total) in problems where the sampling plan is uninformative (selection probabilities depend only on quantities whose values are known at the time of selection, see also the

discussion by Little in the previous chapter). Although there is broad agree-
ment about the definition of likelihood functions for (i), consensus has not been
reached on (ii). We will use simple transparent examples to examine and
compare likelihood functions for (i) and (ii), and to study the probability of
misleading evidence in relation to (ii). We suggest that Bjørnstad's (1996)
assertion that 'Survey sampling under a population model is a field where the
likelihood function has not been defined properly' is mistaken, and we note that
his proposed redefinition of the likelihood function is incompatible with the law
of likelihood.

5.2. THE EVIDENCE IN A SAMPLE FROM A FINITE POPULATION

We begin with the simplest case. Consider a condition (disease, genotype,
behavioural trait, etc.) that is either present or absent in each member of a
population of $N = 393$ individuals. Let x_t be the zero–one indicator of whether
the condition is present in the tth individual. We choose a sample, s, consisting
of $n = 30$ individuals and observe their x-values: x_t, $t \in s$.

5.2.1. Evidence about a probability

If we model the x as realised values of *iid* Bernoulli (θ) random variables,
$X_1, X_2, \ldots, X_{393}$, then our 30 observations constitute evidence about the probability θ. If there are 10 instances of the condition $(x = 1)$ in the sample, then
the evidence about θ is represented by the likelihood function $L(\theta) \propto \theta^{10}$
$(1 - \theta)^{20}$ shown by the solid curve in Figure 5.1, which we have standardised
so that the maximum value is one. The law of likelihood explains how to
interpret this function: the sample constitutes evidence about θ, and for any
two values θ_1 and θ_2 the ratio $L(\theta_1)/L(\theta_2)$ measures the strength of the evidence
supporting θ_1 over θ_2. For example, our sample constitutes very strong evidence supporting $\theta = 1/4$ over $\theta = 3/4$ (the likelihood ratio is nearly 60 000),
weak evidence for $\theta = 1/4$ over $\theta = 1/2$ (likelihood ratio 3.2), and even weaker
evidence supporting $\theta = 1/3$ over $\theta = 1/4$ (likelihood ratio 1.68).

5.2.2. Evidence about a population proportion

The parameter θ is the probability in a conceptual model for the process that
generated the 393 population values $x_1, x_2, \ldots, x_{393}$. It is not the same as the
actual proportion with the condition in this population, which is $\sum_{t=1}^{393} x_t/393$.
The evidence about the *proportion* is represented by a second likelihood function (derived below), shown in the solid dots in Figure 5.1. This function is
discrete, because the only possible values for the proportion are $k/393$, for
$k = 0, 1, 2, \ldots, 393$. Since 10 ones and 20 zeros have been observed, population
proportions corresponding to fewer than 10 ones (0/393, 1/393, $\ldots$, 9/393) and
fewer than 20 zeros (374/393, $\ldots$, 392/393, 1) are incompatible with the sample,
so the likelihood is zero at all of these points.

To facilitate interpretation of this evidence we have supplied some numerical summaries in Figure 5.1, indicating, for example, where the likelihood is maximised and the range of values where the standardised likelihood is at least 1/8. This range, the '1/8 likelihood interval', consists of the values of the proportion that are consistent with the sample in the sense that there is no alternative value that is better supported by a factor of 8 ('moderately strong evidence') or greater.

To clarify further the distinction between the probability and the proportion, let us suppose that the population consists of only $N = 50$ individuals, not 393. Now since the sample consists of 30 of the individuals, there are only 20 whose x-values remain unknown, so there are 21 values for the population proportion that are compatible with the observed data, 10/50, 11/50, ..., 30/50. The likelihoods for these 21 values are shown by the open dots in Figure 5.1. The likelihood function that represents the evidence about the *probability* does not depend on the size of the population that is sampled, but Figure 5.1 makes it clear that the function representing the evidence about the actual *proportion* in that population depends critically on the population size N.

5.2.3. The likelihood function for a population proportion or total

The discrete likelihoods in Figure 5.1 are obtained as follows. In a population of size N, the likelihood ratio that measures the strength of the evidence supporting one value of the population proportion versus another is the factor

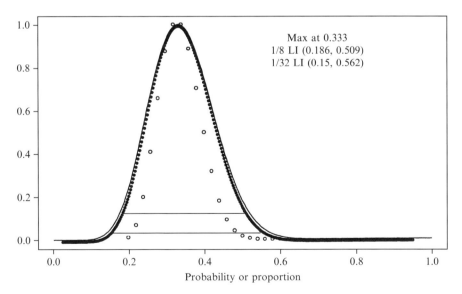

Figure 5.1 Likelihood functions for probability of success (curved line) and for proportion of successes in finite populations under a Bernoulli probability model (population size $N = 393$, black dots; $N = 50$, white dots). The sample of $n = 30$ contains 10 successes.

by which their probability ratio is changed by the observations. Now before the sample is observed, the ratio of the probability that the proportion equals k/N to the probability that it equals j/N is

$$\frac{Pr(T = k)}{Pr(T = j)} = \binom{N}{k} \theta^k (1 - \theta)^{N-k} \bigg/ \binom{N}{j} \theta^j (1 - \theta)^{N-j}$$

where T is the population total $\sum_{t=1}^{N} X_t$.

After observation of a sample of size n in which τ_s have the condition, the probability that the population proportion equals τ_U/N is just the probability that the total for the $N - n$ non-sample individuals is $\tau_U - \tau_s$, so the ratio of the probability that the proportion equals k/N to the probability that it equals j/N is changed to

$$\frac{Pr(T = k|sample)}{Pr(T = j|sample)} = \binom{N-n}{k-\tau_s} \theta^{k-\tau_s} (1-\theta)^{N-n-k+\tau_s} \bigg/ \binom{N-n}{j-\tau_s} \theta^{j-\tau_s} (1-\theta)^{N-n-j+\tau_s}.$$

The likelihood ratio for proportions k/N and j/N is the factor by which the observations change their probability ratio. It is obtained by dividing this last expression by the one before it, and equals

$$\binom{N-n}{k-\tau_s} \binom{N}{j} \bigg/ \binom{N-n}{j-\tau_s} \binom{N}{k}.$$

Therefore the likelihood function is (Royall, 1976)

$$L(\tau_U/N) \propto \binom{N-n}{\tau_U - \tau_s} \bigg/ \binom{N}{\tau_U}.$$

This is the function represented by the dots in Figure 5.1.

An alternative derivation is based more directly on the law of likelihood: an hypothesis asserting that the actual population proportion equals τ_U/N is easily shown to imply (under the Bernoulli trial model) that the probability of observing a particular sample vector consisting of τ_s ones and $n - \tau_s$ zeros is proportional to the hypergeometric probability

$$\binom{\tau_U}{\tau_s} \binom{N - \tau_U}{n - \tau_s} \bigg/ \binom{N}{n}.$$

Thus an alternative representation of the likelihood function is

$$L^*(\tau_U/N) \propto \binom{\tau_U}{\tau_s} \binom{N - \tau_U}{n - \tau_s}.$$

It is easy to show that $L^*(\tau_U/N) \propto L(\tau_U/N)$, so that L and L^* represents the same likelihood function.

It is also easy to show (using Stirling's approximation) that for any fixed sample the likelihood function for the population proportion, $L(\tau_U/N)$, converges to that for the probability, θ, as $N \to \infty$. That is, for any sequences of values of the population total, τ_{1N} and τ_{2N}, for which $\tau_{1N}/N \to \theta_1$ and $\tau_{2N}/N \to \theta_2$ the likelihood ratio is

$$\frac{L(\tau_{1N}/N)}{L(\tau_{2N}/N)} \rightarrow \frac{\theta_1^{\tau_s}(1-\theta_1)^{n-\tau_s}}{\theta_2^{\tau_s}(1-\theta_2)^{n-\tau_s}}.$$

5.2.4. The probability of misleading evidence

According to the law of likelihood, the likelihood ratio measures the strength of the evidence in our sample supporting one value of the population proportion versus another. Next we examine the probability of observing misleading evidence. Suppose that the population total is actually τ_1, so the proportion is $p_1 = \tau_1/N$. For an alternative value, $p_2 = \tau_2/N$, what is the probability of observing a sample that constitutes at least moderately strong evidence supporting p_2 over p_1? That is, what is the probability of observing a sample for which the likelihood ratio $L(p_2)/L(p_1) \geq 8$? As we have just seen, the likelihood ratio is determined entirely by the sample total τ_s, which has a hypergeometric probability distribution

$$Pr(\tau_s = j | T = \tau_1) \binom{\tau_1}{j}\binom{N-\tau_1}{n-j} \bigg/ \binom{N}{n}.$$

Thus we can calculate the probability of observing a sample that represents at least moderately strong evidence in favour of p_2 over the true value p_1. This probability is shown, as a function of p_2, by the heavy lines in Figure 5.2 for a population of $N = 393$ in which the true proportion is $p_1 = 100/393 = 0.254$, or 25.4 per hundred.

Note how the probability of misleading evidence varies as a function of p_2. For alternatives very close to the true value (for example, $p_2 = 101/393$), the probability is zero. This is because no possible sample of 30 observations

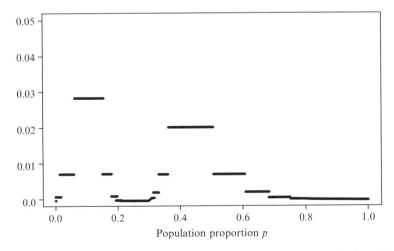

Figure 5.2 Probability of misleading evidence: probability that a sample of $n = 30$ from population of $N = 393$ will produce a likelihood ratio of 8 or more supporting population proportion p over the true value, $100/393 = 0.254$.

represents evidence supporting $p_2 = 101/393$ over $p_1 = 100/393$ by a factor as large as 8. This is true for all values of p_2 from 101/393 through 106/393. For the next larger value, $p_2 = 107/393$, the likelihood ratio supporting p_2 over p_1 can exceed 8, but this happens only when all 30 sample units have the trait, $\tau_s = 30$, and the probability of observing such a sample, when 100 out of the population of 393 actually have the trait, is very small (4×10^{-20}).

As the alternative, p_2, continues to increase, the probability of misleading evidence grows, reaching a maximum at values in an interval that includes $170/393 = 0.42$. For any alternative p_2 within this interval a sample in which 13 or more have the trait ($\tau_s \geq 13$) gives a likelihood ratio supporting p_2 over the true mean, 100/393, by a factor of 8 or more. The probability of observing such a sample is 0.0204. Next, as the alternative moves even farther from the true value, the probability of misleading evidence decreases. For example, at $p_2 = 300/393 = 0.76$ only samples with $\tau_s \geq 17$ give likelihood ratios as large as 8 in favour of p_2, and the probability of observing such a sample is only 0.000 15.

Figure 5.2 shows that when the true population proportion is $p_1 = 100/393$ the probability of misleading evidence (likelihood ratio ≥ 8) does not exceed 0.028 at any alternative, a limit much lower than the universal bound, which is $1/8 = 0.125$. For other values of the true proportion the probability of misleading evidence shows the same behaviour – it equals zero near the true value, rises with increasing distance from that value, reaches a maximum, then decreases.

5.2.5. Evidence about the average count in a finite population

We have used the population size $N = 393$ in the above example for consistency with what follows, where we will examine the evidence in a sample from an actual population consisting of 393 short-stay hospitals (Royall and Cumberland, 1981) under a variety of models. Here the variate x is not a zero-one indicator, but a count – the number of patients discharged from a hospital in one month. We have observed the number of patients discharged from each of the $n = 30$ hospitals in a sample, and are interested in the total number of patients discharged from all 393 hospitals, or, equivalently, the average, $\bar{x}_U = \sum x_t / 393$.

First we examine the evidence under a Poisson model: the counts $x_1, x_2, \ldots, x_{393}$ are modelled as realised values of *iid* Poisson (λ) random variables, $X_1, X_2, \ldots, X_{393}$. Under this model the likelihood function for the actual population mean, $\bar{x}_U$, is proportional to the probability of the sample, given that $\bar{X} = \bar{x}_U$, which for a sample whose mean is $\bar{x}_s$ is easily shown to be

$$L(\bar{x}_U) \propto \binom{N\bar{x}_U}{n\bar{x}_s} \left(\frac{n}{N}\right)^{n\bar{x}_s} \left(1 - \frac{n}{N}\right)^{N\bar{x}_U - n\bar{x}_s} \quad \text{for } \bar{x}_U = (n\bar{x}_s + j)/N, \ j = 0, 1, 2, \ldots.$$

Just as with the proportion and the probability in the first problem that we considered, we must be careful to distinguish between the actual population mean (average), $\bar{x}_U$, and the mean (expected value) in the underlying probability model, $E(X) = \lambda$. And just as in the first problem (i) the likelihood for the finite population mean, $L(\bar{x}_U)$, is free of the model parameter, λ, and (ii) for any given sample, as $N \to \infty$ the likelihood function for the population mean, $L(\bar{x}_U)$, converges to the likelihood function for the expected value λ, which is proportional to $\lambda^{n\bar{x}_s} e^{-n\lambda}$. That is, for any positive values $\bar{x}_1$ and $\bar{x}_2$, $L(\bar{x}_1)/L(\bar{x}_2) \to (\bar{x}_1/\bar{x}_2)^{n\bar{x}_s} e^{-n(\bar{x}_1 - \bar{x}_2)}$.

In our sample of $n = 30$ from the 393 hospitals the mean number of patients discharged per hospital is $\bar{x}_s = 24\,103/30 = 803.4$ and the likelihood function $L(\bar{x}_U)$ is shown by the dashed line in Figure 5.3.

The Poisson model for count data is attractive because it provides a simple explicit likelihood function for the population mean. But the fact that this model has only one parameter makes it too inflexible for most applications, and we will see that it is quite inappropriate for the present one. A more widely useful model is the negative binominal with parameters r and θ, in which the count has the same distribution as the number of failures before the rth success in iid Bernoulli (θ) trials. Within this more general model the Poisson (λ) distribution appears as the limiting case when the expected value $r(1 - \theta)/\theta$ is fixed at λ, while $r \to \infty$.

Under the negative binominal model the likelihood for the population mean $\bar{x}_U$ is free of θ, but it does involve the nuisance parameter r:

$$L(\bar{x}_U, r) \propto P(sample|\bar{X}_U =$$

$$\bar{x}_U; r; \theta) = \frac{\prod_{t \in s} \binom{x_t + r - 1}{r - 1} \binom{N\bar{x}_U - n\bar{x}_s + (N - n)r - 1}{(N - n)r - 1}}{\binom{N\bar{x}_U - Nr - 1}{Nr - 1}}.$$

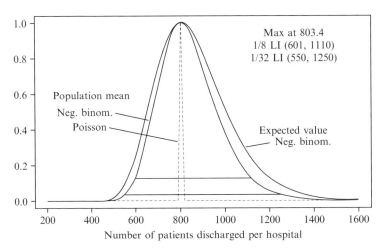

Figure 5.3 Likelihood functions for expected number and population mean number of patients discharged in population $N = 393$ hospitals under negative binomial model. Dashed line shows likelihood for population mean under Poisson model. Sample of $n = 30$ hospitals with mean 803.4 patients/hospital.

The profile likelihood, $L_p(\bar{x}_U) \propto \max_r L(\bar{x}_U, r)$, is easily calculated numerically, and it too is shown in Figure 5.3 (the solid line). For this sample the more general negative binomial model gives a much more conservative evaluation of the evidence about the population mean, $\bar{x}_U$, than the Poisson model.

Both of the likelihood functions in Figure 5.3 are correct. Each one properly represents the evidence about the population mean in this sample under the corresponding model. Which is the more appropriate? Recall that the negative binomial distribution approaches the Poisson as the parameter r approaches infinity. The sample itself constitutes extremely strong evidence supporting small values of r (close to one) over the very large values which characterise an approximate Poisson distribution. This is shown by the profile likelihood function for r, or equivalently, what is easier to draw and interpret, the profile likelihood for $1/\sqrt{r}$, which places the Poisson distribution ($r = \infty$) at the origin. This latter function is shown in Figure 5.4. For comparison, Figure 5.4 also shows the profile likelihood function for $1/\sqrt{r}$ generated by a sample of 30 independent observations from a Poisson probability distribution whose mean equals the hospital sample mean. The evidence for 'extra-Poisson variability' in the sample of hospital discharge counts is very strong indeed, with values of $1/\sqrt{r}$ near one supported over values near zero (the Poisson model) by enormous factors.

Under the previous models (Bernoulli and Poisson), for a fixed sample the likelihood function for the finite population mean converges, as the population size grows, to the likelihood for the expected value. Whether the same results applied to the profile likelihood function under the negative binomial model is an interesting outstanding question.

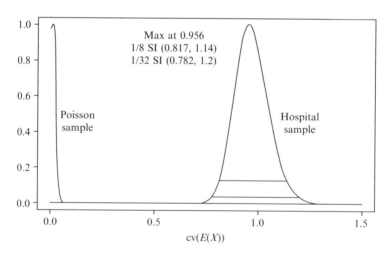

Figure 5.4 Profile likelihood for $1/\sqrt{r}$: negative binomial model. Two samples: (1) numbers of patients discharged from 30 hospitals, and (2) 30 *iid* Poisson variables.

5.2.6.   Evidence about a population mean under a regression model

For each of the 393 hospitals in this population we know the value of a potentially useful covariate – the number of beds, z. For our observed sample of $n = 30$ counts, let us examine the evidence about the population mean number of patients discharged, $\bar{x}_U$, under a model that includes this additional information about hospital size. The nature of the variables, as well as inspection of a scatterplot of the sample, suggests that as a first approximation we consider a proportional regression model (expected number of discharges proportional to the number of beds, z), with variance increasing with z: $E(X) = \beta z$ and $var(X) = \sigma^2 z$. Although a negative binomial model with this mean and variance structure would be more realistic for count data such as these, we will content ourselves for now with the simple and convenient normal regression model.

Under this model the likelihood function for the population mean depends on the nuisance parameter σ, but not the slope β:

$$L(\bar{x}_U, \sigma^2) \propto Pr(sample|\bar{X}_U =$$

$$\bar{x}_U; \beta, \sigma^2) \propto \sigma^{-n} \exp\left[-\frac{1}{2\sigma^2}\left(\sum_{t \in s} \frac{x_t^2}{z_t} + \frac{(N\bar{x}_U - n\bar{x}_s)^2}{N\bar{z}_U} - \frac{(N\bar{x}_U)^2}{N\bar{z}_U}\right)\right].$$

Maximising over the nuisance parameter σ^2 produces the profile likelihood function

$$L_p(\bar{x}_U) \propto \left(1 + \frac{e_s^2}{n-1}\right)$$

where

$$e_s = \frac{\bar{x}_U - (\bar{x}_s/\bar{z}_s)\bar{z}_U}{v_s^{1/2}} \quad \text{with} \quad v_s = \left(\frac{\bar{z}_U(\bar{z}_U - (n/N)\bar{z}_s)}{n(n-1)\bar{z}_s}\right) \sum_{t \in s} \frac{1}{z_t}\left(x_t - \frac{\bar{x}_s}{\bar{z}_s}z_t\right)^2.$$

This profile likelihood function, with n replaced by $(n-1)$ in the exponent (as suggested by Kalbfleisch and Sprott, 1970), is shown in Figure 5.5.

The sample whose evidence is represented in the likelihood functions in Figures 5.3, 5.4, and 5.5 is the 'best-fit' sample described by Royall and Cumberland (1981). It was obtained by ordering 393 hospitals according to the size variable (z), then drawing a centred systematic sample from the ordered list. For this demonstration population the counts (the x) are actually known for all 393 hospitals, so we can find the true population mean and see how well our procedure for making observations and interpreting them as evidence under various probability models has worked in this instance. The total number of discharges in this population is actually $\tau_U = 320\,159$, so the true mean is 814.65. Figures 5.3 and 5.5 show excellent results, with the exception of the clearly inappropriate Poisson model in Figure 5.3, whose overstatement of the evidence resulted in its 1/8 likelihood interval excluding the true value.

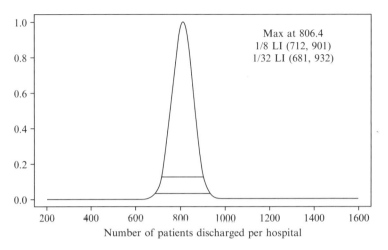

Figure 5.5 Likelihood for population mean number of patients discharged in popula-tion of $N = 393$ hospitals: normal regression model. Sample of $n = 30$ hospitals with mean 803.4 patients/hospital.

5.3. DEFINING THE LIKELIHOOD FUNCTION FOR A FINITE POPULATION

Bjørnstad (1996) proposed a general definition of the 'likelihood function' that, in the case of a finite population total or mean, differs from the one derived from the law of likelihood, which we have adopted in this chapter. In our first example (a population of N under the Bernoulli (θ) probability model) a sample of n units in which the number with the trait is τ_s is evidence about θ. That evidence is represented by the likelihood function

$$L(\theta) = \theta^{\tau_s}(1 - \theta)^{n-\tau_s}.$$

The sample is also evidence about the population total τ_U represented by the likelihood function

$$L(\tau_U) = \binom{\tau_U}{\tau_s}\binom{N - \tau_U}{n - \tau_s}.$$

Bjørnstad's function is the product of $L(\tau_U)$ and a function that is *independent of the sample*, namely the model probability that $T = \tau_U$:

$$L_B(\tau_U; \theta) \propto L(\tau_U)Pr(T = \tau_U; \theta) \propto \binom{\tau_U}{\tau_s}\binom{N - \tau_U}{n - \tau_s}\binom{N}{\tau_U}\theta^{\tau_U}(1 - \theta)^{N-\tau_U}.$$

Consider a population of $N = 100$ and a sample that consists of half of them ($n = 50$). If 20% of the sample have the trait (sample total $\tau_s = 10$), then our observations represent very strong evidence supporting the hypothesis that 20% of the population have the trait (population total $\tau_U = 20$) versus the hypothesis that 50% have it ($\tau_U = 50$): $L(20)/L(50) = 1.88 \times 10^8$. This is what the law of likelihood says, because the hypothesis that $\tau_U = 20$ implies that the probability of our sample is 1.88×10^8 times greater than the probability implied by the hypothesis that $\tau_U = 50$, regardless of the value of θ. Compare this likelihood ratio with the values of the Bjørnstad function:

$$L_B(20; \theta)/L_B(50; \theta) = [(1 - \theta/\theta]^{30}.$$

This is the ratio of the conditional probabilities, $Pr(T = 20|\tau_s = 10; \theta)/Pr(T = 50|\tau_s = 10; \theta)$. It represents, not the evidence in the observations, but a synthesis of that evidence and the probabilities $Pr(T = 20; \theta)$ and $Pr(T = 50; \theta)$ and it is strongly influenced by these 'prior' probabilities. For example, when the parameter θ equals 1/2 the Bjørnstad ratio is $L_B(20; 0.5)/L_B(50; 0.5) = 1$. If this were a likelihood ratio, its value, 1, would signify that when θ is 1/2 the sample represents evidence of no strength at all in support of $\tau_U = 20$ over $\tau_U = 50$, which is quite wrong. The Bjørnstad function's ratio of 1 results from the fact that the very strong evidence in favour of $\tau_U = 20$ in the sample, $L(20)/L(50) = 1.88 \times 10^8$, is the exact reciprocal of the very large probability ratio in favour of $\tau_U = 50$ that is given by the model *independently of the empirical evidence*:

$$Pr(T = 50|\tau_s = 10; \theta)/Pr(T = 20|\tau_s = 10; \theta) = \binom{100}{50} / \binom{100}{20} = 1.88 \times 10^8.$$

In our approach, the likelihood function is derived from the law of likelihood, and it provides a mathematical representation of the evidence in the observations (the empirical evidence) vis-á-vis the population total, under the adopted model. Our goal is analytic – to isolate and examine the empirical evidence per se. The Bjørnstad function represents something different, a synthesis of the empirical evidence with model probabilities. Whatever its virtues and potential uses in estimation and prediction, the function $L_B(\tau_U; \theta)$ does not represent what we seek to measure and communicate, namely, the evidence about τ_U in the sample.

PART B

Categorical Response Data

CHAPTER 6

Introduction to Part B

C. J. Skinner

6.1. INTRODUCTION

This part of the book deals with the analysis of survey data on categorical responses. Chapter 7 by Rao and Thomas provides a review of methods for handling complex sampling schemes for a wide range of methods of categorical data analysis. Chapter 8 by Scott and Wild extends the discussion of logistic regression in Chapter 7 in the specific context of case–control studies, where sampling is stratified by the categories of the binary response variable. In this chapter, we provide an introduction, as background to these two chapters. Our main emphasis will be on the basic methods used in Rao and Thomas's chapter.

Two broad ways of analysing categorical response data may be distinguished:

(i) analysis of tables;
(ii) analysis of unit-level data.

In the first case, the analysis effectively involves two stages. First, a table is constructed from the unit-level data. This will usually involve the cross-classification of two or more categorical variables. The elements of the table will typically consist of estimated proportions. These estimated proportions, together perhaps with associated estimates of the variances and covariances of these estimated proportions, may be considered as 'sufficient statistics', carrying all the relevant information in the data. This tabular information will then be analysed in some way, as the second stage of the analysis.

In the second case, there is no intermediate step of constructing a table. The basic unit-level data are analysed directly. A simple example arises in fitting a logistic regression model with a binary response variable and a single continuous covariate. Because the covariate is continuous it is not possible to construct sufficient statistics for the unit-level data in the form of a standard table. Instead the data are analysed directly. This approach may be viewed as a generalisation of the first. We deal with these two cases separately in the following two sections.

Analysis of Survey Data Edited by R. L. Chambers and C. J. Skinner

6.2. ANALYSIS OF TABULAR DATA

The analysis of tabular data builds most straightforwardly on methods of descriptive surveys. Tables are most commonly formed by cross-classifying categorical variables with the elements of the table consisting of proportions. These may be the (estimated) proportions of the finite population falling into the various cells of the table. Alternatively, the proportions may be domain proportions, either because the table refers only to units falling into a given domain or, for example in the case of a two-way table, because the proportions of interest are row proportions or column proportions. In all these cases, the estimation of either population proportions or domain proportions is a standard topic in descriptive sample surveys (Cochran, 1977, Ch. 3).

6.2.1. One-way classification

Consider first a one-way classification, formed from a single categorical variable, Y, taking possible values $i = 1, \ldots, I$. For a finite population of size N, we let N_i denote the number of units in the population with $Y = i$. Assuming then that the categories of Y are mutually exclusive and exhaustive, we have $\sum N_i = N$. In categorical data analysis it is more common to analyse the population proportions corresponding to the population counts N_i than the population counts themselves. These (finite) population proportions are given by $N_i/N, i = 1, \ldots, I$, and may be interpreted as the probabilities of falling into the different categories of Y for a randomly selected member of the population (Bishop Fienberg and Holland, 1975, p. 9). Instead of considering finite population proportions, it may be of more scientific interest to consider corresponding model probabilities in a superpopulation model. Suppose, for example, that the vector $(N_1, \ldots, N_I)$ is assumed to be the realisation of a multinomial random variable with parameters N and $(\mu_1, \ldots, \mu_I)$, so that the probability μ_i corresponds to the finite population proportion N_i/N. Suppose also, for example, that it is of interest to test whether two categories are equiprobable. Then it might be argued that it would be more appropriate to test the equality of two model probabilities than the equality of two finite population proportions because the latter proportions would not be exactly equal in a finite population except by rare chance (Cochran, 1977, p. 39). Rao and Thomas use the same notation μ_i in Chapter 7 to denote either the finite population proportion N_i/N or the corresponding model probability. It is assumed that the analyst will have specified which one of these parameters is of interest, dependent upon the purpose of the analysis. Note that in either case $\sum \mu_i = 1$.

One reason why it is convenient to use the same notation μ_i for both the finite population and model parameters is that it is natural to use the same point estimator for each. A common general approach in descriptive surveys is to estimate the ith proportion (or probability) by

$$\hat{\mu}_i = \hat{N}_i/\hat{N}, \tag{6.1}$$

where

$$\hat{N}_i = \sum_s w_t I(y_t = i), \quad \hat{N} = \sum_s w_t = \sum_i \hat{N}_i. \tag{6.2}$$

Here, it is assumed that Y is measured for a sample s, with y_t and w_t denoting the value of Y and the value of a survey weight respectively for the tth element of s, and with $I(.)$ denoting the indicator function. The survey weight w_t may be equal to the reciprocal of the sample inclusion probability of sample unit t or may be a more complex weight (Chambers, 1996), reflecting for example poststratification to make use of auxiliary population information or to adjust for nonresponse in some way. Note that the multiplication of w_t by a constant leaves $\hat{\mu}_i$ unchanged. This can be useful in practice for secondary data analysts, since the weights provided to such survey data users are often arbitrarily scaled, for example to force the sum of the weights to equal the sample size. Such scaling has no impact on the estimator $\hat{\mu}_i$.

6.2.2. Multi-way classifications and log–linear models

The one-way classification above, based upon the single categorical variable Y, may be extended to a multi-way classification, formed by cross-classifying two or more categorical variables. Consider, for example, a 2×2 table, formed by cross-classifying two binary variables, A and B, each taking the values 1 or 2. The four cells of the table might then be labelled $(a, b), a = 1, 2; b = 1, 2$. For the purpose of defining log–linear models, as discussed by Rao and Thomas in Chapter 7, these cells may alternatively be labelled just with a single index $i = 1, \dots, 4$. We shall follow the convention, adopted by Rao and Thomas, of letting the index i correspond to the lexicographic ordering of the cells, as illustrated in Table 6.1.

The same approach may be adopted for tables of three or more dimensions. In general, we suppose the cells are labelled $i = 1, \dots, I$ in lexicographic order, where I denotes the number of cells in the table. The cell proportions (or probabilities) μ_i are then gathered together into the $I \times 1$ vector $\mu = (\mu_1, \dots, \mu_I)'$. A model for the table may then be specified by representing μ as a function of an $r \times 1$ vector θ of parameters, that is writing $\mu = \mu(\theta)$. Because the vector μ is subject to the constraint $\sum \mu_i = \mu' 1 = 1$, where 1 is the $I \times 1$ vector of ones, the maximum necessary value of r is $I - 1$. For, in this case

Table 6.1 Alternative labels for cells of 2×2 table.

Bivariate labels ordered lexicographically (a, b)	Single index i
(1,1)	1
(1,2)	2
(2,1)	3
(2,2)	4

we could define a saturated model, with θ as the first $I - 1$ elements of μ and the Ith element of μ expressed in terms of θ using $\mu'1 = 1$.

A *log–linear model* for the table is one which may be expressed as

$$\log(\mu) = u(\theta)1 + X\theta, \tag{6.3}$$

as in Equation (7.1), where $\log(\mu)$ denotes the $I \times 1$ vector of $\log(\mu_i)$, X is a known $I \times r$ matrix and $u(\theta)$ is a normalising factor that ensures $\mu'1 = 1$. Note the key feature of this model that, apart from the normalising factor, the logarithms of the μ_i may be expressed as linear combinations of the elements of the parameter vector θ.

Consider, for example, the model of independence in the 2×2 table discussed above. Denoting the probability of falling into cell (a, b) as $\mu^{AB}_{(a,b)}$, the independence of A and B implies that $\mu^{AB}_{(a,b)}$ may be expressed as a product $\mu^A_{(a)} \times \mu^B_{(b)}$ of marginal probabilities. Taking logarithms, we have the additive property

$$\log(\mu^{AB}_{(a,b)}) = \log(\mu^A_{(a)}) + \log(\mu^B_{(b)}).$$

Since $\mu^A_{(1)} + \mu^A_{(2)} = \mu^B_{(1)} + \mu^B_{(2)} = 1$, the cell probabilities $\mu^{AB}_{(a,b)}$ can be expressed in terms of two parameters, say $\mu^A_{(1)}$ and $\mu^B_{(1)}$. To express the model of independence as a log–linear model, we first reparameterise (e.g. Agresti, 1990, Ch. 5) by transforming from $\mu^A_{(1)}$ and $\mu^B_{(1)}$ to the parameter vector $\theta = (\theta^A, \theta^B)$, where

$$\theta^A = \log(\mu^A_{(1)}) - [\log(\mu^A_{(1)}) + \log(\mu^A_{(2)})]/2,$$
$$\theta^B = \log(\mu^B_{(1)}) - [\log(\mu^B_{(1)}) + \log(\mu^B_{(2)})]/2.$$

Then, setting

$$u(\theta) = [\log(\mu^A_{(1)}) + \log(\mu^A_{(2)})]/2 + [\log(\mu^B_{(1)}) + \log(\mu^B_{(2)})]/2$$

(which is a function of θ through the reparameterisation) and setting $\mu_i = \mu^{AB}_{(a,b)}$ with the correspondence between i and (a, b) in Table 6.1, we may express the logarithms of the cell probabilities in a matrix equation as

$$\begin{pmatrix} \log \mu_1 \\ \log \mu_2 \\ \log \mu_3 \\ \log \mu_4 \end{pmatrix} = u(\theta) \begin{pmatrix} 1 \\ 1 \\ 1 \\ 1 \end{pmatrix} + \begin{pmatrix} 1 & 1 \\ 1 & -1 \\ -1 & 1 \\ -1 & -1 \end{pmatrix} \begin{pmatrix} \theta^A \\ \theta^B \end{pmatrix}.$$

This equation is then a special case of Equation (6.3) above. For further discussion of log–linear models and their interpretation see Agresti (1990, Ch. 5).

Having defined the general log–linear model above, we now consider inference about the parameter vector θ. We discuss point estimation first. In standard treatments of log–linear models, a common approach is to use the maximum likelihood estimator (MLE) under the assumption that the sample cell counts are generated by multinomial sampling. The MLE of θ is then obtained by solving the following likelihood equations for θ:

$$X'\mu(\theta) = X'\hat{\mu},$$

where $\hat{\mu}$ is the (unweighted) vector of sample cell proportions (Agresti, 1990, section 6.4). In the case of complex sampling schemes, the multinomial assumption will usually be unreasonable and the MLE may be inconsistent for θ. Instead, Rao and Thomas (Chapter 7) discuss how θ may be estimated consistently by the *pseudo-MLE* of θ, obtained by solving the same equations, but with $\hat{\mu}$ replaced by $(\hat{\mu}_1, \ldots, \hat{\mu}_I)$, where $\hat{\mu}_i$ is a design-consistent estimator of μ_i such as in (6.1) above.

Let us next consider the estimation of standard errors or, more generally, the variance–covariance matrix of the pseudo-MLE of θ. Under multinomial sampling, the usual estimated covariance matrix of the MLE $\hat{\theta}$ is given by (Agresti, 1990, section 6.4.1; Rao and Thomas, 1988)

$$n^{-2}[X'\Omega(\hat{\theta})X]^{-1}$$

where n is the sample size upon which the table is based and $\Omega(\theta)$ is the covariance matrix of $\hat{\mu}$ under multinomial sampling (with the log–linear model holding):

$$\Omega(\theta) = n^{-1}[D(\mu(\theta)) - \mu(\theta)\mu(\theta)']$$

where $D(\mu)$ denotes the $I \times I$ diagonal matrix with diagonal elements μ_i. Replacing the MLE of θ by its pseudo-MLE in this expression will generally fail to account adequately for the effect of a complex sampling design on standard errors. Instead it is necessary to use an estimator of the variance–covariance matrix of $\hat{\mu}$, which makes appropriate allowance for the actual complex design. Such an estimator is denoted $\hat{\Sigma}$ by Rao and Thomas (Chapter 7). There are a variety of standard methods from descriptive surveys, which may be used to obtain such an estimator – see for example Lehtonen and Pahkinen (1996, Ch. 6) or Wolter (1985).

Given $\hat{\Sigma}$, a suitable estimator for the covariance matrix of $\hat{\theta}$ is (Rao and Thomas, 1989)

$$n^{-2}[X'\Omega(\hat{\theta})X]^{-1}(X\hat{\Sigma}X)[X'\Omega(\hat{\theta})X]^{-1}$$

where $\hat{\theta}$ is now the pseudo-MLE of θ. Rao and Thomas (1988) provide a numerical illustration of how the standard errors obtained from the diagonal elements of this matrix may differ from the usual standard errors based upon multinomial sampling assumptions.

Rao and Thomas (Chapter 7) focus on methods for testing hypotheses about θ. Many hypotheses of interest in contingency tables may be expressed as a hypothesis about θ in a log–linear model. Rao and Thomas consider the impact of complex sampling schemes on the distribution of standard Pearson and likelihood ratio test statistics and discuss the construction of Rao–Scott adjustments to these standard procedures. As in the construction of appropriate standard errors, these adjustments involve the use of the $\hat{\Sigma}$ matrix. Rao and Thomas also consider alternative test procedures, such as the Wald test, which also make use of $\hat{\Sigma}$, and make comparisons between the different methods.

6.2.3. Logistic models for domain proportions

A log–linear model for a cross-classification treats the cross-classifying variables symmetrically. In many applications, however, one of these variables will be considered as the response and others as the explanatory variables (also called factors). The simplest case involves the cross-classification of a binary response variable Y by a single categorical variable X. Let the categories of Y be labelled as 0 and 1, the categories (also called levels) of X as $1, \ldots, I$ and let N_{ji} be the number of population units with $X = i$ and $Y = j$. In order to study how Y depends upon X, it is natural to compare the proportions N_{1i}/N_i, where $N_i = N_{0i} + N_{1i}$, for $i = 1, \ldots, I$. For example, suppose Y denotes smoking status (1 if smoker, 0 if non-smoker) and X denotes age group. Then N_{1i}/N_i is the proportion of people in age group i who smoke. If these proportions are equal then there is no dependence of smoking status on age group. On the other hand, if the proportions vary then we may wish to model how the N_{1i}/N_i depend upon i.

In contingency table terminology, if Y and X define the rows and columns of a two-way table then the N_{1i}/N_i are *column proportions*, as compared with the N_{ji}/N ($N = \sum N_i$), used to define a log–linear model, which are *cell proportions*. In survey sampling terminology, the subsets of the population defined by the categories of X are referred to as *domains* and the N_{1i}/N_i are referred to as *domain proportions*. See Section 7.4, where the notation $\mu_{1i} = N_{1i}/N_i$ is used.

In a more general setting, there may be several categorical explanatory variables (factors). For example, we may be interested in how smoking status depends upon age group, gender and social class. In this case it is still possible to represent the cells in the cross-classification of these factors by a single index $i = 1, \ldots, I$, by using a lexicographic ordering, as in Table 6.1. For example, if age group, gender and social class have five, two and five categories (levels) respectively, then $I = 5 \times 2 \times 5 = 50$ domains i are required. In this case, each domain i refers to a cell in the cross-classification of the explanatory factors, i.e. a specific combination of the levels of the factors.

The dependence of a binary response variable on one or more explanatory variables may then be represented by modelling the dependence of the μ_{1i} on the factors defining the domains. Just as μ_i could refer to a finite population proportion or model probability, depending upon the scientific context, μ_{1i} may refer to either a domain proportion N_{1i}/N_i or a corresponding model probability. For example, it might be assumed that N_{1i} is generated by a binomial superpopulation model (conditional on N_i): $N_{1i}|N_i \sim Bin(N_i, \mu_{1i})$. A common model for the dependence of μ_{1i} on the explanatory factors is a logistic regression model, which is specified by Rao and Thomas in Section 7.4 as

$$\log \left[\mu_{1i}/(1 - \mu_{1i}) \right] = x_i'\theta, \tag{6.4}$$

where x_i is an $m \times 1$ vector of known constants, which may depend upon the factor levels defining domain i, and θ is an $m \times 1$ vector of regression parameters. In general $m \leq I$ and the model is saturated if $m = I$. For example, suppose that there are $I = 4$ domains defined by the cross-classification of two binary factors. A saturated model in which μ_{1i} may vary freely according to the levels of these two factors may be represented by this model with $m = 4$ parameters. On the other hand a model involving main effects of the two factors but no interaction (on the logistic scale) may be represented with a θ vector of dimension $m = 3$. See Agresti (1990, section 4.3.2) for further discussion of such models.

Rao and Thomas discuss how θ may be estimated by pseudo-MLE. This requires the use of consistent estimates of both the μ_{1i} and the proportions N_i/N. These may be obtained from standard descriptive survey estimates of the N_{1i} and the N_i, as in (6.2). Rao and Thomas discuss appropriate methods for testing hypotheses about θ, which require an estimated covariance matrix for the vector of estimated μ_{1i}. These approaches and methods for estimating the standard errors of the pseudo-MLEs are discussed further in Roberts, Rao and Kumar (1987) and are extended, for example to polytomous Y, in Rao, Kumar and Roberts (1989).

6.3. ANALYSIS OF UNIT-LEVEL DATA

The approach to logistic regression described in the previous section can become impractical in many survey settings. First, survey data often include a large number of variables which an analyst may wish to consider using as explanatory variables and this may result in small or zero sample sizes in many of the domains formed by cross-classifying these explanatory variables. This may lead in particular to unstable $\hat{\mu}_{1i}$ and problems in estimating the covariance matrix of the $\hat{\mu}_{1i}$. Second, some potential explanatory variables may be continuous and it may be undesirable to convert these arbitrarily to categorical variables. It may therefore be preferable to undertake regression modelling at the unit level.

6.3.1. Logistic regression

The basic unit-level data consist of the values (x_t, y_t) for units t in the sample s, where x_t is the vector of values of the explanatory variables and y_t is the value (0 or 1) of the response variable. A conventional logistic regression model assumes that y_t is generated by a model, for fixed x_t, in such a way that the probability that $y_t = 1$, denoted $\mu_t(\theta) = Pr(y_t = 1) = E(y_t)$, obeys

$$\log[\mu_t(\theta)/(1 - \mu_t(\theta))] = x_t'\theta.$$

This is essentially the same model as in (6.4), except that $Pr(y_t = 1)$ can now vary from unit to unit, whereas before the same probability applied to all units within a given domain.

Under the conventional assumption that the sample y_t values independently follow this model, the MLE of θ is obtained by solving

$$\sum_{t \in s} u_t(\theta) = 0 \tag{6.5}$$

where $u_t(\theta) = [y_t - \mu_t(\theta)]x_t$ is the component of the score function for unit t. For a complex sampling scheme, this assumption is invalid and the MLE may be inconsistent. Instead, Rao and Thomas consider the pseudo-MLE obtained from solving the weighted version of these score equations:

$$\sum_{t\in s} w_{ts} u_t(\theta) = 0$$

where w_{ts} is the survey weight for unit t. For standard Horvitz–Thompson weights, i.e. proportionate to the reciprocals of the inclusion probabilities, the pseudo-MLE will be design consistent for the *census parameter* θ_U (as discussed in Section 1.2), the value of θ solving the corresponding finite population score equations:

$$\sum_{1}^{N} u_t(\theta) = 0. \tag{6.5}$$

If the model is true at the population level then this census parameter will be model consistent for the model parameter θ and the pseudo-MLE may be viewed as a consistent estimator (with respect to both the design and the model) of θ. If the model is not true, only holding approximately, then the pseudo-MLE is still consistent for the census parameter. The extent to which this is a sensible parameter to make inference about is discussed further below in Section 6.3.2 and by Scott and Wild in Chapter 8.

Rao and Thomas discuss the estimation of the covariance matrix of the pseudo-MLE of θ as well as procedures for testing hypotheses about θ. One advantage of the approach to fitting logistic regression models discussed in the previous section is that goodness-of-fit tests are directly generated by comparing the survey estimates $\hat{\mu}_{1i}$ of the domain proportions with the proportions $\mu_{1i}(\hat{\theta})$ predicted by the model. Such goodness-of-fit tests are not directly generated when the model is fitted to unit-level data and Rao and Thomas only discuss testing hypotheses about θ, treating the logistic model as given. See Korn and Graubard (1999, section 3.6) for further discussion of the assessment of the goodness of fit of logistic regression models with survey data.

Properties of pseudo-MLE are explored further by Scott and Wild (Chapter 8). They consider the specific kinds of sampling design that are used in case–control studies. A basic feature of these studies is the stratification of the sample by Y. In a case–control study, the binary response variable Y classifies individuals into two groups, the *cases* which possess some condition and for whom $Y = 1$, and the *controls* which do not possess the condition and for whom $Y = 0$. Scott and Wild give an example, where the cases are children with meningitis and the controls are children without meningitis. The aim is to study how Y depends on a vector of explanatory variables via a logistic regression model, as above. Sampling typically involves independent samples from the cases and from the controls, often with very different sampling fractions if the cases are rare. The sampling fraction for rare cases may be high, even 100%, whereas the sampling fraction for controls may be much lower. In Scott and Wild's example the ratio is 400 to 1.

Pseudo-MLE, weighting inversely by the probabilities of selection, provides one approach to consistent estimation of θ in case–control studies, provided these probabilities are available, at least up to a scale factor. Alternatively, as Scott and Wild discuss, a special property of case–control sampling and the logistic regression model is that the MLE, obtained from solving (6.5) above and ignoring the unequal probabilities of selection, provides consistent estimates of the elements of θ other than the intercept (if the model holds).

Scott and Wild refer to the pseudo-MLE as the *design-weighted* estimator and the MLE which ignores the sample selection as the *sample-weighted* estimator. They provide a detailed discussion of the relative merits of these two estimators for case–control studies. An introduction to this discussion is provided in the following section.

6.3.2. Some issues in weighting

The question of whether or not to use survey weights when fitting regression models, such as the logistic regression model above, has been widely discussed in the literature – see e.g. Pfeffermann (1993, 1996) and Korn and Graubard (1999, Ch. 4). Two key issues usually feature in these discussions. First, there is the issue of bias and efficiency – in certain senses the use of survey weights may reduce bias but also reduce efficiency. Second, there is the issue of the impact of model misspecification.

If the specified model is assumed correct and if sampling depends only upon the explanatory variables (i.e. the survey weights are unrelated to the response variable conditional upon the values of the explanatory variables), then it may be argued that the sampling will not induce bias in the unweighted estimators of the regression coefficients. The survey-weighted estimator will, however, typically have a greater variance than the unweighted estimator and thus the latter estimator is preferred in this case (Pfeffermann, 1996). The loss of efficiency of the survey-weighted estimator will usually be of more concern, the greater the variation in the survey weights.

On the other hand, if sampling depends upon factors which may be related to the response variable, even after conditioning on the explanatory variables, then sampling may induce bias in the unweighted estimator. The use of survey weights provides a simple method to protect against this potential bias. There may remain an issue of trade-off in this case between bias and efficiency. The case–control setting with logistic regression is unusual, since the unweighted estimator does not suffer from bias, other than in the intercept, despite sampling being directly related to the response variable.

It is desirable to use diagnostic methods when specifying regression models, so that the selected model fits the data well. Nevertheless, however much care we devote to specifying a model well, it will usually be unreasonable to assume that the model is entirely correct. It is therefore desirable to protect against the possibility of some misspecification. A potential advantage of the survey-weighted estimator is that, even under misspecification, it is consistent for a clearly defined census parameter θ_U, which solves (6.6) in the case of logistic regression. This census parameter may be interpreted as defining a logistic regression model which best approximates the actual model in a certain sense in the population. Scott and Wild refer to this as the 'whole population' approximation. In the same way, the unweighted estimator will be design consistent for another value of θ, the one which solves

$$\sum_1^N \pi_t u_t(\theta) = 0, \tag{6.7}$$

where π_t is the inclusion probability of unit t. This parameter defines a logistic regression model which approximates the actual model in a different sense. As a result, the issue of whether survey weighting reduces bias is replaced by the question of which approximation is more scientifically sensible. A conceptual attraction of the whole population approximation is that it does not depend upon the sampling scheme. As indicated by (6.7), the second approximation depends upon the π_t, which seems unattractive since it seems conceptually unnatural for the parameter of interest to depend upon the sampling scheme. There may, however, be other countervailing arguments. Consider, for example, a longitudinal survey to study the impact of birthweight on educational attainment at age 7. The survey designer might deliberately oversample babies of low birthweight and undersample babies of normal birthweight in order to achieve a good spread of birthweights for fitting the regression relationship. It may therefore be scientifically preferable to fit a model which best approximates the true relationship across the sample distribution of birthweights rather than across the population distribution. In the latter case the approximation will be dominated by the fit of the model for babies of normal birthweight (cf. Korn and Graubard, 1999, section 4.3). Further issues, relating specifically to binary response models, are raised by Scott and Wild. In their application they show that the whole population approximation only provides a good approximation to the actual regression relationship for values in the tail of the covariate distribution, whereas the second approximation provides a good approximation for values of the covariate closer to the centre of the population distribution. They conclude that the second approximation 'represents what is happening in the population better than' the whole population approximation and that the unweighted MLE is more robust to model misspecification.

Issues of weighting and alternative approaches to correcting for nonignorable sampling schemes are discussed further in Sections 9.3 and 11.6 and in Chapter 12.

CHAPTER 7

Analysis of Categorical Response Data from Complex Surveys: an Appraisal and Update

J. N. K. Rao and D. R. Thomas

7.1. INTRODUCTION

Statistics texts and software packages used by researchers in the social and health sciences and in other applied fields describe and implement classical methods of analysis based on the assumption of independent identically distributed data. In most cases, there is little discussion of the analysis difficulties posed by data collected from complex sample surveys involving clustering, stratification and unequal probability selection. However, it has now been well documented that applying classical statistical methods to such data without making allowance for the survey design features can lead to erroneous inferences. In particular, ignoring the survey design can lead to serious underestimation of standard errors of parameter estimates and associated confidence interval coverage rates, as well as inflated test levels and misleading model diagnostics. Rapid progress has been made over the last 20 years or so in the development of methods that account for the survey design and as a result provide valid inferences, at least for large samples.

This chapter will focus on the complex survey analysis of cross-classified count data using log–linear models, and the analysis of binary or polytomous response data using logistic regression models. In Chapter 4 of the book edited by Skinner, Holt and Smith (1989), abbreviated SHS, we provided an overview of methods for analyzing cross-classified count data. The methods reviewed included Wald tests, Rao–Scott first-and second-order corrections to classical chi-squared and likelihood ratio tests, as well as residual analysis to detect the

Analysis of Survey Data Edited by R. L. Chambers and C. J. Skinner
© 2003 John Wiley & Sons, Ltd

nature of deviations from the null hypothesis. Other accounts of these developments in survey data analysis have been published. For example, the book by Lehtonen and Pahkinen (1995) builds on an earlier review by the authors (Rao and Thomas, 1988), and the recent book by Lohr (1999) includes a description of methods for analyzing regression data and count data from complex samples. The objective of this chapter is to update our 1989 review and to present an appraisal of the current status of methods for analyzing survey data with categorical variables.

To make this chapter self-contained, we begin with an overview of Rao–Scott and Wald tests for log–linear models, together with a description of some approaches not covered in Chapter 4 of SHS. These will include the jackknife procedures developed by Fay (1985) and the Bonferroni procedures applied to complex survey data by Thomas (1989), Korn and Graubard (1990) and Thomas, Singh and Roberts (1996). These developments are based on large-sample theory, and it cannot be assumed that asymptotic properties will always hold for the 'effective' sample sizes encountered in practice. Therefore, recent simulation studies will be reviewed in Section 7.3 in order to determine which classes of test procedures should be recommended in practice. In Sections 7.4 and 7.5 we present a discussion of logistic regression methods for binary or polytomous response variables, including recent general approaches based on quasi- score tests (Rao, Scott and Skinner, 1998). To conclude, we discuss some applications of complex survey methodology to other types of problems, including data subject to misclassification, toxicology data on litters of experimental animals, and marketing research data in which individuals can respond to more than one item on a list. We describe how methods designed for complex surveys can be extended and/or applied to provide convenient test procedures.

7.2. FITTING AND TESTING LOG–LINEAR MODELS

7.2.1. Distribution of the Pearson and likelihood ratio statistics

Consider a multi-way cross-classification of categorical variables for a sample (or domain) of size n, in which the cells in the table are numbered lexicographically as $i = 1, \ldots, I$, where I is the total number of cells. Let μ denote an $I \times 1$ vector of finite population cell proportions, or cell probabilities under a superpopulation model, and let $\hat{\mu}$ denote a corresponding design-consistent estimator of μ. For any survey design, a consistent estimate of the elements of μ is given by $\hat{\mu}_i = \hat{N}_i/\hat{N}$, where the $\hat{N}_i$ represent weighted-up survey counts, i.e., consistent estimates of the population cell counts N_i, and $\hat{N} = \sum \hat{N}_i$; published tables often report weighted-up counts. A log–linear model on the cell proportions (or probabilities) $\mu = \mu(\theta)$ can be expressed as

$$\log(\mu) = u(\theta)1 + X\theta, \tag{7.1}$$

where $\log(\mu)$ is the I-vector of log cell proportions. Here X is a known $I \times r$ matrix of full rank $r \leq I - 1$, such that $X'\underset{\sim}{1} = \underset{\sim}{0}$, where $\underset{\sim}{0}$ and $\underset{\sim}{1}$ are I-vectors of zeros and ones, respectively, θ is an r-vector of parameters and $u(\theta)$ is a normalizing factor that ensures that $\mu'1 = 1$.

For the case of simple random sampling with replacement (frequently referred to in this context as multinomial sampling), likelihood equations can be solved to yield the maximum likelihood estimates (MLEs) of θ. For general designs, however, this approach cannot be implemented due to the difficulty of defining appropriate likelihoods. Rao and Scott (1984) obtained instead a 'pseudo-MLE' of θ by solving the estimating equations

$$X'\mu(\hat{\theta}) = X'\hat{\mu}, \tag{7.2}$$

namely the multinomial likelihood equations in which unweighted cell proportions have been replaced by $\hat{\mu}$. Consistency of $\hat{\mu}$ under the complex survey design ensures the consistency of $\hat{\theta}$ as an estimator of θ, and consequently ensures the consistency of $\mu(\hat{\theta})$ as an estimator of μ under the log–linear model (7.1).

A general Pearson statistic for testing the goodness of fit of model (7.1) can be defined as

$$X_P^2 = n \sum_{i=1}^{I} (\hat{\mu}_i - \mu_i(\hat{\theta}))^2 / \mu_i(\hat{\theta}), \tag{7.3}$$

and a statistic corresponding to the likelihood ratio is defined as

$$X_{LR}^2 = 2n \sum_{i=1}^{I} \hat{\mu}_i \log[\hat{\mu}_i/\mu_i(\hat{\theta})]. \tag{7.4}$$

Rao and Scott (1984) showed that for general designs, X_P^2 and X_{LR}^2 are asymptotically distributed as a weighted sum, $\delta_1 W_1 + \ldots + \delta_{I-r-1} W_{I-r-1}$, of $I - r - 1$ independent χ_1^2 random variables, W_i. The weights, δ_i, are eigenvalues of a general design effects matrix, given by

$$\Delta = (\tilde{Z}'\Omega\tilde{Z})^{-1}(\tilde{Z}'\Sigma\tilde{Z}), \tag{7.5}$$

where

$$\tilde{Z} = [I - X(X'\Omega X)^{-1}X'\Omega]Z, \tag{7.6}$$

and Z is any matrix of dimension $I \times (I - r - 1)$ such that $(1, X, Z)$ is non-singular and that satisfies $Z'1 = 0$. The matrix Ω is defined by $\Omega = n^{-1}[D(\mu) - \mu\mu']$, where $D(\mu) = \mathrm{diag}(\mu)$, the $I \times I$ diagonal matrix with diagonal elements $\mu_1, \ldots, \mu_I$. Thus Ω has the form of a multinomial covariance matrix, corresponding to the covariance matrix of $\hat{\mu}$ under a simple random sampling design. In this case, however, $\hat{\mu}$ is a vector of consistent estimates of cell proportions under a complex survey design, with asymptotic covariance matrix under the complex survey design denoted by $\Sigma = Var(\hat{\mu})$. When the actual design is simple random, $\Sigma = \Omega$, in which case Δ reduces to the identity matrix so that all $\delta_i = 1$, and both X_P^2 and X_{LR}^2 recover their classical χ_{I-r-1}^2 asymptotic distributions. As illustrated in Chapter 4 of SHS using data from the Canada Health Survey, the generalized design effects, δ_i, obtained from (7.5) can differ markedly from one when the survey design is complex.

The result (7.5) can also be extended to tests on nested models. Rao and Scott (1984) considered the case when $X\theta$ of Equation (7.3) is partitioned as $X_1\theta_1 + X_2\theta_2$, with X_1 and X_2 of full column rank r_1 and r_2, respectively, with $r_1 + r_2 = r \leq I$. The Pearson statistic for testing the nested hypothesis $H_0(2|1): \theta_2 = \underset{\sim}{0}$ is given by

$$X_P^2(2|1) = n \sum_{i=1}^{I} (\mu_i(\hat{\theta}) - \mu_i(\hat{\theta}_1))^2 / \mu_i(\hat{\theta}_1), \qquad (7.7)$$

where $\pi_i(\hat{\theta}_1)$ is the pseudo-MLEs under the reduced model, $X_1\theta_1$, obtained from the pseudo-likelihood equations $X_1'\mu(\hat{\theta}_1) = X_1'\mu$. A general form of the likelihood ratio statistic can also be defined, namely

$$X_{LR}^2(2|1) = 2n \sum_{i=1}^{I} \mu_i(\hat{\theta}) \log[\mu_i(\hat{\theta})/\mu_i(\hat{\theta}_1)]. \qquad (7.8)$$

The asymptotic distributions of $X_P^2(2|1)$ and $X_{LR}^2(2|1)$ are weighted sums of r_2 independent χ_1^2 variables. The design effects matrix defining the weights has the form of Equation (7.5), with $\tilde{Z}$ replaced by $\tilde{X}_2$, where $\tilde{X}_2$ is defined as in Equation (7.6), with X replaced by X_1 and Z replaced by X_2.

7.2.2. Rao–Scott procedures

Under multinomial sampling, tests of model fit and tests of nested hypotheses are obtained by referring the relevant Pearson or likelihood ratio statistic to $\chi_d^2(\alpha)$, the upper $100\alpha\%$ point of a chi-squared distribution on d degrees of freedom, where $d = I - r - 1$ for the test of model fit and $d = r_2$ for the test of a nested hypothesis. Under complex sampling, alternatives to the classical testing strategy are required to deal with the design-induced distortion of the asymptotic distribution of the Pearson and likelihood ratio statistics. The adjusted chi-squared procedures of Rao and Scott (1981, 1984) are described in this section, while Sections 7.2.3, 7.2.4 and 7.2.5 consider Wald tests and their variants, tests based on the Bonferroni inequality, and Fay's jackknifed tests, respectively.

First-order Rao–Scott procedures

The first-order Rao–Scott strategy is based on the observation that $X_P^2/\delta.$ and $X_{LR}^2/\delta.$ (where $\delta.$ is the mean of the generalized design effects, δ_i) have the same first moment as χ_d^2, the asymptotic distribution of the Pearson statistic under multinomial sampling. Thus a first-order Rao–Scott test of model fit refers

$$X_P^2(\hat{\delta}.) = X_P^2/\hat{\delta}. \quad \text{or} \quad X_{LR}^2(\hat{\delta}.) = X_{LR}^2/\hat{\delta}. \qquad (7.9)$$

to the upper $100\alpha\%$ point of χ_d^2, where $d = I - r - 1$, $\hat{\delta}$ is a consistent estimator of $\delta.$ given by

$$\hat{\delta}. = \text{trace}(\hat{\Delta})/d. \tag{7.10}$$

Here $\hat{\Delta}$ is obtained from Equation (7.5), with Ω replaced by its estimate, $\hat{\Omega}$, given by $\hat{\Omega} = n^{-1}[\text{diag}(\mu(\hat{\theta})) - \mu(\hat{\theta})'\mu(\hat{\theta})]$, and with Σ replaced by $\hat{\Sigma}$, the consistent estimator of the covariance matrix of $\hat{\mu}$ under the actual design (or superpopulation model). The tests based on $X_P^2(\hat{\delta}.)$ and $X_{LR}^2(\hat{\delta}.)$ are asymptotically correct when the individual design effects, δ_i, are equal, and simulation evidence to be discussed later confirms that they work well in practice when the variation among the δ_i is small. The second-order Rao–Scott procedure outlined below is designed for cases when the variation among the design effects is expected to be appreciable.

One of the main advantages of the first-order procedures is that they do not always require the full estimate, $\hat{\Sigma}$, of the covariance matrix. The trace function of Equation (7.10) can often be expressed in closed form as a function of the estimated design effects of specific cell and marginal probabilities, which may be available from published data. Closed form expressions are given in section 4.3 of SHS for a simple goodness-of-fit test on a one-way array, the test of homogeneity of vectors of proportions, the test of independence in a two-way table, as well as for specific tests on log–linear models for multi-way tables for which closed form MLEs exist. In the latter case, $\hat{\delta}.$ depends only on the estimated cell and marginal design effects. Moreover, if the cell and marginal design effects are all equal to D, then $\hat{\delta}.$ reduces to D provided the model is correct (Nguyen and Alexander, 1989). For testing association in an $A \times B$ contingency table, Shin (1994) studied the case where only the estimated marginal row and column design effects, $D_a(1)$ and $D_b(2)$, are available, and proposed an approximation to $\hat{\delta}.$, denoted $\hat{\delta}.^{\text{min}}$, that depends only on $D_a(1)$ and $D_b(2), a = 1, \ldots, A; \ b = 1, \ldots, B$. For $\hat{\delta}.$, we have (under independence) the closed form expression

$$(A-1)(B-1)\hat{\delta}. = \sum_a \sum_b (1 - \hat{\mu}_a.\hat{\mu}_{.b})\hat{D}_{ab} - \sum_a (1 - \hat{\mu}_a.)\hat{D}_a(1)$$
$$- \sum_b (1 - \hat{\mu}_{.b})\hat{D}_b(2), \tag{7.11}$$

while Shin's (1994) approximation takes the form

$$(A-1)(B-1)\hat{\delta}. = \sum_a \sum_b (1 - \hat{\mu}_a.\hat{\mu}_{.b})\min\{\hat{D}_a(1), \hat{D}_b(2)\}$$
$$- \sum_a (1 - \hat{\mu}_a.)\hat{D}_a(1) - \sum_b (1 - \hat{\mu}_{.b})\hat{D}_b(2), \tag{7.12}$$

where $\hat{\mu}_{ab}, \hat{\mu}_a.$ and $\hat{\mu}_{.b}$ are the estimated cell, row and column proportions, and $\hat{D}_{ab}$ is the estimated cell design effect of $\hat{\mu}_{ab}$. This approximation turned out to be quite close to the true $\hat{\delta}.$ and performed better than the approximation $\min\{\sum_a \hat{D}_a(1)/A, \sum_b \hat{D}_b(2)/B\}$, proposed earlier by Holt, Scott and Ewings (1980) when applied to some survey data obtained from a longitudinal study of drinking behaviour among US women living in non-institutional settings.

Second-order Rao–Scott procedures
When an estimate $\hat{\Sigma}$ of the full covariance matrix $\Sigma = V(\hat{\mu})$ is available, a more accurate correction to X_P^2 or X_{LR}^2 can be obtained by matching the first two moments of the weighted sum of χ_1^2 random variables to a single chi-squared distribution. Unlike the first-order correction, the second-order correction accounts for variation among the generalized design effects, $\hat{\delta}_i$. For a test of model fit, it is implemented by referring

$$X_P^2(\hat{\delta}., \hat{a}) = \frac{X_P^2(\hat{\delta}.)}{1 + \hat{a}^2} \quad \text{or} \quad X_{LR}^2(\hat{\delta}., \hat{a}) = \frac{X_{LR}^2(\hat{\delta}.)}{1 + \hat{a}^2} \tag{7.13}$$

to the upper $100\alpha\%$ point of a χ_{d*}^2 random variable with $d^* = d/(1 + \hat{a}^2)$ fractional degrees of freedom, where $\hat{a}$ is a measure of the variation among the design effects, given by the expressions

$$\hat{a}^2 = d \operatorname{trace}(\hat{\Delta}^2)/[\operatorname{tr}(\hat{\Delta})]^2 - 1 = \sum_{i=1}^{d} (\hat{\delta}_i - \hat{\delta}.)^2/d\hat{\delta}.^2. \tag{7.14}$$

From Equation (7.14), it can be seen that $\hat{a}$ has the form of a coefficient of variation. The simulation results of Thomas and Rao (1987) for simple goodness of fit showed that second-order corrected tests do a better job of controlling Type I error when the variation among design effects is appreciable. The simulation results for tests of independence in two-way tables discussed in Section 7.4 confirm this conclusion.

F-based variants of the Rao–Scott procedures
For a simple goodness-of-fit test, a heuristic large-sample argument was given by Thomas and Rao (1987) to justify referring F-based versions of the first-and second-order Rao–Scott tests to F-distributions. An extension of this argument to the log–linear case yields the following variants.

F-based first-order tests   Refer $FX_P^2(\hat{\delta}.)$ to an F-distribution on d and dv degrees of freedom, where

$$FX_P^2(\hat{\delta}.) = X_P^2(\hat{\delta}.)/d, \tag{7.15}$$

and v is the degrees of freedom for estimating the covariance matrix Σ. The statistic $FX_{LR}^2(\hat{\delta}.)$ is defined similarly.

F-based second-order tests   Denoted $FX_P^2(\hat{\delta}., \hat{a})$ and $FX_{LR}^2(\hat{\delta}., \hat{a})$, these second-order procedures are similar to the first-order F-based tests in that they use the same test statistics, namely $X_P^2(\hat{\delta}.)/d$ or $X_{LR}^2(\hat{\delta}.)/d$, but refer them instead to an F-distribution on d^* and d^*v degrees of freedom, again with $d^* = d/(1 + \hat{a}^2)$.

Rao–Scott procedures for nested hypotheses
For testing nested hypotheses on log–linear models, first-order and second-order Rao–Scott procedures, similar to (7.9) and (7.13), are reported in

section 4.4 of SHS. By analogy with the above strategy, F-based versions of these tests can also be defined.

7.2.3. Wald tests of model fit and their variants

Log–linear form
Wald tests, like the second-order Rao–Scott procedures, cannot be implemented without an estimate, $\hat{\Sigma}$, of the full covariance matrix of $\hat{\mu}$ under the actual survey design. Wald tests can be constructed for all of the cases considered in Chapter 4 of SHS, and in theory Wald tests provide a general solution to the problem of the analysis of survey data. In practice, however, they have some serious limitations as will be discussed. As an example, we consider a Wald test of goodness of fit of the log–linear model (7.1). The saturated model corresponding to model (7.1) can be written as

$$\log(\mu) = u(\theta)1 + X\theta + Z\theta_Z, \tag{7.16}$$

where Z is an $I \times (I - r - 1)$ matrix as defined earlier (with the added constraint that $Z'X = 0$), and θ_Z is a vector of $(I - r - 1)$ parameters. The goodness-of-fit hypothesis can be expressed as $H_0: \theta_Z = 0$, which is equivalent to $H_0: \varphi = Z' \log(\mu) = 0$. The corresponding log–linear Wald statistic can therefore be written as

$$X_W^2(LL) = \hat{\varphi}'[\hat{V}(\hat{\varphi})]^{-1}\hat{\varphi}', \tag{7.17}$$

where $\hat{\varphi} = Z' \log(\hat{\mu})$ and the estimated covariance matrix of $\hat{\varphi}$ is given by

$$\hat{V}(\hat{\varphi}) = Z'D(\hat{\mu})^{-1}\hat{\Sigma}D(\hat{\mu})^{-1}Z, \tag{7.18}$$

where $D(\hat{\mu}) = \text{diag}(\hat{\mu})$. An asymptotically correct Wald test is obtained by referring X_W^2 to the upper α-point of a χ_{I-r-1}^2 random variable.

Residual form
For a test of independence on an $A \times B$ table, the null hypothesis of independence can be expressed in residual form as

$$H_0: h_{ab} = \mu_{ab} - \mu_{a+}\mu_{+b} = 0; \quad a = 1, \ldots, (A-1), b = 1, \ldots, (B-1). \tag{7.19}$$

Rao and Thomas (1988) described a corresponding residual form of the Wald statistic, given by

$$X_W^2(R) = \hat{h}'[\hat{V}(\hat{h})]^{-1}\hat{h}, \tag{7.20}$$

where the $(A-1)(B-1)$ elements of $\hat{h}$ are defined by the estimated residuals $\hat{\mu}_{ab} - \hat{\mu}_{a+}\hat{\mu}_{+b}$, and $\hat{V}(\hat{h})$ is a consistent estimate of the covariance matrix of $\hat{h}$ under the general survey design. Residual forms of the Wald test can be constructed for testing the fit of any log–linear model, by using the expressions given by Rao and Scott (1984) for the variances of estimated residuals for log–linear models of the form (7.1). Thus $X_W^2(LL)$ and $X_W^2(R)$ are direct competitors.

As discussed in Section 7.3 on simulation studies, both versions of the Wald test are very susceptible to serious Type I error inflation in practice even for moderate-sized tables, and should be used in practice only for small tables and large values of v, the degrees of freedom for estimating Σ.

F-based Wald procedures
Thomas and Rao (1987) described an F-based version of the Wald test that exhibits better control of test levels at realistic values of v. For a Wald test of model fit, this procedure consists of referring

$$FX_W^2 = \frac{(v - d + 1)}{dv} X_W^2 \qquad (7.21)$$

to an F-distribution on d and $(v - d + 1)$ degrees of freedom, where $d = I - r - 1$ for the test of fit of model (7.1). The FX_W^2 test can be justified under the heuristic assumption that $\hat{\Sigma}$ has a Wishart distribution. F-based versions of $X_W^2(LL)$ and $X_W^2(R)$ will be denoted $FX_W^2(LL)$ and $FX_W^2(R)$, respectively. This simple strategy for taking account of the variation in $\hat{\Sigma}$ improves the Type I error control of both procedures.

Wald tests on the parameter vector
Scott, Rao and Thomas (1990) developed a unified weighted least squares approach to estimating θ in the log–linear model (7.1) and to developing Wald tests on θ. They used the following log–linear model:

$$\log (\hat{\mu}) = u\underline{1} + X\theta + \delta \qquad (7.22)$$

where δ is the error vector with asymptotic mean $\underline{0}$ and singular asymptotic covariance matrix $D(\mu)^{-1}\Sigma D(\mu)^{-1}$. By appealing to the optimal theory for linear models having singular covariance matrices, Scott, Rao and Thomas (1990) obtained an asymptotically best linear unbiased estimator of θ and a consistent estimator of its asymptotic covariance matrix, together with Wald tests of general linear hypotheses on the model parameter vector θ.

7.2.4.   Tests based on the Bonferroni inequality

Log–linear form
Thomas and Rao's (1987) simulation results show that the Type I error control of a Wald test rapidly worsens as the dimension, d, of the test increases. Since the hypothesis of goodness of fit of log–linear model (7.1), $H_0: \varphi = \underline{0}$, can be replaced by d simultaneous hypotheses, $\varphi_i = 0, i = 1, \ldots, d$, it is natural to consider whether or not d simultaneous tests might provide better control than the Wald test, or even the F-based Wald tests. A log–linear Bonferroni procedure, $Bf(LL)$, with asymptotic error rate bounded above by the nominal test level, α, is given by the rule

$$\text{Reject } H_0 \text{ if: } |\hat{\varphi}_i| > t_{\alpha/2d}^v (\hat{v}_{\varphi_i})^{1/2}, \text{ for any } i, i = 1, \ldots, d, \qquad (7.23)$$

where $\hat{\varphi}_i$ is the survey estimate of φ_i, $\hat{v}_{\varphi_i}$ denotes the estimated variance of $\hat{\varphi}_i$, given by the ith diagonal element of the covariance matrix (7.18), and $t^v_{\alpha/2d}$ represents the upper $100(\alpha/2d)\%$ point of a t-distribution on v degrees of freedom, namely the degrees of freedom for estimating the error variance v_{φ_i}. The use of a t-distribution in place of the asymptotic standard normal distribution is a heuristic device that was shown to yield much better control of overall error rates by Thomas (1989), in his study of simultaneous confidence intervals for proportions under cluster sampling.

Residual form
The hypothesis of goodness of fit of log–linear model (7.1) can also be expressed in the residual form $H_0 \colon h = \underline{0}$. Since H_0 can be replaced by the d simultaneous hypotheses $h_i = 0$, $i = 1, \ldots, d$, a residual form of the Bonferroni test is given by

$$\text{Reject } H_0 \text{ if } |\hat{h}_i| > t^v_{\alpha/2d}(\hat{v}_{h_i})^{1/2}, \text{for any } i, \ i = 1, \ldots, d, \qquad (7.24)$$

where $\hat{v}_{h_i}$ denotes the estimated variance of the residual $\hat{h}_i$.

Caveats Wald procedures based on the log–linear form of the hypothesis ($H_0 \colon \varphi = \underline{0}$) are invariant to the choice of Z, provided it is of full column rank and satisfies $Z'1 = Z'X = \underline{0}$. Similarly, Wald procedures based on the residual form ($H_0 \colon h = \underline{0}$) of the independence hypothesis are invariant to the choice of the residuals to be dropped from the specification of the null hypothesis, a total of $A + B - 1$ in the case of a two-way $A \times B$ table. Unfortunately, these invariances do not hold for the Bonferroni procedures. The residual procedure, $Bf(R)$, can be modified to include all residuals, e.g., AB residuals in the case of a two-way table, but in this case there is a risk that the Bonferroni procedure will become overly conservative. An alternative to the log–linear Bonferroni procedure, $BF(LL)$, can be derived by using a Scheffe-type argument to get simultaneous confidence intervals on the φ_i. By inverting these intervals, a convenient testing procedure of the form (7.23) can then be obtained, with the critical constant $t^v_{\alpha/2d}$ replaced by $[(dv/(v - d + 1))F_{d,(v-d+1)}(\alpha)]^{1/2}$. Unfortunately, it can be shown that this procedure is very conservative. As an example, for a 3×3 table and large v, the asymptotic upper bound on the Type I error of this procedure corresponding to a nominal 5% α-level is 0.2%, far too conservative in practice.

The only practical approach is for the analyst to recognize that the Bonferroni procedure $Bf(LL)$ of Equation (7.23) yields results that are specific to the choice of Z. This is quite acceptable in many cases, since the choice of Z is frequently determined by the nature of the data. For example, orthogonal polynomial contrasts are often used to construct design matrices for ordinal categorical variables. Design matrices obtained by dropping a column from an identity matrix are usually recommended when the categories of a categorical variable represent comparisons to a control category.

7.2.5. Fay's jackknifed tests

In a series of papers, Fay (1979, 1984, 1985) proposed and developed jack-knifed versions of X_P^2 and X_{LR}^2, which can be denoted X_{PJ} and X_{LRJ}, respectively. The jackknifed tests are related to the Rao–Scott corrected tests, and can be regarded as an alternative method of removing, or at least reducing, the distortion in the complex survey distributions of X_P^2 and X_{LR}^2 characterized by the weights δ_i, $i = 1, \dots, I - r - 1$. Though more computationally intensive than other methods, Fay's procedures have the decided advantage of simplicity. Irrespective of the analytical model to be tested, and subject to some mild regularity conditions, all that is required is an algorithm for computing X_P^2 or X_{LR}^2. Software for implementing the jackknife approach is available and the simulation studies described in Section 7.3 have shown that jackknifed tests are competitive from the point of view of control of test levels, and of power. A brief outline of the procedure will be given here, together with an example.

The form of X_{PJ} and X_{LRJ}

It was noted in Section 7.2.1 that consistent estimators of the I finite population proportions in a general cross-tabulation can be obtained as ratios of the population count estimates, written in vector form as $\hat{N} = (\hat{N}_1, \dots, \hat{N}_I)'$. Fay (1985) considered the class of sample replication methods represented by $\hat{N} + W^{(j,k)}$, which satisfy the condition $\sum_k W^{(j,k)} = 0$ and for which an estimator, $V^*(\hat{N})$, of the covariance matrix of $\hat{N}$ is given by

$$V^*(\hat{N}) = \sum_j b_j \sum_k W^{(j,k)} W^{(j,k)'}. \tag{7.25}$$

A number of replication schemes can be represented this way, including jackknife replicates for stratified cluster sampling, with $j = 1, \dots, J$ strata and n_j independently sampled clusters in statum j. In this case, the $W^{(j,k)}$ are given by

$$W^{(j,k)} = \left[\sum_{k'} Z^{(j,k')} - n_j Z^{(j,k)} \right] / (n_j - 1) \tag{7.26}$$

where it is assumed that $\hat{N}$ can be represented in the form $\hat{N} = \sum_j \sum_k Z^{(j,k)}$, with $Z^{(j,k)}$ that are i.i.d. within strata. In this case, $b_j = (n_j - 1)/n_j$.

Let $X_P^2(\hat{N})$ and $X_P^2(\hat{N} + W^{(j,k)})$ represent general chi-squared statistics based on the full sample and the (j, k)th replicate, respectively. The form of these statistics is given by Equation (7.3) except that the sample size n is replaced by the sum of the cell estimates in $\hat{N}$ and $\hat{N} + W^{(j,k)}$, respectively. Fay's (1985) jackknife statistic, X_{PJ}, is then defined as

$$X_{PJ} = [(X_P^2(\hat{N}))^{1/2} - (K^+)^{1/2}]/[V/(8X_P^2(\hat{N}))]^{1/2}, \tag{7.27}$$

where

$$K = \sum_j b_j \sum_k P_{jk}, \tag{7.28}$$

$$V = \sum_j b_j \sum_k P_{jk}^2, \tag{7.29}$$

and

$$P_{jk} = X_P^2(\hat{N} + W^{(j,k)}) - X_P^2(\hat{N}). \tag{7.30}$$

In Equation (7.27), $K^+ = K$ for K positive, and $K^+ = 0$ otherwise. An equivalent expression for X_{LRJ}, the jackknifed version of X_{LR}^2, is obtained by replacing X_P^2 by X_{LR}^2 in Equations (7.27) through (7.30). Similarly, jackknifed statistics $X_{PJ}(2|1)$ and $X_{LRJ}(2|1)$ for testing a nested hypothesis can be obtained by replacing X_P^2 in the above equations by $X_P^2(2|1)$ and $X_{LRJ}^2(2|1)$, respectively.

The distribution of X_{PJ} and X_{LRJ}
Fay (1985) showed that X_{PJ} and X_{LRJ} are both asymptotically distributed as a *function* of weighted sums of $d = I$-r-1 independent χ_1^2 variables. The weights themselves are functions of the δ_i featured in the asymptotic distributions of X_P^2 and X_{LR}^2. When the δ_i are all equal, the asymptotic distribution simplifies to

$$X_{PJ} \text{ and } X_{LRJ} \sim 2^{1/2}[(\chi_d^2)^{1/2} - d^{1/2}]. \tag{7.31}$$

Numerical investigations and simulation experiments have shown that (7.31) is a good approximation to the distribution of the jackknifed statistics even when the δ_i are not equal. In practice, therefore, the jackknife test procedure consists of rejecting the null hypothesis whenever X_{PJ} (or X_{LRJ}) exceeds the upper $100\,\alpha\%$ critical value determined from (7.31).

Example 1
The following example is taken from the program documentation of PC-CPLX (Fay, 1988), a program written to analyze log–linear models using the jackknife methodology. The data come from an experimental analysis of the response rates of school teachers to a survey; thus the response variable (RESPONSE) was binary (*response/no response*). Two variables represent experimental treatments designed to improve response, namely **PAY**, a binary variable with levels *paid/unpaid*, and FOLLOWUP, with levels *phone/mail*. Other variables entered into the log–linear model analysis were RGRADE (*elementary/secondary/combined*), and STATUS (*public/private*). The schools from which the teachers were sampled were treated as clusters, and the 10 participating states were treated as separate strata for the public schools. All private schools were aggregated into an eleventh stratum. The reported analysis involves 2630 teachers; state strata contained an average of 20 clusters (schools) while the private school stratum contained 70 clusters. Thus the average cluster size was close to 10. Differential survey weights were not included in the analysis, so that the design effect can be attributed to the clustering alone.

The following three hierarchical log–linear models were fitted, specified below in terms of the highest order interactions that they contain:

Model 1 {PAY × FOLLOWUP × RGRADE × STATUS}, {RESPONSE × RGRADE × STATUS}

Model 2 {PAY × FOLLOWUP × RGRADE × STATUS}, {RESPONSE × RGRADE × STATUS} {RESPONSE × PAY}

Model 3 {PAY × FOLLOWUP × RGRADE × STATUS}, {RESPONSE × RGRADE × STATUS} {RESPONSE × PAY}, {RESPONSE × FOLLOWUP}

Table 7.1 shows the results of two comparisons. The first compares model 3 to model 1, to test the joint relationship of RESPONSE to the two experimental variables PAY and FOLLOWUP. The second compares model 3 to model 2, to test the relationship between RESPONSE and FOLLOWUP alone. Results are shown for the jackknifed likelihood and first-and second-order Rao–Scott procedures for testing nested hypotheses. F based versions of the Rao–Scott procedures are omitted in view of the large number of degrees of freedom for variance estimation (over 200) in this survey.

It can be seen from Table 7.1 that the p-values obtained from the uncorrected X_P^2 and X_{LR}^2 procedures are very low, in contrast to the p-values generated by the Fay and Rao–Scott procedures, which account for the effect of the stratified cluster design. The values of $\hat{\delta}$. for the two model comparisons are 4.32 and 4.10, for model 3 versus model 1, and model 2 versus model 1, respectively. This is indicative of a strong clustering effect. Thus the strong 'evidence' for the joint effect of the RESPONSE by PAY and RESPONSE by FOLLOWUP interactions provided by the uncorrected tests is erroneous. The comparison of model 3 to model 2 shows that, when the effect of the design is accounted for, there is marginal evidence for a RESPONSE by FOLLOWUP interaction alone. The p-values generated by Fay's jackknifed likelihood test (denoted $X_{LRJ}(2|1)$ for nested hypotheses) and the Rao–Scott tests are quite similar. In this example, the value of $\hat{a}$ for the two degree of freedom test of model 3 versus model 1 is very low at 0.09, so that the second-order Rao–Scott procedure, $X_{LR}^2(2|1)/\hat{\delta}.(1 + \hat{a}^2)$, differs very little from the first-order procedure. The nested comparison of model 3 to model 2 involves only one degree of freedom; thus the first-order Rao–Scott test is asymptotically valid in this case, and the second-order test is not applicable.

Table 7.1   Tests of nested log–linear models for the teacher response study.

| Statistic | Model 3–model 1 | | Model 3–model 2 | |
	Value	Estimated p-value ($d = 2$)	Value	Estimated p-value ($d = 1$)
$X_P^2(2\|1)$	11.80	0.003	11.35	<0.001
$X_{LR}^2(2\|1)$	11.86	0.003	11.41	<0.001
$X_{LRJ}(2\|1)$	0.27	0.276	0.85	0.109
$X_P^2(2\|1)/\hat{\delta}.$	2.73	0.255	2.76	0.097
$X_P^2(2\|1)/\hat{\delta}.(1 + \hat{a}^2)$	2.71	0.258	n.a.	n.a.

7.3. FINITE SAMPLE STUDIES

The techniques reported in Section 7.2 rely for their justification on asymptotic results. Thus simulation studies of the various complex survey test procedures that have been proposed are essential before definitive recommendations can be made. Early simulation studies featuring tests on categorical data from complex surveys were reported by Brier (1980), Fellegi (1980), Koehler and Wilson (1986) and others, but these studies were limited in scope and designed only to examine the specific procedures developed by these authors. The first study designed to examine a number of competing procedures was that of Thomas and Rao (1987), who focused on testing the simple hypothesis of goodness of fit for a vector of cell proportions, i.e., $H_0: \mu = \mu_0$, under a mixture Dirichlet multinomial model of two-stage cluster sampling. They found that the Rao–Scott family of tests generally provided good control of Type I error and adequate power. For large values of a, the coefficient of variation of the design effects, δ_i, second-order corrections were best. First-order corrections were somewhat liberal in this case, $FX_P^2(\hat{\delta}.)$ providing better control than $X_P^2(\hat{\delta}.)$. Fay's jackknifed statistic, X_{PJ}, also performed well, but tended to be somewhat liberal for small numbers of clusters. They also found that while the Wald test of goodness of fit was extremely liberal and should not be used, the F-based version was a much better performer, providing adequate control of Type I error except for small numbers of clusters.

More recently, Thomas, Singh and Roberts (1996) undertook a large simulation study of tests of independence in two-way tables, under cluster sampling, and included in their comparison a number of test procedures not included in the Thomas and Rao (1987) study. An independent replication of this study, which used a different method of simulating the cluster sampling, has been reported (in Spanish) by Servy, Hachuel and Wojdyla (1998). The results of both studies will be reviewed in this section.

7.3.1. Independence tests under cluster sampling

Thomas, Singh and Roberts (1996) developed most of their conclusions by studying a 3×3 table, and then verified their results on a 3×4 and a 4×4 table. They developed a flexible model of cluster sampling, based on a modified logistic normal distribution that enabled them to study a variety of clustering situations under a partially controlled experimental design. All cluster sizes were set equal to 20, and sample sizes consisting of from 15 to 70 clusters were simulated. Servy, Hachuel and Wojdyla (1998) set all cluster sizes equal to 5, and examined 3×3 tables only.

Test statistics
With the exception of Fay's statistic, studied by Thomas, Singh and Roberts (1996) but not by Servy, Hachuel and Wojdyla (1998), all versions of the procedures discussed in Section 7.2 were examined together with the following:

(i)   Conservative versions of the F-based Rao–Scott tests, denoted $F^*X_P^2(\hat{\delta}.)$ and $F^*X_P^2(\hat{\delta}.)$, in which $FX_P^2(\hat{\delta}.)$ and $FX_{LR}^2(\hat{\delta}.)$ are referred to F-distributions on $d = (A - 1)(B - 1)$ and $v = L - 1$ degrees of freedom, where A and B are the row and column dimensions of the table, respectively, and L is the number of clusters.

(ii) A heuristic modification of a Wald statistic suggested by Morel (1989), designed to improve test performance for small numbers of clusters. It is denoted $X_W^2(L; M)$. Details are given in the technical report by Thomas, Singh and Roberts (1995).

(iii) A family of modified Wald statistics proposed by Singh (1985). These are variants obtained by first expressing the Wald statistic, having $d = (A - 1)(B - 1)$ degrees of freedom, in terms of its spectral decomposition, and then discarding $d - T$ of the smallest eigenvalues (and corresponding eigenvectors) that account for no more than ε of the original eigenvalue sum. The preset constant ε is the range 0.01 to 0.1. The test procedure consists of referring the statistics, denoted $Q^T(LL)$ and $Q^T(R)$, to the upper $100\alpha\%$ point of a chi-squared distribution on T degrees of freedom. F-based versions, $FQ^T(LL)$ and $FQ^T(R)$, can be defined as in Equation (7.21), with v replaced by T.

(iv) A second family of modified Wald statistics designed to increase the stability of Wald tests by adding to each eigenvalue of the kernel covariance matrix a small quantity that tends to zero as $L \to \infty$. The small quantity is defined as k times the estimated standard error of the corresponding eigenvalue, where k is selected in the range 0.25 to 1.0. Under suitable assumptions, this results in a modified Wald statistic that is a scalar multiple of the original. These statistics are denoted $X_W^2(LL; EV1)$ and $X_W^2(R; EV1)$, and are referred to chi-squared distributions on v degrees of freedom. F-based versions, denoted $FX_W^2(LL; EV1)$ and $FX_W^2(R; EV1)$, can be obtained as before.

Further details are given in the technical report by Thomas, Singh and Roberts (1995).

7.3.2. Simulation results

Type I error control was estimated using empirical significance levels (ESLs), namely the proportion of simulation trials leading to rejection of the (true) null hypothesis of independence. Similarly, the empirical powers (EPs) of the various test procedures were estimated as the proportion of trials leading to rejection of the null, when the simulations were generated under the alternative hypothesis. The results discussed below relate to the 5% nominal α-level.

  Though the main recommendations of both studies are very similar, there are some interesting differences. Servy, Hachuel and Wojdyla (1998) found that for almost all procedures, their ESL estimates of Type I error were lower than those of Thomas, Singh and Roberts (1996) and as a result they focused their recommendations on the relative powers of the competing procedures. In contrast, Thomas, Singh and Roberts (1996) put primary emphasis on Type I error control. Some general conclusions are listed below, followed by a ranking of the performance of the viable test procedures. Differences between the two studies are specifically noted.

General conclusions

- The standard Wald procedures, $X_W^2(R)$ and $X_W^2(LL)$, yield seriously inflated ESLs, a finding that is consistent with the earlier study by Thomas and Rao (1987). Further, their variants $Q^T(.)$ and $X_W^2(.; EV1)$ are also seriously inflated. These procedures are not viable in practice.
- There are no differences of practical importance between ESLs for tests based on X_P^2 and tests based on X_{LR}^2. Only the former are listed.
- ESLs for X_{LRJ} are lower than for X_{PJ}, the reverse of Thomas and Rao's (1987) finding for the goodness-of-fit test. The difference is small, however, less than one percentage point. For samples consisting of small numbers of clusters, ESLs for both X_{PJ} and X_{LRJ} are unacceptably inflated ($> 9\%$ for $L = 15$), confirming the earlier finding of Thomas and Rao (1987).
- Thomas, Singh and Roberts (1996) found that the F-based log–linear Wald procedure, $FX_W^2(LL)$, and the log–linear Bonferroni procedure, $Bf(LL)$, provide better control of Type I error than their residual based versions, $FX_W^2(R)$ and $Bf(R)$. Servy, Hachuel and Wojdyla (1998) found the reverse to be true.
- Though they provide good control of Type I error, the powers of the residual statistics $Bf(R)$ and $FQ^T(R)$ are very low. They should not be used.
- Servy, Hachuel and Wojdyla (1998) noted erratic Type I error control for several test procedures for samples containing 15 clusters, the smallest number of clusters considered.

Rankings of the viable procedures

Table 7.2 gives rankings (based on Thomas, Singh and Roberts, 1996) of the test procedures providing adequate Type I error control. The best-performing tests appear at the top of their respective columns. Adequate control is defined as an ESL between 3 % and 7 %, for the nominal 5 % test level. Similarly, only statistics exhibiting reasonable power levels are included.

ESLs in Table 7.2 are averages over three sets of cluster numbers ($L = 15, 30$ and 70) and all values of δ. and other study parameters (see Thomas, Singh and Roberts 1996), for two different groupings of a, the study parameter that has greatest impact on Type I error control. The power rankings shown in Table 7.2 are average EP values over $L = 30$ and 70 and over all other study parameters, for one specific form of the alternative hypothesis.

In their technical report, Servy, Hachuel and Wojdyla (1998) also provided a ranked list of 'best' procedures. Their list contains essentially the same procedures as Table 7.2, with two exceptions. First, the settings of the parameters ε and k featured in the Singh and eigenvalue-adjusted procedures were different,

Table 7.2   Rankings of test performance, based on Type I error and power.

Average ESLs		Average EPs
$0.3 \leq a \leq 0.6$	$0.7 \leq a \leq 1.0$	
$(4.8\% \leq \mathrm{ESL} \leq 7.0\%)$	$(5.4\% \leq \mathrm{ESL} \leq 7.0\%)$	$(55\% \leq \mathrm{EP} \leq 73\%)$
$FX_P^2(\hat{\delta}., \hat{a})$	$FX_W^2(LL; EV1),\ k = 0.5$	$Bf(LL)$
$F^*X_P^2(\hat{\delta}.)$	$Bf(LL)$	$FQ(T)(LL),\ \varepsilon = 0.05$
$FX_W^2(LL; EV1),\ k = 0.5$	$F^*X_P^2(\hat{\delta}.)$	$FX_W^2(LL)$
$Bf(LL)$	$FQ(T)(LL),\ \varepsilon = 0.05$	X_{LRJ}
$X_P^2(\hat{\delta}., \hat{a})$	$FX_P^2(\hat{\delta}., \hat{a})$	$FX_W^2(LL; EV1),\ k = 0.5$
$FQ(T)(LL),\ \varepsilon = 0.05$	$X_P^2(\hat{\delta}., \hat{a})$	$FX_P^2(\hat{\delta}.)$
$FX_W^2(LL)$	$F^*X_P^2(\hat{\delta}.)$	$F^*X_P^2(\hat{\delta}.)$
$X_P^2(\hat{\delta}.)X_{LRJ^a}$		$FX_P^2(\hat{\delta}., \hat{a})$

[a] For $L > 30$, to avoid error rate inflation for small L.

and second, Servy, Hachuel and Wojdyla (1998) included the Morell modified Wald procedure, $X_W^2(LL; M)$, in their list. They gave the second-order Rao–Scott statistic, $X_P^2(\hat{\delta}., \hat{a})$, a higher power ranking than did Thomas, Singh and Roberts (1996), but an important point on which both studies agreed is that the log–linear Bonferroni procedure, $Bf(LL)$, is the most powerful procedure overall. Thomas, Singh and Roberts (1996) also noted that $Bf(LL)$ provides the highest power and most consistent control of Type I error over tables of varying size (3×3, 3×4 and 4×4).

7.3.3.   Discussion and final recommendations

The benefits of Bonferroni procedures in the analysis of categorical data from complex surveys were noted by Thomas (1989), who showed that Bonferroni simultaneous confidence intervals for population proportions, coupled with log or logit transformations, provide better coverage properties than competing procedures. This is consistent with the preeminence of $Bf(LL)$ for tests of independence. Nevertheless, it is important to bear in mind the caveat that the log–linear Bonferroni procedure is not invariant to the choice of basis for the interaction terms in the log–linear model.

The runner-up to $Bf(LL)$ is the Singh procedure, $FQ^T(LL)$, which was highly rated in both studies with respect to power and Type I error control. However, some practitioners might be reluctant to use this procedure because of the difficulty of selecting the value of ε. For example, Thomas, Singh and Roberts (1996) recommend $\varepsilon = 0.05$, while Servy, Hachuel and Wojdyla (1998) recommend $\varepsilon = 0.1$. Similar comments apply to the adjusted eigenvalue procedure, $FX_W^2(LL; EV1)$. Nevertheless, an advantage of $FX_W^2(LL; EV1)$ is that it is a simple multiple of the Wald procedure $FX_W^2(LL)$, and it is reasonable to recommend it as a stable improvement on $FX_W^2(LL)$.

If the uncertainties of the above procedures are to be avoided, the choice of test comes down to the Rao–Scott family, Fay's jackknifed tests, or the F-based Wald test $FX^2_W(LL)$. Table 7.2 can provide the necessary guidance, depending on the degree of variation among design effects. There is a choice of Rao–Scott procedures available whatever the variation among design effects. If full survey information is not available, then the first-order Rao–Scott tests might be the only option. Fay's jackknife procedures are viable alternatives when full survey information is available, provided the number of clusters is not small. These jackknifed tests are natural procedures to choose when survey variance estimation is based on a replication strategy. Finally, $FX^2_W(LL)$, the F-based Wald test based on the log–linear representation of the hypothesis, provides adequate control and relatively high power provided that the variation in design effects is not extreme. It should be noted that in both studies, all procedures derived from the F-based Wald test exhibited low power for small numbers of clusters, so some caution is required if these procedures are to be used.

7.4. ANALYSIS OF DOMAIN RESPONSE PROPORTIONS

Logistic regression models are commonly used to analyze the variation of subpopulation (or domain) proportions associated with a binary response variable. Suppose that the population of interest consists of I domains corresponding to the levels of one or more factors. Let $\hat{N}_i$ and $\hat{N}_{1i}$ ($i = 1, \ldots, I$) be survey estimates of the ith domain size, N_i, and the ith domain total, N_{1i}, of a binary $(0, 1)$ response variable, and let $\sum N_i = N$. A consistent estimate of the domain response proportions $\mu_{1i} = N_{1i}/N_i$ is denoted by $\hat{\mu}_{1i} = \hat{N}_{1i}/\hat{N}_i$. The asymptotic covariance matrix of $\hat{\mu}_1 = (\hat{\mu}_{11}, \ldots, \hat{\mu}_{1I})'$, denoted Σ_1, is consistently estimated by $\hat{\Sigma}_1$.

A logistic regression model on the response proportions μ_{1i} is given by

$$\log[\mu_{1i}/(1 - \mu_{1i})] = x'_i \theta, \qquad (7.32)$$

where x_i is an m-vector of known constants derived from the factor levels with $x_{i1} = 1$, and θ is an m-vector of regression parameters.

The pseudo-MLE $\hat{\theta}$ is obtained by solving the estimating equations specified by Roberts, Rao and Kumar (1987), namely

$$X'_1 D(\hat{\pi}) \mu_1(\hat{\theta}) = X'_1 D(\hat{\pi}) \hat{\mu}_1, \qquad (7.33)$$

where $X'_1 = (x_1, \ldots, x_I)$, $D(\hat{\pi}) = \text{diag}(\hat{\pi}_1, \ldots, \hat{\pi}_I)$ with $\hat{\pi}_i = \hat{N}_i/\hat{N}$ and $\hat{N} = \sum \hat{N}_i$, and where $\mu_1(\hat{\theta})$ is the m-vector with elements $\mu_{1i}(\hat{\theta})$. Equations (7.33) are obtained from the likelihood equations under independent binomial sampling by replacing n_i/n by $\hat{\pi}_i$ and n_{1i}/n_i by the ratio estimator $\hat{\mu}_{1i}$, where n_i and n_{1i} are the sample size and the sample total of the binary response variable from the ith domain ($\sum n_i = n$). A general Pearson statistic for testing the goodness of fit of the model (7.32) is given by

$$X_{P1}^2 = n \sum_{i=1}^{I} \hat{\pi}_i [\hat{\mu}_{1i} - \mu_{1i}(\hat{\theta})]^2 / [\mu_{1i}(\hat{\theta})(1 - \mu_{1i}(\hat{\theta}))] \qquad (7.34)$$

and a statistic corresponding to the likelihood ratio is given by

$$X_{LR1}^2 = 2n \sum_{i=1}^{I} \hat{\pi}_i \{ \hat{\mu}_{1i} \log [\hat{\mu}_{1i}/\mu_{1i}(\hat{\theta})] + (1 - \hat{\mu}_{1i}) \log [(1 - \hat{\mu}_{1i})/(1 - \mu_{1i}(\hat{\theta}))] \}.$$

$$(7.35)$$

Roberts, Rao and Kumar (1987) showed that both X_{P1}^2 and X_{LR1}^2 are asymptotically distributed under the model as a weighted sum, $\sum \delta_{1i} W_i$, of independent χ_1^2 variables W_i, where the weights δ_{1i}, $i = 1, \ldots, I - m$, are eigenvalues of a generalized design effects matrix, Δ_1, given by

$$\Delta_1 = (\tilde{Z}_1' \Omega_1 \tilde{Z}_1)^{-1} (\tilde{Z}_1' \Sigma_1 \tilde{Z}_1), \qquad (7.36)$$

where

$$\tilde{Z}_1' = I - D(\pi)^{-1} D(\mu_1) D(1 - \mu_1) Z_1 [Z_1' D(\mu_1) D(1 - \mu_1) Z_1]^{-1} Z_1' D(\pi), \quad (7.37)$$

Z_1 is any $I \times (I - m)$ full rank matrix that satisfies $Z_1' X_1 = 0$, and $\Omega_1 = n^{-1} D(\pi)^{-1} D(\mu_1) D(1 - \mu_1)$ with $D(a)$ denoting the diagonal matrix with diagonal elements a_i, where $a = (a_1, \ldots, a_I)^I$. Under independent binomial sampling, $\Delta_1 = I$ and hence $\delta_{1i} = 1$ for all i so that both X_{P1}^2 and X_{LR1}^2 are asymptotically χ_{I-m}^2. A first-order Rao–Scott correction refers

$$X_{P1}^2(\hat{\delta}_{1\cdot}) = X_{P1}^2/\hat{\delta}_{1\cdot}. \quad \text{or} \quad X_{LR1}^2(\hat{\delta}_{1\cdot}) = X_{LR1}^2/\hat{\delta}_{1\cdot}. \qquad (7.38)$$

to the upper $100\alpha\%$ point of χ_{I-m}^2, where

$$\hat{\delta}_{1\cdot} = \text{trace}(\hat{\Delta}_1)/(I - m), \qquad (7.39)$$

and $\hat{\Delta}_1$ is given by (7.36) with μ_1 replaced by $\mu_1(\hat{\theta})$, π by $\hat{\pi}$, and Σ_1 by $\hat{\Sigma}_1$. Similarly, a second-order Rao–Scott correction refers

$$X_{P1}^2(\hat{\delta}_{1\cdot}, \hat{a}_1) = \frac{X_{P1}^2(\hat{\delta}_{1\cdot})}{1 + \hat{a}_1^2} \quad \text{or} \quad X_{LR1}^2(\hat{\delta}_{1\cdot}, \hat{a}_1) = \frac{X_{LR1}^2(\hat{\delta}_{1\cdot})}{1 + \hat{a}_1^2} \qquad (7.40)$$

to the upper $100\alpha\%$ point of a χ^2 with degrees of freedom $(I - m)/(1 + \hat{a}_1^2)$, where $(1 + \hat{a}_1^2)/(I - m) = tr(\hat{\Delta}_1^2)/[tr(\hat{\Delta}_1)]^2$. A simple conservative test, depending only on the domain design effects $\hat{D}_{1i} = \hat{\sigma}_{1ii}/[n\hat{\pi}_i \hat{\mu}_{1i}(1 - \hat{\mu}_{1i})]$, is obtained by using $X_{P1}^2(\hat{\delta}_1^*)$ or $X_{LR1}^2(\hat{\delta}_1^*)$, where $\hat{\delta}_1^* = [I/(I - m)]\hat{D}_1$ is an upper bound on $\hat{\delta}_{1\cdot}$ given by $\hat{D}_1 = I^{-1}\Sigma\hat{D}_{1i}$. The modified first-order corrections $X_{P1}^2(\hat{\delta}_1^*)$ and $X_{LR1}^2(\hat{\delta}_1^*)$ can be readily implemented from published tables provided the estimated proportions $\hat{\mu}_{1i}$ and their design effects $\hat{D}_{1i}$ are known.

Roberts, Rao and Kumar (1987) also developed first-order and second-order corrections for nested hypotheses as well as model diagnostics to detect any outlying domain response proportions and influential points in the factor space, taking account of the design features. They also obtained a linearization estimator of $Var(\hat{\theta})$ and the standard errors of 'smoothed' estimates $\mu_{1i}(\hat{\theta})$.

Example 2

Roberts, Rao and Kumar (1987) and Rao, Kumar and Roberts (1989) applied the previous methods to data from the October 1980 Canadian Labour Force Survey. The sample consisted of males aged 15–64 who were in the labour force and not full-time students. Two factors, age and education, were chosen to explain the variation in employment rates via logistic regression models. Age group levels were formed by dividing the interval [15, 64] into 10 groups with the jth age group being the interval $[10 + 5j, 14 + 5j]$ for $j = 1, \ldots, 10$ and then using the mid-point of each interval, $A_j = 12 + 5j$, as the value of age for all persons in that age group. Similarly, the levels of education, E_k, were formed by assigning to each person a value based on the median years of schooling resulting in the following six levels: 7, 10, 12, 13, 14 and 16. The resulting age by education cross-classification provided a two-way table of $I = 60$ estimated cell proportions or employment rates, $\hat{\mu}_{1jk}$, $j = 1, \ldots, 10$; $k = 1, \ldots, 6$.

A logistic regression model involving linear and quadratic age effects and a linear education effect provided a good fit to the two-way table of estimated employment rates, namely:

$$\log[\hat{\mu}_{1jk}/(1 - \hat{\mu}_{1jk})] = -3.10 + 0.211A_j - 0.002\,18A_j^2 + 0.151E_k.$$

The following values were obtained for testing the goodness of fit of the above model: $X_{P1}^2 = 99.8$, $X_{LR1}^2 = 102.5$, $X_{P1}^2(\hat{\delta}_1., \hat{a}_1) = 23.4$ and $X_{LR1}^2(\hat{\delta}_1., \hat{a}_1) = 23.9$. If the sample design is ignored and the value of X_{P1}^2 or X_{LR1}^2 is referred to the upper 5% point of χ^2 with $I - m = 56$ degrees of freedom, the model with linear education effect and linear and quadratic effects could be rejected. On the other hand, the value of $X_{P1}^2(\hat{\delta}_1., \hat{a}_1)$ or $X_{LR1}^2(\hat{\delta}_1., \hat{a}_1)$ when referred to the upper 5% point of a χ^2 with $56/(1 + \hat{a}_1^2)$ degrees of freedom is not significant. The simple conservative test $X_{P1}^2(\hat{\delta}_1^*.)$ or $X_{LR1}^2(\hat{\delta}_1^*.)$, depending only on the domain design effects, is also not significant, and is close to the first-order correction $X_{P1}^2(\hat{\delta}_1^*.)$ or $X_{LR1}^2(\hat{\delta}_1^*.)$; $\hat{\delta}_1^*. = 2.04$ compared to $\delta_1. = 1.88$.

Roberts, Rao and Kumar (1987) also applied their model diagnostic tools to the Labour Force Survey data. On the whole, their investigation suggested that the impact of cells flagged by the diagnostics is not large enough to warrant their deletion.

Example 3

Fisher (1994) also applied the previous methods to investigate whether the use of personal computers during interviewing by telephone (treatment) versus in-person on-paper interviewing (control) has an effect on labour force estimates. For this purpose, he used split panel data from the US Current Population Survey (CPS) obtained by randomly splitting the CPS sample into two panels and then administering the 'treatment' to respondents in one of the panels and the 'control' to those in the other panel. Three other binary factors, sex, race and ethnicity, were also included. A logistic regression model containing only the main effects of the four factors fitted the four-way table of estimated unemployment rates well. A test of the nested hypothesis that the 'treatment' main effect was absent given the model containing all four main effects was rejected, suggesting that the use of a computer during interviewing does have an effect on labour force estimates.

Rao, Kumar and Roberts (1989) studied several extensions of logistic regression. They extended the previous results to Box–Cox-type models involving power transformations of domain odds ratios, and illustrated their use on data from the Canadian Labour Force Survey. The Box–Cox approach would be useful in those cases where it could lead to additive models on the transformed scale while the logistic regression model would not provide as good a fit without interaction terms. Methods for testing equality of parameters in two logistic regression models, corresponding to consecutive time periods, were also developed and applied to data from the Canadian Labour Force Survey. Finally, they studied a general class of polytomous response models and developed Rao–Scott adjusted Pearson and likelihood tests which they applied to data from the Canada Health Survey (1978–9).

In this section we have discussed methods for analyzing domain response proportions. We turn next to logistic regression analysis of unit-specific sample data.

7.5.   LOGISTIC REGRESSION WITH A BINARY RESPONSE VARIABLE

Suppose that (x_t, y_t) denote the values of a vector of explanatory variables, x, and a binary response variable, y, associated with the tth population unit, $t = 1, \ldots, N$. We assume that for a given x_t, y_t is generated from a model with mean $E(y_t) = \mu_t(\theta) = g(x_t, \theta)$ and 'working' variance $var(y_t) = v_{0t} = v_0(\mu_t)$. Specifically, we assume a logistic regression model so that

$$\log [\mu_t(\theta)/(1 - \mu_t(\theta))] = x_t'\theta \qquad (7.41)$$

and $v_0(\mu_t) = \mu_t(1 - \mu_t)$. Our interest is in estimating the parameter vector θ and in testing hypotheses on θ using the sample data $\{(x_t, y_t), t \in s\}$. A pseudo-MLE $\hat{\theta}$ of θ is obtained by solving

$$\hat{T}(\theta) = \sum_{t \in s} w_{ts} u_t(\theta) = \underset{\sim}{0}, \qquad (7.42)$$

where the w_{ts} are survey weights that may depend on s, e.g., when the design weights, w_t, are adjusted for post-stratification, and (see section 3.4.5 of SHS)

$$u_t(\theta) = [y_t - \mu_t(\theta)]x_t. \qquad (7.43)$$

The estimator $\hat{\theta}$ is a consistent estimator of the census parameter θ obtained as the solution of the estimating equations

$$T(\theta) = \sum_{1}^{N} u_t(\theta) = \underset{\sim}{0}.$$

Following Binder (1983), a linearization estimator of the covariance matrix $V(\hat{\theta})$ of $\hat{\theta}$ is given by

$$\hat{V}_L(\hat{\theta}) = [J(\hat{\theta})]^{-1} \hat{V}(\hat{T})[J(\hat{\theta})]^{-1} \qquad (7.44)$$

where $J(\hat{\theta}) = -\sum_{t \in s} w_{ts} \partial u_t(\theta)/\partial\theta'$ is the observed information matrix and $\hat{V}(\hat{T})$ is the estimated covariance matrix of the estimated total $\hat{T}(\theta)$ evaluated at $\theta = \hat{\theta}$. The estimator $\hat{V}(\hat{T})$ should take account of post-stratification and other adjustments. It is straightforward to obtain a resampling estimator of $V(\hat{\theta})$ that takes account of post-stratification adjustment. For stratified multi-stage sampling, a jackknife estimator of $V(\hat{\theta})$ is given by

$$\hat{V}_J(\hat{\theta}) = \sum_{j=1}^{J} \frac{n_j - 1}{n_j} \sum_{k=1}^{n_j} (\hat{\theta}_{(jk)} - \hat{\theta})(\hat{\theta}_{(jk)} - \hat{\theta})' \qquad (7.45)$$

where J is the number of strata, n_j is the number of sampled primary clusters from stratum j and $\hat{\theta}_{(jk)}$ is obtained in the same manner as $\hat{\theta}$ when the data from the (jk)th sample cluster are deleted, but using jackknife survey weights $w_{ts(jk)}$ (see Rao, Scott and Skinner, 1998). Bull and Pederson (1987) extended (7.44) to the case of a polytomous response variable, but without allowing for post-stratification adjustment as in Binder (1983). Again it is straightforward to define a jackknife variance estimator for this case.

Wald tests of hypotheses on θ are readily obtained using either $\hat{V}_L(\hat{\theta})$ or $\hat{V}_J(\hat{\theta})$. For example, suppose $\theta = (\theta'_1, \theta'_2)$, where θ_2 is $r_2 \times 1$ and θ_2 is $r_1 \times 1$, with $r_1 + r_2 = r$, and we are interested in testing the nested hypothesis $H_0: \theta_2 = \theta_{20}$. Then the Wald statistic

$$X_W^2 = (\hat{\theta}_2 - \hat{\theta}_{20})'[\hat{V}(\hat{\theta}_2)]^{-1}(\hat{\theta}_2 - \hat{\theta}_{20}) \qquad (7.46)$$

is asymptotically $\chi^2_{r_2}$ under H_0, where $\hat{V}(\hat{\theta})$ may be taken as either $\hat{V}_L(\hat{\theta})$ or $\hat{V}_J(\hat{\theta})$. A drawback of Wald tests is that they are not invariant to non-linear transformations of θ. Further, one has to fit the full model (7.1) before testing H_0, which could be a disadvantage if the full model contains a large number, r, of parameters. This would be the case with a factorial structure of explanatory variables containing a large number of potential interactions, θ_2, if we wished to test for the absence of interaction terms, i.e., $H_0: \theta_2 = 0$. Rao, Scott and Skinner (1998) proposed quasi-score tests to circumvent the problems associated with Wald tests. These tests are invariant to non-linear transformations of θ and we need only fit the simpler model under H_0, which is a considerable advantage if the dimension of θ is large, as noted above.

Let $\tilde{\theta} = (\tilde{\theta}'_1, \theta'_{20})$ be the solution of $\hat{T}_1(\tilde{\theta}) = 0$, where $\hat{T}(\theta) = [\hat{T}_1(\theta)', \hat{T}_2(\theta)']'$ is partitioned in the same manner as θ. Then a quasi-score test of $H_0: \theta_2 = \theta_{20}$ is based on the statistic

$$X_{QS}^2 = \tilde{T}'_2[\hat{V}(\hat{T}_2)]^{-1}\tilde{T}_2, \qquad (7.47)$$

where $\tilde{T}_2 = \hat{T}_2(\tilde{\theta})$ and $\hat{V}(\hat{T}_2)$ is either a linearization estimator or a jackknife estimator.

The linearization estimator, $\hat{V}_L(\hat{T}_2)$, is the estimated covariance matrix of the estimated total

$$T_2^*(\theta) = \sum_{t \in s} w_{ts}\{u_{2t}(\theta) - Bu_{1t}(\theta)\}$$

evaluated at $\theta = \tilde{\theta}$ and $B = J_{21}(\tilde{\theta})[J_{11}(\tilde{\theta})]^{-1}$, where

$$J(\theta) = \begin{pmatrix} J_{11}(\theta) & J_{12}(\theta) \\ J_{21}(\theta) & J_{22}(\theta) \end{pmatrix}$$

corresponds to the partition of θ as $(\theta_1', \theta_2')'$. Again, $\hat{V}_L(\hat{T}_2)$ should take account of post-stratification and other adjustments. The jackknife estimator $\hat{V}_J(\hat{T}_2)$ is similar to (7.45), with $\hat{\theta}$ and $\hat{\theta}_{(jk)}$ changed to $\hat{T}_2$ and $\hat{T}_{2(jk)}$, where $\hat{T}_{2(jk)}$ is obtained in the same manner as $\hat{T}_2$, i.e., when the data from the (jk)th sample cluster are deleted, but using the jackknife survey weights $w_{ts(jk)}$. Under H_0, X_{QS}^2 is asymptotically $\chi_{r_2}^2$ so that X_{QS}^2 provides a valid test of H_0.

A difficulty with both X_W^2 and X_{QS}^2 is that when the effective degrees of freedom for estimating $V(\hat{\theta}_2)$ or $V(\hat{T}_2)$ are not large, the tests become unstable when the dimension of θ_2 is large because of instability of $[\hat{V}(\hat{\theta}_2)]^{-1}$ and $[\hat{V}(\hat{T}_2)]^{-1}$. To circumvent the degrees of freedom problem, Rao, Scott and Skinner (1998) proposed alternative tests including an *F*-version of the Wald test (see also Morel, 1989), and Rao–Scott corrections to naive Wald or score tests that ignore survey design features as in the case of X_P^2 and X_{LR}^2 of Section 7.2, as well as Bonferroni tests. We refer the reader to Rao, Scott and Skinner (1998) for details.

It may be noted that with unit-level data $\{(x_t, y_t), t \in s\}$ we can only test nested hypotheses on θ given the model (7.41), unlike the case of domain proportions which permits testing of model fit as well as nested hypotheses given the model.

7.6. SOME EXTENSIONS AND APPLICATIONS

In this section we briefly mention some recent applications and extensions of Rao–Scott and related methodology.

7.6.1. Classification errors

Rao and Thomas (1991) studied chi-squared tests in the presence of classification errors which often occur with survey data. Following the important work of Mote and Anderson (1965) for multinomial sampling, they considered the cases of known and unknown misclassification probabilities for simple goodness of fit. In the latter case two models were studied: (i) misclassification rate is the same for all categories; (ii) for ordered categories, misclassification only occurs between adjacent categories, at a constant rate. First- and second-order Rao–Scott corrections to Pearson statistics that account for the survey design features were obtained. A model-free approach using two-phase sampling was also developed. In two-phase sampling, error-prone measurements are made on a large first-phase sample selected according to a specified design and error-free measurements are then made on a smaller subsample selected according to another specified design (typically SRS or stratified SRS). Rao–Scott corrected Pearson statistics were proposed under double sampling for both the goodness-of-fit test and the tests of independence in a two-way table. Rao and Thomas (1991) also extended Assakul and Proctor's (1967) method of testing of independence in a two-way table with known misclassification probabilities to general survey designs. They developed Rao–Scott corrected tests using a general methodology that can also be used for testing the fit of log–linear models on multi-way contingency tables. More recently, Skinner (1998) extended the methods of Section 7.4 to the case of longitudinal survey data subject to classification errors.

7.6.2. Biostatistical applications

Cluster-correlated binary response data often occur in biostatistical applications; for example, toxicological experiments designed to assess the teratogenic effects of chemicals often involve animal litters as experimental units. Several methods that take account of intracluster correlations have been proposed but most of these methods assume specific models for the intracluster correlation, e.g., the beta-binomial model. Rao and Scott (1992) developed a simple method, based on conceptual design effects and effective sample size, that can be applied to problems involving independent groups of clustered binary data with group-specific covariates. It assumes no specific models for the intracluster correlation in the spirit of Zeger and Liang (1986). The method can be readily implemented using any standard computer program for the analysis of independent binary data after a small amount of pre-processing.

Let n_{ij} denote the number of units in the jth cluster of the ith group, $i = 1, \ldots, I$, and let y_{ij} denote the number of 'successes' among the n_{ij} units, with $\sum_j y_{ij} = y_i$ and $\sum_j n_{ij} = n_i$. Treating y_i as independent binomial $B(n_i, p_i)$ leads to erroneous inferences due to the clustering, where p_i is the success probability in the ith group. Denoting the design effect of the ith estimated proportion, $\hat{p}_i = y_i/n_i$, by D_i, and the effective sample size by $\tilde{n}_i = n_i/D_i$, the Rao–Scott (1992) method simply treats $\tilde{y}_i = y_i/D_i$ as independent binomial $B(\tilde{n}_1, p_i)$ which leads to asymptotically correct inferences. The method was applied to a variety of biostatistical problems; in particular, testing homogeneity of proportions, estimating dose response models and testing for trends in proportions, computing the Mantel–Haenszel chi-squared test statistic for independence in a series of 2×2 tables and estimating the common odds ratio and its variance when the independence hypothesis is rejected. Obuchowski (1998) extended the method to comparisons of correlated proportions.

Rao and Scott (1999a) proposed a simple method for analyzing grouped count data exhibiting overdispersion relative to a Poisson model. This method is similar to the previous method for clustered binary data.

7.6.3. Multiple response tables

In marketing research surveys, individuals are often presented with questions that allow them to respond to more than one of the items on a list, i.e., multiple responses may be observed from a single respondent. Standard methods cannot be applied to tables of aggregated multiple response data because of the multiple response effect, similar to a clustering effect in which each independent respondent plays the role of a cluster. Decady and Thomas (1999, 2000) adapted the first-order Rao–Scott procedure to such data and showed that it leads to simple approximate methods based on easily computed, adjusted chi-squared tests of simple goodness of fit and homogeneity of response probabilities. These adjusted tests can be calculated from the table of aggregate multiple response counts alone, i.e., they do not require knowledge of the correlations between the aggregate multiple response counts. This is not true in general; for example, the test of equality of proportions will require either the full dataset or an expanded table of counts in which each of the multiple response items is treated as a binary variable. Nevertheless, the first-order Rao–Scott approach is still considerably easier to apply than the bootstrap approach recently proposed by Loughin and Scherer (1998).

CHAPTER 8

Fitting Logistic Regression Models in Case–Control Studies with Complex Sampling

Alastair Scott and Chris Wild

8.1. INTRODUCTION

In this chapter we deal with the problem of fitting logistic regression models in surveys of a very special type, namely population-based case–control studies in which the controls (and sometimes the cases) are obtained through a complex multi-stage survey. Although such surveys are special, they are also very common. For example, we are currently involved in the analysis of a population-based case–control study funded by the New Zealand Ministry of Health and the Health Research Council looking at meningitis in young children. The study population consists of all children under the age of 9 in the Auckland region. There were about 250 cases of meningitis over the three-year duration of the study and all of these are included. In addition, a sample of controls was drawn from the remaining children in the study population by a complex multi-stage design. At the first stage, a sample of 300 census mesh blocks (each containing roughly 50 households) was drawn with probability proportional to the number of houses in the block. Then a systematic sample of 20 households was selected from each chosen mesh block and children from these households were selected for the study with varying probabilities that depend on age and ethnicity as in Table 8.1. These probabilities were chosen to match the expected frequencies among the cases. Cluster sample sizes varied from one to six and a total of approximately 250 controls was achieved. This corresponds to a sampling fraction of about 1 in 400, so that cases are sampled at a rate that is 400 times that for controls.

Analysis of Survey Data Edited by R. L. Chambers and C. J. Skinner
© 2003 John Wiley & Sons, Ltd

Table 8.1   Selection probabilities.

	Maori	Pacific Islander	Other
≤ 1 year	0.29	0.70	0.10
≤ 3 years	0.15	0.50	0.07
≤ 5 years	0.15	0.31	0.04
≤ 8 years	0.15	0.17	0.04

A similar population-based case–control study of the effects of artificial sweeteners on bladder cancer is described in Graubard, Fears and Gail (1989). Again all cases occurring during the study were included, but the control sample was drawn using Waksberg's two-stage telephone sampling technique. A sample of exchanges, each containing 100 telephone numbers, is chosen at the first stage and then a simple random sample of households is drawn from each of the selected exchanges.

Complex sampling may also be used in the selection of cases. For example, we have recently helped with the analysis of a study in which cases (patients who had suffered a mild stroke) were selected through a two-stage design with doctors' practices as primary sampling units. However, this is much less common than with controls.

As we said at the outset, these studies are a very special sort of survey but they share the two key features that make the analysis of survey data distinctive. The first feature is the lack of independence. In our example, we would expect substantial intracluster correlation because of unmeasured socio-economic variables, factors affecting living conditions (e.g. mould on walls of house), environmental exposures and so on. Ignoring this will lead to standard errors that are too small and confidence intervals that are too narrow. The other distinctive feature is the use of unequal selection probabilities. In case–control studies the selection probabilities can be extremely unbalanced and they are based directly on the values of the response variable, so that we have informative sampling at its most extreme.

In recent years, efficient semi-parametric procedures have been developed for handling the variable selection probabilities (see Scott and Wild (2001) for a survey of recent work). 'Semi-parametric' here means that a parametric model is specified for the response as a function of potential explanatory variables, but that the joint distribution of the explanatory variables is left completely free. This is important because there are usually many potential explanatory variables (more than 100 in some of the studies in which we are involved) and it would be impossible to model their joint behaviour, which is of little interest in its own right. However, the problem of clustering is almost completely ignored, in spite of the fact that a large number of studies use multi-stage sampling. The paper by Graubard, Fears and Gail (1989) is one of the few that even discuss the problem. Most analyses simply ignore the problem and use a program designed for simple (or perhaps stratified) random sampling of cases and controls. This chapter is an attempt to remedy this neglect.

In the next section we give a summary of standard results for simple case–control studies, where cases and controls are each selected by simple random sampling or by stratified random sampling. In Section 8.3, we extend these results to cover cases when the controls (and possibly the cases) are selected using an arbitrary probability sampling design. We investigate the properties of these methods through simulation studies in Section 8.4. The robustness of the methods in situations when the assumed model does not hold exactly is explored in Section 8.5, and some possible alternative approaches are sketched in the final section.

8.2. SIMPLE CASE–CONTROL STUDIES

We shall take it for granted throughout this chapter that our purpose is to make inferences about the parameters of a superpopulation model since interest is centred in the underlying process that produces the cases and not on the composition of a particular finite population at a particular point in time. Suppose that the finite population consists of values $\{(x_t, y_t), t = 1, \ldots, N\}$, where y_t is a binary response and x_t is the vector of potential explanatory variables. The x_t can be generated by any mechanism, but the y_t are assumed to be conditionally independent given the x_t and $\Pr(Y_t = y|\{x_j\}) = \Pr(Y_t = y|x_t)$. These assumptions underlie all standard methods for the analysis of case–control data from population-based studies. For simplicity, we will work with the logistic regression model, logit $\{\Pr(Y = 1|x; \beta)\} = x'\beta$ (see Sections 6.3.1 and 7.5), but similar results can be obtained for any binary regression model. We denote $\Pr(Y = 1|x; \beta)$ by $p_1(x; \beta)$ and often shorten it to $p_1(x)$. We also set $p_0(x) = 1 - p_1(x)$.

If we had data on the whole population, that is what we would analyse, treating it as a random sample from the process that produces cases and controls. The finite population provides score (first derivative of the log-likelihood) equations

$$\frac{1}{N}\sum_{t=1}^{N} x_t\{y_t - p_1(x_t; \beta)\} = \frac{N_1}{N}\sum_{t:y_t=1}\frac{x_t p_0(x_t)}{N_1} - \frac{N_0}{N}\sum_{t:y_t=0}\frac{x_t p_1(x_t)}{N_0} = 0, \qquad (8.1)$$

where N_1 is the number of cases in the finite population and N_0 is the number of controls. As $N \to \infty$, Equations (8.1) converge to

$$E[X\{Y - p_1(X; \beta)\}] = W_1 E_1\{Xp_0(X)\} - W_0 E_0\{Xp_1(X)\} = 0, \qquad (8.2)$$

where $W_i = \Pr(Y = i)$ and $E_i\{\}$ denotes the expected value over the distribution of X conditional on $Y = i$ ($i = 0, 1$). Under the model, Equations (8.2) have solution β.

Standard logistic regression applies if our data comes from a simple random sample of size n from the finite population. For rare events, as is typical in biostatistics and epidemiology, enormous efficiency gains can be obtained by stratifying on the value of Y and drawing a random sample of size n_i from the stratum defined by $Y = i$ ($i = 0, 1$) with $n_1 \approx n_0$.

Since the estimating equations (8.2) simply involve population means for the case and control subpopulations, they can be estimated directly from case–control data using the corresponding sample means. This leads to a design-weighted (pseudo-MLE, Horvitz Thompson) estimator, $\hat{\beta}_{HT}$, that satisfies

$$W_1 \sum_{\text{cases}} \frac{x_t p_0(x_t)}{n_1} - W_0 \sum_{\text{controls}} \frac{x_t p_1(x_t)}{n_0} = 0, \tag{8.3}$$

as in Sections 6.3.1 and 7.5. This method can be used when the W_i are known or can be estimated consistently (by the corresponding population proportions, for example). More efficient estimates, however, are obtained using the semi-parametric maximum likelihood estimator, $\hat{\beta}_{ML}$ say. This is essentially obtained by ignoring the sampling scheme (see Prentice and Pyke, 1979) and solving the prospective score equations (i.e. those that would be appropriate if we had a simple random sample from the whole population)

$$\frac{n_1}{n} \sum_{\text{cases}} \frac{x_t p_0(x_t)}{n_1} - \frac{n_0}{n} \sum_{\text{controls}} \frac{x_t p_1(x_t)}{n_0} = 0. \tag{8.4}$$

If $n_i/n \to \rho_i$ as $n = n_1 + n_0 \to \infty$, these equations tend to

$$\rho_1 E_1\{X p_0(X)\} - \rho_0 E_0\{X p_1(X)\} = 0, \tag{8.5}$$

which have solution $\beta + k e_1$, where $k = \log[\rho_1 W_0/(\rho_0 W_1)]$ and $e_1 = (1, 0, \ldots, 0)'$. We see that only the intercept term is affected by the case–control sampling. The intercept term can be corrected simply by using k as an offset in the model, but if we are only interested in the coefficients of the risk factors, we do not even need to know the relative stratum weights. More generally, Scott and Wild (1986) show that, for any positive λ_1 and λ_0, equations of the form

$$\lambda_1 E_1\{X p_0(X; b)\} - \lambda_0 E_0\{X p_1(X; b)\} = 0 \tag{8.6}$$

have the unique solution $b = \beta + k_\lambda e_1$ with $k_\lambda = \log[\lambda_1 W_0/(\lambda_0 W_1)]$, provided that the model contains a constant term (i.e. the first component of x is 1). This can be seen directly by expanding (8.6). Suppose, for simplicity, that X is continuous with joint density function $f(x)$. Then, noting that the conditional density of X given that $Y = i$ is $f(x|Y = i) = p_i(x; \beta)f(x)/W_i$, (8.6) is equivalent to

$$\int \frac{x e^{x'\beta + k_\lambda} f(x)}{(1 + e^{x'b})(1 + e^{x'\beta})} dx = \int \frac{x e^{x'b} f(x)}{(1 + e^{x'b})(1 + e^{x'\beta})} dx.$$

The result then follows immediately.

The increase in efficiency of $\hat{\beta}_{ML}$ (or, more generally, estimates based upon (8.6)) over $\hat{\beta}_W$ can be modest when, as in Scott and Wild (1989), we have only one explanatory variable and the ratio of the sampling rates between cases and controls is not too extreme. However, the differences become bigger as cases become rarer in the population and the sampling rates become less balanced, and also when we have more than one covariate. We have seen 50% efficiency gains in some simple case–control problems. In more complex stratified samples, design weighting can be very inefficient indeed (see Lawless, Kalbfleisch and Wild 1999).

Most case–control studies incorporate stratification on other variables, such as age and ethnicity as in our motivating example. This is one aspect of complex survey design that is taken into account in standard case–control methodology. A common assumption is that relative risk is constant across these strata but that absolute levels of risk vary from stratum to stratum, leading to a model of the form

$$\text{logit}\{P(Y = 1|x, \text{stratum } h; \beta)\} = \beta_{0h} + x_1'\beta_1 \qquad (8.7)$$

for observations in the hth stratum, where x now contains dummy variables for all the strata as well as the measured covariates, x_1 say. There is no problem adapting the design-weighted (pseudo-MLE, Horvitz–Thompson) approach to this problem; we simply add a dummy variable for each stratum to x and replace the sample means in (8.3) by their stratified equivalents. The maximum likelihood solution for $\hat{\beta}_1$ is even simpler, assuming simple random sampling of cases and controls within each stratum; we simply fit the model in (8.7) as if we had a simple random sample from the whole population, ignoring the stratification completely. If we want estimates of the β_{0h} we need to adjust the output by an additive constant depending on the relative sampling fractions of cases and controls in the hth stratum, but if we are interested only in β_1, we do not even need to know these sampling fractions.

In some studies, we want to model the stratum constants as functions of the variables in x_1. For example, if the population is stratified by age, we might still want to model the effect of age by some smooth function. Again, adapting the design-weighted approach is completely straightforward and only requires the specification of the appropriate sampling weights. Extending the maximum likelihood approach to this case is considerably more difficult, however, and a fully efficient solution has only been obtained relatively recently (for details see Scott and Wild, 1997; Breslow and Holubkov, 1997).

8.3.   CASE–CONTROL STUDIES WITH COMPLEX SAMPLING

Now consider situations in which controls (and possibly cases) are obtained using a complex sampling plan involving multi-stage sampling. Our only assumption is that the population is stratified into cases and controls and that samples of cases and controls are drawn independently. Note that this assumption does not hold exactly in our motivating example since a control could be drawn from the same cluster as a case. However, cases are sufficiently rare for this possibility to be ignorable in the analysis. (In fact, it occurred once in the study.) One common option is simply to ignore the sample design (see Graubard, Fears and Gail, 1989, for examples) and use a standard logistic regression program, just as if we had a simple (or stratified) case–control study. Underestimates of variance are clearly a worry with this strategy; the coverage frequencies of nominally 95% confidence intervals dropped to about 80% in some of our simulations. In fact, estimates obtained in this way need not even be consistent. To obtain consistency, we need $(1/n_0) \sum_{\text{controls}} x_t p_1(x_t)$ to converge to $E_0\{Xp_1(X)\}$ (with a similar condition for cases), which will be true for self-weighting designs but not true in general.

The estimating equations (8.2) involve separate means for the case and control subpopulations. Both these terms can be estimated using standard survey sampling techniques for estimating means from complex samples. Variances and covariances of the estimated means, which are required for sandwich estimates of $\text{Cov}(\hat{\beta})$, can also be obtained by standard survey sampling techniques. Such analyses can be carried out routinely using specialist sampling packages such as SUDAAN or with general packages such as SAS or Stata that include a survey option. This approach is a direct generalisation of the design-weighted (Horvitz–Thompson) approach to the analysis of simple case–control data which, as we have seen, can be quite inefficient.

An alternative is to apply standard sampling methods to (8.6), with appropriate choices for λ_1 and λ_0, rather than to (8.2). This leads to an estimator, $\hat{\beta}_\lambda$ say, that satisfies

$$\hat{S}_\lambda(\beta) = \lambda_1 \hat{\mu}_1(x; \beta) - \lambda_0 \hat{\mu}_0(x; \beta) = 0, \tag{8.8}$$

where $\hat{\mu}_1$ is the sample estimate of the stratum mean $E_1\{(Xp_0(X))\}$ etc. The covariance matrix of $\hat{\beta}_\lambda$ can then be obtained by standard linearisation arguments. This leads to an estimated covariance matrix

$$\widehat{\text{Cov}}(\hat{\beta}_\lambda) \approx I_\lambda^{-1}(\hat{\beta}_\lambda)\widehat{\text{Cov}}(\hat{S}_\lambda)I_\lambda^{-1}(\hat{\beta}_\lambda), \tag{8.8}$$

where $I_\lambda(\beta) = (\partial \hat{S}_\lambda/\partial \beta')$ and $\widehat{\text{Cov}}(\hat{\beta}_\lambda)$ is the standard survey estimate of $\text{Cov}(\hat{S}_\lambda)$, just as in Equation (7.44). (Note that $\hat{S}$ is just a linear combination of two estimated means.) All of this can also be carried out straightforwardly in SUDAAN or similar packages simply by specifying the appropriate weights (i.e. by scaling the case weights and control weights separately so that the sum of the case weights is proportional to λ_1 and the sum of the control weights is proportional to λ_0).

We still have to decide on appropriate values for λ_1 and λ_0. For simple random samples of cases and controls, maximum likelihood, in which we set $\lambda_i = n_i/n$ for $i = 0, 1$, is the most efficient strategy (Cosslett, 1981). For more complex schemes, using the sample proportions will no longer be fully efficient and we might expect weights based on some form of equivalent sample sizes to perform better. We have carried out some simulations to investigate this. Our limited experience so far suggests that setting $\lambda_i = n_i/n$ for $i = 0, 1$ leads to only a moderate loss of efficiency unless design effects are extreme.

We have not said anything specific about stratification beyond the basic division into cases and controls so far in this discussion. Implicitly, we are assuming that stratification on other variables such as age or ethnicity is handled by specifying the appropriate weights when we form the sample estimators $\hat{\mu}_i(i = 0, 1)$. One advantage of this approach is that, if we want to model the stratum constants in terms of other variables, then this is taken care of automatically. There is another approach to fitting model (8.6), which is to ignore the different selection probabilities of units in different strata. This mimics the maximum likelihood method of the previous section more closely and usually leads to more efficient estimates of the coefficients in β_1. However, further adjustments are necessary if we wish to model the stratum constants. We explore the difference in efficiency between the two approaches in our simulations in the next section. Note that we can again implement the method simply in a package like SUDAAN by choosing the weights appropriately.

8.4. EFFICIENCY

In this section, we describe the results of some of the simulations that we have done to test and compare the methods discussed previously. The simulations reported here use two explanatory variables, a continuous variable $X_1 \sim$ normal (0, 1) and an independent binary variable $X_2 \sim$ binomial (1, 0.5). We generated data from a linear logistic model with $\beta_1 = \beta_2 = 0.5$. With these values, an increase of one standard deviation in X_1 corresponds to a 65% increase in risk, and the effect of exposure ($X_2 = 1$) is also a 65% increase in risk. Values of X_2 were kept constant within a cluster and values of X_1 were highly correlated within a cluster ($\rho > 0.9$). We used an intercept of $\beta_0 = -6.1$ which produces a population proportion of approximately 1 case in every 300 individuals, a little less extreme than the 1 in 400 in the meningitis study. The number of cases in every simulation study is 200. The number of control clusters is always fixed. The number of controls in each study averages 200, but the actual number is random in some studies (when 'ethnicities' are subsampled at different rates – see later). All simulations used 1000 generated datasets.

The first row of Table 8.2 shows results for a population in which 60% of the clusters are of size 2 and the remaining 40% of size 4. We see from Table 8.2 that, under these conditions, the sample-weighted estimator of β_1 is 60% more efficient than the population-weighted estimator, whereas for β_2 there is a 20% increase in efficiency. It is possible that the gain in efficiency is smaller for the

Table 8.2   Relative efficiencies (cf. design weighting).

	β_1		β_2	
Stratified sampling	Sample weighted	Ethnic weighted	Sample weighted	Ethnic weighted
None				
2s and 4s	160		120	
of 'ethnic groups'				
2s and 4s (whole)	149	150	118	122
2s and 4s (random)	152	145	119	114
4s and 8s (whole)	186	210	141	156
4s and 8s (random)	189	177	127	122

binary variable because it represents a smaller change in risk. We also investigated the effect of using a standard logistic regression program ignoring clustering. Members of the same cluster are extremely closely related here, so we would expect the coverage frequency of nominally 95 % confidence intervals to drop well below 95 % if the clustering was ignored. We found that coverage frequencies for β_1 and β_2 dropped to 80 % and 82 % respectively when clustering was ignored. When we correct for clustering in both the sample-weighted and population-weighted variance estimates, the coverage frequencies were close to 95 %.

In Section 8.3, we discussed the possible benefits of using weights that reflected 'effective sample' size in place of sample weights. We repeated the preceding simulation (using the clusters of size 2 and 4) down-weighting the controls by a factor of 3 (which is roughly equal to the average design effect). This produced a 5 % increase in the efficiency of $\widehat{\beta}_1$ and only a 1 % increase in the efficiency of $\widehat{\beta}_2$ compared with sample weighting.

The rows of Table 8.2 headed 'Stratified sampling of "ethnic groups" ' show the results of an attempt to mimic (on a smaller scale) the stratified sampling of controls within clusters in the meningitis study. We generated our population so that 60 % of the individuals belong to ethnic group 1, and 20 % belong to each of groups 2 and 3. This was done in two ways. Where '(random)' has been appended, the ethnic groups have been generated independently of cluster. Where '(whole)' has been appended, all members of a cluster have the same ethnicity. All members of a sampled cluster in groups 2 and 3 were retained, while members of group 1 were retained with probability 0.33. We varied the sizes of the clusters, with either 60 % twos and 40 % fours, or 60 % fours and 40 % eights, before subsampling. The subsampling (random removal of group 1 individuals) left cluster sample sizes ranging from one to the maximum cluster size. The models now also include dummy variables for ethnic group and we can estimate these in two ways. The columns headed 'Ethnic weighted' show results when cases have sample weights, whereas controls have weights which are the sample weights times 1.8 for group 1 or 0.6 for groups 2 or 3. This reconstitutes relative population weightings so that contrasts in ethnic coefficients are correctly estimated.

In this latter set of simulations the increase in efficiency for sample weighting over population weighting is even more striking. Reweighting to account for different sampling rates in the different ethnic groups has led to similar efficiencies. This is important as the type of 'ethnic weighting' we have done in the control group would allow us to model group differences rather than being forced to fit a complete set of dummy variables. The coverages for β_2 are good for all methods that take account of the clustering. The coverages for β_1 dropped a little, however, in the '4s and 8s' clusters with design weighting. We can interpret this as meaning that, with more clustering and additional parameters to estimate, our effective sample sizes have dropped and, with smaller sample sizes, asymptotic approximations are less accurate.

8.5. ROBUSTNESS

In the previous section we considered efficiency when we were fitting the correct model. In this section, we look at what happens when the model is misspecified. In this case, the estimator $\hat{\beta}_\lambda$ converges in probability to b_λ, the solution to (8.6), i.e.

$$\lambda_1 E_1\{Xp_0(X)\} - \lambda_0 E_0\{Xp_1(X)\} = 0.$$

If the true model is logistic with parameter β, then $b_\lambda = \beta + k_\lambda e_1$. Only the intercept is affected by the choice of weightings λ_i used in the estimating equations. It is the remaining regression coefficients that are of primary interest, however, since they determine the relative risk of becoming a case associated with a change in the values of explanatory variables. These remaining coefficients are unaffected by the choice of weightings.

By contrast, if the model is misspecified, every element of b_λ depends upon the weightings used. Choosing design weights ($\lambda_i = W_i$) leads to B, the quantity we would be estimating if we fitted the model to the whole population. Since all models are almost certainly misspecified to some extent, many survey statisticians (see Kish and Frankel, 1974) would suggest that a reasonable aim is to estimate B. We adopted this perspective uncritically in Scott and Wild (1986, 1989), suggesting as a consequence that, although maximum likelihood estimation was more efficient, design weighting was more robust because it alone led to consistent estimates of B in the presence of model misspecification. This has been quoted widely by both biostatisticians and survey samplers. However, a more detailed appraisal of the nature of B suggests that using sample weights ($\lambda_i = n_i/n$), which is maximum likelihood under the model, may often be more robust as well as being more efficient.

Table 1 of Scott and Wild (1989) showed results from fitting a linear logistic regression model when the true model was quadratic with logit $\{\Pr(Y = 1|x)\} = \beta_0 + \beta_1 x + \beta_2 x^2$. Any model can be expanded in a Taylor series on a logistic scale and a quadratic approximation should give a reasonable idea of what happens when the working logistic model is not too badly misspecified. Two population curves were investigated, with the extent of the curvature being chosen so that an analyst would fail to detect the curvature about 50% of the time with samples of size $n_1 = n_2 = 200$. Scott and Wild (1989) assumed a standard normal(0, 1) distribution for the covariate X and the population curves logit $\{\Pr(Y = 1|x)\} = \beta_0 + 4x - 0.6x^2$ with $\beta_0 = -5.8$, plotted in Figure 8.1(a), and logit $\{\Pr(Y = 1|x)\} = \beta_0 + 2x + 0.3x^2$ with $\beta_0 = -5.1$, plotted in Figure 8.1(b). We will refer to the former curve as the *negative quadratic* and the latter as the *positive quadratic*. Both models result in about 1 case in every 20 individuals (i.e. $W_1 \approx 0.05$). By reducing the value of β_0 we reduce the value of W_1. Figure 8.1(c) plots the negative quadratic curve for $\beta_0 = -9.21$ and Figure 8.1(d) plots the positive quadratic curve for $\beta_0 = -9.44$. These values correspond to 1 case in every 400 individuals. For each population curve,

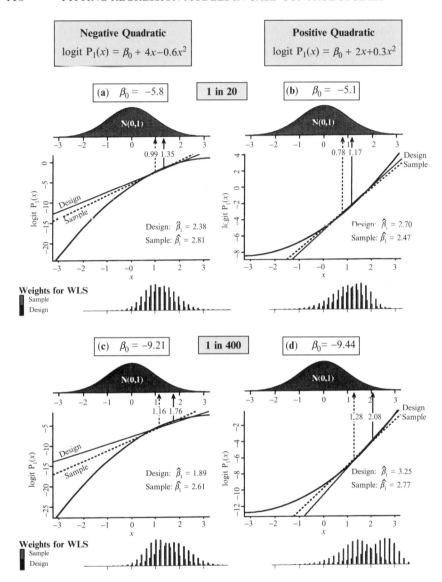

Figure 8.1   Approximations to population curve.

we have plotted two linear approximations. The solid line corresponds to B, the 'whole-population approximation'. The dashed-line approximation corresponds to using sample weights (maximum likelihood).

The slope of the logit curve tells us about the relative risk of becoming a case associated with a change in x. Using vertical arrows, we have indicated the position at which the slope of the linear approximation agrees with the slope of the curve. The results are clearest in the more extreme situation of 1 case in 400 (cf. the meningitis study). We see that B (design weights) is telling us about changes in risk only in the extreme tail of the covariate distribution, while the sample-weighted maximum likelihood estimator is telling us about the effects of changes in x closer to the centre of the population. In other words, the sample-weighted estimator represents what is happening in the population better than the design-weighted one. The same thing is happening with 1 case in 20 but to a lesser extent. This should not be surprising as the relative weightings are less severe here.

Why does the whole-population approximation provide a worse description of the effects of changes in x on risk? To gain some insight we use a discrete approximation to a normal distribution of X and look at the asymptotic weights employed in a weighted least squares linear fit to the true logit at x, namely $\log\{\Pr(Y = 1|x)/\Pr(Y = 0|x)\}$ (weighted least squares with the appropriate weights are asymptotically equivalent to maximum likelihood when the model is true). Note that weighted least squares use weights that are inversely proportional to variances and recall that the asymptotic variance of the logit is proportional to $1/n(x)p_1(x)p_0(x)$, where $f(x)$ denotes the frequency of x in the population. Thus, the weight at x corresponding to design weighting ($\lambda_i = W_i$) is proportional to $f(x)p_1(x)p_0(x)$. Similarly, when $n_1 = n_0$, the weight corresponding to sample weighting ($\lambda_i = n_i/n$) is proportional to $\{f(x|0) + f(x|1)\}$ $\tilde{p}_1(x)\tilde{p}_0(x)$, where $\tilde{p}_i(x)$ is obtained by replacing β with b_λ in $p_i(x)$ (i.e. $\tilde{p}_1(x)$ represents the smoothed sample proportion, and $p_1(x)$ the fitted population proportion, of cases at x). These weights are plotted at the bottom of each graph in Figure 8.1 with sample–control weights in grey and population weights in black. We see that as cases become rarer (going from 1 in 20 to 1 in 400), the weights used move away from the centre of the X-distribution with the design weights moving more quickly out into extreme regions that the sample weights, and therefore approximating the logit curve only for extreme (and rare) high-risk individuals.

We have argued above that, for misspecified models, using sample weights often estimates a more relevant quantity than using design weights. The weights also offer some insight into the efficiency losses, and slower asymptotic convergence of the design-weighted estimates.

Figure 8.2 relates the weights to $f(x|1)$, the X-distribution among cases, and $f(x|0)$, the X-distribution among controls. These densities were obtained for the more extreme (1 in 400) models in Figure 8.1, which are curved, but almost identical behaviour is observed when we look at fitting correctly specified linear models. In the following we assume a case–control sample of $n_1 = n_0$ for simplicity. Under case–control sampling, the sample odds at x_j are of the form n_{j1}/n_{j0} where n_{j1} is the number of cases in the sample at x_j. Multiplication by W_1/W_0 converts this to an estimate of the population odds at x_j. Clearly, these quantities are most reliably estimated in X-regions where there are substantial numbers of both cases and controls in the population, i.e. where $f(x|1)$ and $f(x|0)$ overlap. We see that this is the region from which the sample-weighted estimates draw most heavily. In situations where x has a strong effect and cases are rare, as shown in Figure 8.2, the design-weighted estimate puts its large weights where cases are most common. These weights can

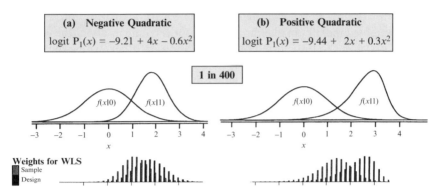

Figure 8.2   Comparing weights to marginal densities.

be shown to be proportional to $f(x|1)\mathrm{Pr}(Y=0|x)$. Thus, in situations where cases are rare across the support of X, the weight profile is almost identical to $f(x|1)$ as in Figure 8.2. Heavy weight is being given to unreliable sample odds estimates formed at x_j-values where cases are common in the sample but controls are rare. One would expect this to lead to precisely the sort of behaviour we have seen over many simulations, namely less efficient estimates and slower convergence to asymptotic behaviour.

These arguments lead us to the conclusion that design-weighted estimates are both less efficient and no more robust that maximum likelihood sample-weighted estimates.

8.6.   OTHER APPROACHES

One alternative approach is to modify the subsampling method suggested by Hinkins, Oh and Scheuren (1997). Instead of trying to develop specialist programs to handle the complex sample, we modify the sample so that it can be processed with standard programs. In our case, this means producing a subsample of controls (cases) that looks like a simple random sample from the population of controls (cases). Then we have simple case–control data which can be analysed using a standard logistic regression program. Can we produce such a subsample? The answer is 'Yes' for most designs (see Hinkins, Oh and Scheuren, 1997) but the resulting subsample size is usually much smaller than the original. Obviously this results in a substantial loss of efficiency, since we are throwing away many (sometimes most) of the observations. We increase power by repeating the process independently many times and averaging the estimates over the repetitions. Note that the repetitions, although conditionally independent, are not unconditionally independent when averaged over the distribution of the original sample. In spite of this, it is relatively simple to estimate standard errors and derive other properties of the final estimates. Details are given in Rao and Scott (1999b).

We have done some simulations to investigate the relative efficiency of this method. Preliminary results suggest that the method works well in relatively simple situations. With a small number of strata and small cluster sizes we observed efficiencies comparable with those obtained using the sample-weighted method of Section 8.3. Unfortunately, this method broke down completely when we attempted to apply it to the example discussed in the introduction. With a maximum cluster sample size of six, a simple random subsample of controls can contain at most $n_0/6$ units. This meant that our subsamples only contained about 40 controls. As a consequence, some of the age×ethnic strata were empty in some of the subsamples and the logistic regression program often failed to converge.

Another approach would be to build a mixed logistic model for $Pr(Y_t = y|x_t)$ containing random cluster effects. In principle, fitting such a model is covered by the methods in Scott and Wild (2001). In practice, this requires numerical integration at each step of an iterative procedure, but software to make this feasible has not yet been developed. Work is proceeding on developing such software. It is important to note that the meaning of the regression coefficients is different with this approach. In the rest of that paper, we are dealing with marginal (or population-averaged) models, which are appropriate when we are interested in the effects of changes in the explanatory variables on the whole population rather than on specific individuals. Random effects models are subject specific and are appropriate when we are interested in the effects of changes in the explanatory variables on specific individuals. This is a huge and important topic, but it is peripheral to the main points of this chapter and we do not pursue it further here. The interested reader is referred to the comprehensive discussions in Neuhaus, Kalbfleisch and Hauck (1991) or Diggle, Liang and Zeger (1994) for more details.

ACKNOWLEDGEMENTS

We wish to thank Joanna Stewart and Nick Garrett for introducing us to the problem and our research assistant Jolie Hutchinson who did most of the programming.

Continuous and General Response Data

CHAPTER 9

Introduction to Part C

R. L. Chambers

The three chapters that make up Part C of this book discuss methods for fitting a regression model using sample survey data. In all three the emphasis is on situations where the sampling procedure may be informative, i.e. where the distribution of the population values of the sample inclusion indicators depends on the response variable in the regression model. This model may be parametrically specified, as in Chapter 12 (Pfeffermann and Sverchkov) or may be nonparametric, as in Chapter 10 (Bellhouse, Goia and Stafford) and Chapter 11 (Chambers, Dorfman and Sverchkov).

To focus our discussion of the basic issues that arise in this situation, consider a population made up of N units, with scalar variables Y and X defined for these units (the extension to multivariate X is straightforward). A sample of n units is taken and their values of Y and X are observed. The aim is to use these data to model the regression of Y on X in the population. Let $g_U(x)$ denote this population regression function. That is, $g_U(x)$ is the expected value of Y given $X = x$ under the superpopulation joint distribution of Y and X. If a parametric specification exists for $g_U(x)$, the problem reduces to estimating the values of the parameters underlying this specification. More generally, such a specification may not exist, and the concern then is with estimating $g_U(x)$ directly.

9.1. THE DESIGN-BASED APPROACH

The traditional design-based approach to this problem weights the sample data in inverse proportion to their sample inclusion probabilities when fitting $g_U(x)$. In particular, if an estimation procedure for $g_U(x)$ can be defined when the values of Y and X for the entire population are known, then its extension to the general unequal probability sampling situation is simply accomplished by replacing all (unknown) population-based quantities in this procedure by appropriate estimates based on the sample data. If the population-level estimation procedure is maximum likelihood this represents an application of the so-called

Analysis of Survey Data Edited by R. L. Chambers and C. J. Skinner
© 2003 John Wiley & Sons, Ltd

pseudo-likelihood approach (see Chapter 2). In Chapter 10, Bellhouse, Goia and Stafford apply this idea to nonparametric estimation of both the population density $f_U(y)$ of Y and the population regression function $g_U(x)$.

Nonparametric density estimation and nonparametric regression estimation are two well-established areas of statistical methodology. See Silverman (1986) and Härdle (1990). In both areas, kernel-based methods of estimation are well known and widely available in statistical software. These methods were originally developed for simple random sampling from infinite populations. For example, given a population consisting of N independent and identically distributed observations $y_1, y_2, \ldots, y_N$ from a population distribution with density $f_U(y)$, the kernel estimate of this density is

$$\hat{f}_U(y) = \frac{1}{Nh} \sum_{t=1}^{N} K\left(\frac{y_t - y}{h}\right) \tag{9.1}$$

where K is its kernel function (usually defined as a zero-symmetric unimodal density function) and $h > 0$ is its bandwidth. The bandwidth h is the most important determinant of the efficiency of $\hat{f}_U(y)$. In particular, the bias of (9.1) increases and its variance decreases as h increases. Asymptotic results indicate that the optimal value of h (in terms of minimising the asymptotic mean squared error of $\hat{f}_U(y)$) is proportional to the inverse of the fifth root of N.

Modifying this estimator for the case where data from a complex sampling scheme are available (rather than the population values) is quite straightforward from a design-based viewpoint. One just replaces $\hat{f}_U(y)$ by

$$\hat{f}_s(y) = \frac{1}{Nh} \sum_{s} w_t K\left(\frac{y_t - y}{h}\right) \tag{9.2}$$

where w_t is the sample weight associated with sampled unit t. More generally, Bellhouse, Goia and Stafford define a binned version of $\hat{f}_s(y)$ in Chapter 10, and develop asymptotic design-based theory for this estimator. Their estimator (10.1) reduces to (9.2) when the bins correspond to the distinct Y-values observed on the sample. Since larger bins are essentially equivalent to increasing bandwidth, there is a strong relationship between the bin size b and the bandwidth h, and Bellhouse and Stafford (1999) show that, analogous to the random sampling case, a good choice is $b = 1.25h$. In Section 10.4 Bellhouse, Goia and Stafford suggest a bootstrap algorithm for correcting the bias in (9.2).

Extension of kernel density estimation to nonparametric estimation of the population regression function $g_U(x)$ is straightforward. The population-based kernel estimator of this function is

$$\hat{g}_U(x) = \sum_{k=0}^{p} b_{kx} x^k \tag{9.3}$$

where the coefficients $\{b_{kx}\}$ are the solution to the local polynomial estimating equation

$$\sum_{t=1}^{N} K\left(\frac{x_t - x}{h}\right) \left(y_t - \sum_{k=0}^{p} b_{kx} x_t^k\right) (1 \quad x_t \ldots x_t^p) = 0. \tag{9.4}$$

That is, this approach estimates $g_U(x)$ via a weighted least squares fit of a polynomial regression function of order p to the population data, with the weight for unit t in the population determined by its 'distance' $K(h^{-1}(x_t - x))$ from x. A popular choice is $p = 1$, in which case the solution to (9.4) is often referred to as the *local linear regression* of Y on X.

With unequally weighted sample survey data (9.3) and (9.4) are replaced by

$$\hat{g}_s(x) = \sum_{k=0}^{p} b_{kxs} x^k \tag{9.5}$$

and

$$\sum_{s} w_t K\left(\frac{x_t - x}{h}\right) \left(y_t - \sum_{k=0}^{p} b_{kxs} x_t^k\right) (1 \quad x_t \ldots x_t^p) = 0. \tag{9.6}$$

Again, binned versions of these formulae are easily derived (see Section 10.5). In this case binning is carried out on X, with Y-values in (9.6) replaced by their sample weighted average values within each bin. Since the estimator (9.5) is just a weighted sum of these weighted averages, its variance (conditional on the estimated bin proportions for X) can be estimated via a sandwich-type estimator. See Equation (10.10). Whether this variance is the right one for (9.5) is a moot point. In general both the estimated bin proportions as well as the kernel weights $K(h^{-1}(x_t - x))$ are random variables and their contribution to the overall variability of (9.5) is ignored in (10.10).

9.2. THE SAMPLE DISTRIBUTION APPROACH

This approach underpins the development in Chapter 11 (Chambers, Dorfman and Sverchkov) and Chapter 12 (Pfeffermann and Sverchkov). See Section 2.3 for a discussion of the basic idea. In particular, given knowledge of the method used to select the sample and specification of the population distribution of Y, one can use Bayes' theorem to infer the marginal distribution of this variable for any sampled unit. This sample distribution then forms the basis for inference about the population distribution.

Suppose the population values of Y can be assumed to be independent and identically distributed observations from a distribution with density $f_U(y)$. This density function may, or may not, be parametrically specified. Let I denote the indicator random variable defined by whether a generic population unit is selected into the sample or not and define Δ_y to be a small interval centred around y. Then

$$Pr(Y \in \Delta_y | I = 1) = \frac{Pr(I = 1 | Y \in \Delta_y) Pr(Y \in \Delta_y)}{Pr(I = 1)}.$$

The sample density $f_s(y)$ for Y is obtained by letting Δ_y shrink to the value y, leading to

$$f_s(y) = \frac{Pr(I = 1|Y = y)f_U(y)}{Pr(I = 1)}. \tag{9.7}$$

When the random variables I and Y are independent of one another, the sample and population densities of Y are identical. In such cases one can apply standard inferential methods to the sample data. In general, however, these two random variables will not be independent, and inference then needs to be based on the distribution of the *sample* data.

In Chapter 11 Chambers, Dorfman and Sverchkov extend this sample-based approach to nonparametric estimation of the population regression function $g_U(x)$. The development in this chapter uses relationships between population and sample moments that hold under this approach to develop 'plug in' and 'estimating equation' based estimators for $g_U(x)$ given four data scenarios corresponding to common types of sample data availability. These estimators are then contrasted with the simple inverse π-weighted approach to estimation of $g_U(x)$, as described, for example, in Chapter 10. Under the sample-based framework, this π-weighted approach is inconsistent in general. However, it can be made consistent via a 'twicing' argument, which adjusts the estimate by subtracting a consistent estimate of its bias.

Finally, in Chapter 12 Pfeffermann and Sverchkov extend this idea to parametric estimation. In particular, they focus on the important case where population values are independently distributed and follow a generalised linear model (GLM). The conditional version of (9.7) appropriate to this situation is

$$f_s(y|x) = \frac{Pr(I = 1|Y = y, X = x)f_U(y|x)}{Pr(I = 1|X = x)} \tag{9.8}$$

where $f_U(y|x)$ denotes the value at y of the population conditional density of Y given $X = x$. In order to use this relationship, it is necessary to model the conditional probability that a population unit is sampled given its values for Y and X. Typically, we characterise sample inclusion for a population unit in terms of its sample inclusion probability π. In addition to the standard situation where this probability depends on the population values z_U of an auxiliary variable Z, it can also depend on the population values y_U of Y and x_U of X. In general therefore $\pi = \pi(y_U, x_U, z_U) = E(I = 1|y_U, x_U, z_U)$ so the values of π in the population correspond to realised values of random variables. Using an iterated expectation argument, Pfeffermann and Sverchkov (Equation (12.1)) show that $Pr(I = 1|Y = y, X = x) = E_U(\pi|Y = y, X = x)$, where E_U denotes expectation with respect to the population distribution of relevant quantities. It follows that (9.8) can be equivalently expressed as

$$f_s(y|x) = \frac{E_U(\pi|Y = y, X = x)}{E_U(\pi|X = x)} f_U(y|x). \tag{9.9}$$

In order to use this relationship to estimate the parameters of the GLM defining $f_U(y|x)$, Pfeffermann and Sverchkov make the further assumption that the sample data values can be treated as realisations of independent random variables (although not true in general, asymptotic results presented in Pfeffermann, Krieger and Rinott (1998) justify this assumption as a first-order approximation). This allows the use of (9.9) to define a sample-based likelihood for the parameters of $f_U(y|x)$ that depends on both the GLM specification for this conditional density as well as the ratio $E_U(\pi|Y = y, X = x)/E_U(\pi|X = x)$.

Let β denote the (unknown) vector-valued parameter characterising the conditional population distribution of Y given X. The sample likelihood estimating equations for β defined by (9.9) are

$$\sum_s d\log f_U(y_t|x_t)/d\beta + \sum_s d\log E_U(\pi_t|y_t, x_t)/d\beta$$

$$- \sum_s d\log E_U(\pi_t|x_t)/d\beta = 0 \qquad (9.10)$$

where these authors note that (i) the conditional expectation $E_U(\pi_t|y_t, x_t)$ will be 'marginally' dependent on β (and so the second term of (9.10) can be ignored), and (ii) the conditional expectation $E_U(\pi_t|x_t)$ in the third term of (9.10) can be replaced by an empirical estimate based on the sample data (in effect leading to a sample-based *quasi-* likelihood approach).

In order to show how $E_U(\pi_t|x_t)$ can be estimated, Pfeffermann and Sverchkov use a crucial identity (see Equation (12.4)) linking population and sample-based moments. From this follows

$$E_U(\pi_t|x_t) = 1/E_s(\pi_t^{-1}|x_t) \qquad (9.11)$$

where the expectation operator $E_s()$ in the denominator on the right hand side of (9.11) denotes expectation conditional on being selected into the sample. It can be estimated by the simple expedient of fitting an appropriate regression model for the sample π_t^{-1} values in terms of the sample x_t values. See Section 12.3.

The sample-based likelihood approach is not the only way one can use (9.9) to develop an estimating equation for β. Pfeffermann and Sverchkov observe that when the population-level 'parameter equations'

$$\sum_{t=1}^{N} E(d\log f(y_t|x_t)/d\beta \,|x_t) = 0$$

are replaced by their sample equivalent and (12.4) applied, then these equations can be expressed as

$$\sum_s E_s(d\log f_s(y_t|x_t)/d\beta \,|x_t) = \sum_s \frac{E_s(\pi_t^{-1}d\log f_U(y_t|x_t)/d\beta \,|x_t)}{E_s(\pi_t^{-1}|x_t)} = 0.$$

They therefore suggest β can also be estimated by solving

$$\sum_s q_t(x_t)d\log f_U(y_t|x_t)/d\beta = 0 \qquad (9.12)$$

where the weights $q_t(x_t)$ are defined as

$$q_t(x_t) = \frac{\pi_t^{-1}}{E_s(\pi_t^{-1}|x_t)}. \tag{9.13}$$

As before, the conditional expectation in the denominator above is unknown, and is estimated from the sample data.

Pfeffermann and Sverchkov refer to the conditional weights (9.13) as adjusted weights and argue that they are superior to the standard weights π_t^{-1} that underpin the pseudo-likelihood approach to estimation of β. Their argument is based on the observation that the pseudo-likelihood approach to estimating β corresponds to replacing the conditional expectation in the denominator (9.13) by an unconditional expectation, and hence includes effects due to sample variation in values of X.

These authors also suggest that variance estimation for estimators obtained by solving the sample-based likelihood estimating equations (9.10) can be defined by assuming that the left hand sides of these equations define a score function and then calculating the inverse of the expected information matrix corresponding to this function. For estimators defined by an estimating equation like (9.12), they advocate either sandwich-type variance estimation (assuming the weights in these equations are treated as fixed constants) or bootstrap methods. See Section 12.4.

9.3. WHEN TO WEIGHT?

In Section 6.3.2 Skinner discusses whether survey weights are relevant for parameteric inference. Here we address the related problem of deciding whether to use the sample distribution-based methods explored in Part C of this book. These methods are significantly more complex than standard methods. However, they are applicable where the sampling method is informative. That is, where sample inclusion probabilities depend on the values of the variable Y. Both Chapter 11 (Chambers, Dorfman and Sverchkov) and Chapter 12 (Pfeffermann and Sverchkov) contain numerical results which illustrate the behaviour of the different approaches under informative sampling.

These results indicate that more complicated sample distribution-based estimators should be used instead of standard unweighted estimators if the sampling procedure is informative. They also show that use of sample distribution-based estimators when the sampling method is noninformative can lead to substantial loss of efficiency. Since in practice it is unlikely that a secondary analyst of the survey data will have enough background information about the sampling method to be sure about whether it is informative or noninformative, the question then becomes one of deciding, on the basis of the evidence in the available sample data, whether they have been obtained via an informative sampling method.

Two basic approaches to resolving this problem are presented in what follows. In Chapter 11, Chambers, Dorfman and Sverchkov consider this problem from the viewpoint of nonparametric regression estimation. They suggest two test methods. The first corresponds to the test statistic approach, with a Wald-type statistic used to test whether there is a significant difference in the fits of the estimated regression functions generated by the standard and sample distribution-based approaches at a number of values of X. Their second (and more promising) approach is to test the hypothesis that the coefficients of a polynomial approximation to the difference between the population and sample regression functions are all zero. In Chapter 12, Pfeffermann and Sverchkov tackle the problem by developing a Hotelling-type test statistic for the expected difference between the estimating functions defining the standard and sample distribution-based estimators. As they point out, 'the sampling process can be ignored for inference if the corresponding parameter equations are equivalent'. Numerical results for these testing methods presented in Chapters 11 and 12 are promising.

CHAPTER 10

Graphical Displays of Complex Survey Data through Kernel Smoothing

D. R. Bellhouse, C. M. Goia and J. E. Stafford

10.1. INTRODUCTION

Many surveys contain measurements made on a continuous scale. Due to the precision at which the data are recorded for the survey file and the size of the sample, there will be multiple observations at many of the distinct values. This feature of large-scale survey data has been exploited by Hartley and Rao (1968, 1969) in their scale-load approach to the estimation of finite population parameters. Here we exploit this same feature of the data to obtain kernel density estimates for a continuous variable y and to graphically examine relationships between a variate y and covariate x through local polynomial regression. In the case of independent and identically distributed random variables, kernel density estimation in various contexts is discussed, for example, by Silverman (1986), Jones (1989), Scott (1992) and Simonoff (1996). Local polynomial regression techniques are described in Härdle (1990), Wand and Jones (1995), Fan and Gijbels (1996), Simonoff (1996) and Eubank (1999). Here we examine how these kernel smoothing techniques may be applied in the survey context. Further, we examine the effect of the complex design on estimates and second-order moments. With particular reference to research in complex surveys, Buskirk (1999) has examined density estimation techniques using a model-based approach and Bellhouse and Stafford (1999) have examined the same techniques under both design- and model-based approaches. Further, Korn and Graubard (1998b) provided a technique for local regression smoothing based on data at the micro level, but did not provide any properties for their procedures. Bellhouse and Stafford (2001) have provided a design-based approach to local polynomial regression. Bellhouse and Stafford (1999, 2001) have analyzed data from the Ontario Health Survey. Chesher (1997) has

Analysis of Survey Data Edited by R. L. Chambers and C. J. Skinner
© 2003 John Wiley & Sons, Ltd

applied nonparametric regression techniques to model data from the British National Food Survey and Härdle (1990) has done the same for the British Family Expenditure Survey. In these latter cases it does not appear that the survey weights were used in the analysis.

One method of displaying the distribution of continuous data, univariate or bivariate, is through the histogram or binning of the data. This is particularly useful when dealing with large amounts of data, as is the case in large-scale surveys. In view of the multiple observations per distinct value and the large size of the sample that are both characteristic of survey data, we can bin to the precision of the data, at least for histograms on one variable. On their own, univariate histograms are usually easy to interpret. However, interpretation can be difficult when comparing several histograms, especially when they are superimposed. Similarly, bivariate histograms, even on their own, can be difficult to interpret. We overcome these difficulties in interpretation by smoothing the binned data through kernel density estimation. What is gained by increased interpretability through smoothing is balanced by an increase in the bias of the density estimate of the variable under study. The basic techniques of smoothing binned data are described in Section 10.2 and applications of the methodology to the 1990 Ontario Health Survey are presented in Section 10.3. A discussion of bias adjustments to the density estimate is given in Section 10.4.

Rather than binning to the precision of the data in y, we could instead bin on a covariate x. For each distinct value of x, we compute an estimate of the average response in y. We can then investigate the relationship between y and x through local polynomial regression. The properties of the estimated relationship are examined in Section 10.5 and applications to the Ontario Health Survey are discussed in Section 10.6.

10.2. BASIC METHODOLOGY FOR HISTOGRAMS AND SMOOTHED BINNED DATA

For density estimation from a finite population it is necessary to define the function to be estimated. We take the point of view that the quantity of interest is a model function obtained through asymptotic arguments based on a nested sequence of finite populations of size N_ℓ, $l = 1, 2, 3, \ldots$ with $N_\ell \to \infty$ as $l \to \infty$. For a given population l we can define the finite population distribution function $F_{N_l}(y)$ as the empirical distribution function for that finite population. Then as $l \to \infty$ we assume $F_{N_l}(y) \to F(y)$ where $F(y)$ is a smooth function. The function of interest is $f(y) = F'(y)$ rather than the empirical distribution function for the given finite population.

For independent and identically distributed observations on a single variable $f(y)$ is the model density function. One avenue for estimation of $f(y)$ is the construction of a histogram. This is a straightforward procedure. The range of the data is divided into I nonoverlapping intervals or bins, and either the number or the sample proportion of the observations that fall into each bin is calculated. This sample proportion for the ith bin, where $i = 1, \ldots, I$, is an unbiased estimate of p_i, the probability that an observation taken from the underlying probability distribution falls in the ith bin. The same sample proportion divided by the bin size b can be used as an estimate of $f(y)$ at all y-values within the bin. The methodology can be easily extended to two variables. Bins are constructed for each variable, I for the first variable and J for the second, yielding a cross-classification of IJ bins in total. In this case p_{ij} is the probability that an observation taken from the underlying bivariate distribution falls in the (i,j)th bin, where $i = 1, \ldots, I$ and $j = 1, \ldots, J$.

A minor modification is needed to obtain binned data for large-scale surveys. In this case survey weights are attached to each of the sample units. These are used to obtain the survey estimates $\hat{p}_i$ of p_i and $\hat{p}_{ij}$ of p_{ij}, where p_i and p_{ij} are now finite population bin proportions. Within the appropriate bin interval, the histogram estimate of the univariate density function is $\hat{p}_i/b$ for any y in the ith bin. Similarly, $\hat{p}_{ij}$ divided by the bin area provides a bivariate density estimate. For simplicity of presentation we assume throughout that the bin sizes b are equal for all bins. This assumption can be relaxed.

We denote the raw survey data by $\{y_s, w_s\}$ where y_s is the vector of observations and w_s is the vector of survey weights. If binning is carried out then the data are reduced to $\{m, \hat{p}\}$, where $\mathbf{m} = (m_1, m_2, \ldots, m_I)$ is the vector of bin midpoints and $\hat{p} = (\hat{p}_1, \hat{p}_2, \ldots, \hat{p}_I)$ is the vector of survey estimates of bin proportions. When binning has been carried out to the level of precision of the data, then, for the purpose of calculating density estimates, there is no loss in information in going from $\{y_s, w_s\}$ to $\{m, \hat{p}\}$.

Visual comparison of density estimates is much easier once the histogram has been smoothed. The smoothed histogram, examined by Bellhouse and Stafford (1999), is also a density estimate. For complex surveys based on the data $\{m, \hat{p}\}$, the smoothed histogram for a single variable y is given by

$$\hat{f}(y) = \frac{1}{h} \sum_{i=1}^{I} \hat{p}_i K\left(\frac{y - m_i}{h}\right). \tag{10.1}$$

In (10.1), $K(x)$ is the kernel evaluated at the point x and h is the bandwidth or smoothing parameter to be chosen. The kernel estimator at a value y is a weighted sum of the estimated bin proportions with higher weights placed on the bins that are closer to the value y. The kernel $K(x)$ is typically chosen to be a probability density function with mean value 0 and finite higher moments, in particular

$$\sigma^2 = \int t^2 K(t) \mathrm{d}t < \infty. \tag{10.2}$$

In the examples that we provide we have chosen the kernel to be the standard normal distribution. The estimate in (10.1) depends on both the bin size b and the window width h. In practical terms for complex surveys, the smallest value that b can take will be the precision to which the data have been recorded. For the variables body mass index and desired body mass index in the Ontario

Health Survey, discussed in Section 10.3, the smallest value for b is 0.1. As noted already, there is no loss in information between $\{y_s, w_s\}$ and $\{m, \hat{p}\}$ when the level of binning is at the precision of the data. As a result, kernel density estimation from the raw data while taking into account the survey weights is equivalent to (10.1) with the smallest possible bin size.

Quantile estimates can be obtained from $\hat{f}(y)$ using methodology given in Bellhouse and Stafford (1999). Denote the estimated αth quantile by $\hat{q}_\alpha$. The estimate may be obtained as the solution to the equation

$$\hat{F}(\hat{q}_\alpha) - \alpha = \int_{-\infty}^{\hat{q}_\alpha} \hat{f}(x)\mathrm{d}x - \alpha = 0.$$

The solution is obtained through Newton–Raphson iteration by setting

$$\hat{q}_\alpha^i = \hat{q}_\alpha^{i-1} - \hat{F}(\hat{q}_\alpha^{i-1})/f(\hat{q}_\alpha^{i-1})$$

for $i = 1, 2, 3, \ldots$ until convergence is achieved. If the kernel is a standard normal density, then $\hat{F}(\hat{q}_\alpha^i)$ has the same functional form as $\hat{f}(\hat{q}_\alpha^i)$ with the density replaced by the standard normal cumulative distribution function. An initial estimate $\hat{q}_\alpha^0$ may be obtained through a histogram estimate. Bellhouse (2000) has shown that the estimated variance of $\hat{q}_\alpha$ may be expressed as

$$\hat{V}(\hat{q}_\alpha) = \hat{V}(\hat{F}(\hat{q}_\alpha))/[\hat{f}(\hat{q}_\alpha)]^2,$$

where $\hat{V}(\hat{F}(\hat{q}_\alpha))$, the estimated variance of the proportion of the observations less than or equal to $\hat{q}_\alpha$, can be obtained as the survey variance estimate of a proportion. In a limited empirical study carried out by Bellhouse (2000), this variance estimate has compared favourably to the Woodruff (1952) procedure.

The population may consist of a number D of disjoint subpopulations or domains, such as gender or age groups. Then a histogram or a smoothed histogram $\hat{f}_d(y)$ may be calculated for each domain d, $d = 1, \ldots, D$. By super-imposing plots of $\hat{f}_d(y)$ for $d = 1, \ldots, D$ we can make visual comparisons of the domains. We could also make visual comparisons of different variables for the same domain. This provides a much less cluttered view than superimposing histograms, and is visually more effective than trying to compare separate plots.

For two variables, x and y, the smoothed two-dimensional (2-D) histogram is given by

$$\hat{f}(x, y) = \frac{1}{h_1 h_2} \sum_{i=1}^{I} \sum_{j=1}^{J} \hat{p}_{ij} K\left(\frac{x - m_{1i}}{h_1}, \frac{y - m_{2j}}{h_2}\right), \tag{10.3}$$

where $K(x, y)$ is the kernel evaluated at the point (x, y), where h_1 and h_2 are the bandwidths in the x direction and y direction respectively, and where m_{1i} and m_{2j} are the midpoints of the ith and jth bins defined for x and y respectively. Again, $K(t, s)$ is usually a density function; we choose $K(x, y)$ to be the product of two standard normal densities. A 3-D plot of (10.3) may be easier to interpret than the associated histogram. Additionally contour plots based on

(10.3) may also be obtained, thus providing more easily accessible information about the relationship between y_1 and y_2.

For a given value of y, the finite population mean square error of a density estimator $\hat{f}(y)$ of $f(y)$ is $E(\hat{f}(y) - f(y))^2$ where the operator E denotes expectation with respect to the sampling design. A measure of global performance of $\hat{f}(y)$ is its integrated mean square error given by $\int E(\hat{f}(y) - f(y))^2 dy$, where the integral is taken over the range of values of y. The limiting value of this measure, taken in the increasing sequence of finite populations and as the bandwidth and bin size decrease, is the asymptotic integrated mean square error (AIMSE). In the case of independent and identically distributed random variables Silverman (1986) and Scott (1992), among others, provide formal definitions of the AIMSE. We examine the AIMSE under complex sampling designs for two reasons. The first is to see the effect of the design on this global measure in comparison to the standard case. The second is that the ideal window size h may be obtained through the AIMSE.

Under the finite population expectation operator, Bellhouse and Stafford (1999) have obtained

$$\text{AIMSE}\,(\hat{f}(y)) = \bar{d}\,\frac{R(K)}{nh} - \bar{d}'\,\frac{R(f)}{n} - (\bar{d} - \bar{d}')\frac{bR(f)R(K)}{n}$$
$$+ \frac{1}{4}\left(\sigma^2 h^2 + \frac{b^2}{12}\right)^2 R(f'') \tag{10.4}$$

as the AIMSE for the smoothed histogram estimator of $f(y)$ in the survey context, where σ^2 is given in (10.2). In (10.4) higher order terms in $1/n$, h and b are ignored. Further in (10.4): $\bar{d} = \sum_{i=1}^{I} d_i/k$, where $d_i = nvar(p_i)/[p_i(1 - p_i)]$ is the design effect for the ith bin; $\bar{d}' = \sum_{i=1}^{I}\sum_{i=1}^{I} d_{ij}/k^2$, where $d_{ij} = -n\,\text{cov}$ $(\hat{p}_i, \hat{p}_j)/[p_i p_j]$ is the covariance design effect when $i \neq j$; and $R(\phi) = \int \phi(x)^2 dx$. When $\bar{d} = \bar{d}' = 1$, (10.4) reduces to that obtained by Jones (1989) in the standard case. The first three terms in (10.4) comprise the asymptotic integrated variance and the last term is the asymptotic integrated square bias. It may be noted that the complex design has no effect on the bias but changes the variance by factors related to the design effects in the survey. The same is true for the AIMSE of the histogram estimate of the density function. The bias term in (10.4) increases with the bin size b. Smoothing the histogram generally leads to an increase in bias of the estimate of the density function, i.e. $\hat{f}(y)$ has greater bias than $\hat{p}_i/b$.

For smoothed histograms the bin size b and the window width h are related. Jones (1989) sets $b = ah$ in (10.4) for the case when $\bar{d} = \bar{d}' = 1$ and the term $R(f)/n$ is negligible. This yields an AIMSE depending on a, denoted by AIMSE_a. Jones then examined the ratio $\text{AIMSE}_a/\text{AIMSE}_0$, where AIMSE_0 is the value for the ideal kernel density estimate. Jones found that the relationship $b = 1.25\,h$ was reasonable. Using the same approach but without the assumption that $\bar{d} = \bar{d}' = 1$, Bellhouse and Stafford (1999) have obtained the same relationship for b and h. Consequently, the ideal window size does not change when going from the standard case to complex surveys.

10.3. SMOOTHED HISTOGRAMS FROM THE ONTARIO HEALTH SURVEY

We illustrate our techniques with results from the Ontario Health Survey. This survey was carried out in 1990 in order to measure the health status of the population and to collect data relating to the risk factors of major causes of morbidity and mortality in the Province of Ontario. A total sample size of 61 239 people was obtained by stratified two-stage cluster sampling. Each public health unit in the province, which is further divided into rural and urban areas, formed the strata, a total of 86 in number. The first-stage units within a stratum were enumeration areas. These enumeration areas, taken from the 1986 Census of Canada, are the smallest geographical units from which census counts can be obtained automatically. An average of 46 enumeration areas was chosen within each stratum. At the second stage of sampling, within an enumeration area dwellings were selected, approximately 15 from an urban enumeration area and 20 from a rural enumeration area. Information was collected on members of the household within the dwelling, which constituted the cluster. See Ontario Ministry of Health (1992) for a detailed description of the survey.

Several health characteristics were measured in the Ontario Health Survey. We will focus on two continuous variables from the survey, body mass index (BMI) and desired body mass index (DBMI). For this survey the measurements were taken to three digits of accuracy and vary between 7.0 and 45.0. The BMI is a measure of weight status and is calculated from the weight in kilograms divided by the square of the height in meters. The DBMI is the same measure with actual weight replaced by desired weight. The index is not applicable to adolescents, adults over 65 years of age and pregnant or breast feeding women (see, for example, US Department of Health and Human Services, 1990). Consequently, there is a reduction in sample size when only the applicable cases are analyzed. A total of 44 457 responses was obtained for BMI and 41 939 for DBMI. With over 40 000 responses and approximately 380 distinct values for the response, there can be multiple observations at each distinct value. Other variables that will be examined in the analysis that follows are age, sex and the smoking status of the respondent.

We constructed density estimates $\hat{f}(y)$ for each of the domains based on gender, age and smoking status for both BMI and DBMI. The ease of interpretation of superimposed smoothed histograms may be seen in Figures 10.1, 10.2 and 10.3.

Figure 10.1 shows the estimated distributions of BMI and DBMI for females alone. The superimposed images are distributions for two variables on a single domain. It is easily seen that for women BMI has a greater mean and variance than DBMI. There is also greater skewness in the BMI distribution than in the DBMI distribution. Inspection of the graph also lends some support to the notion that women generally want to lose weight whatever their current weight may be.

For a single variable we can also superimpose distributions obtained from several domains. We calculated the density estimates for BMI based on

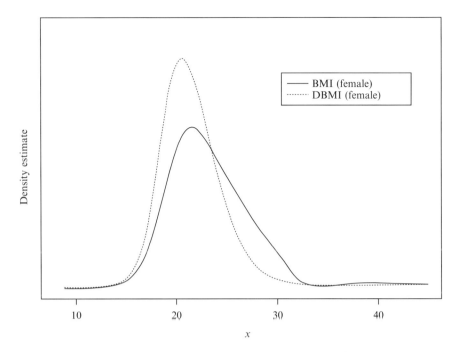

Figure 10.1 Comparison of the distributions of BMI and DBMI for females.

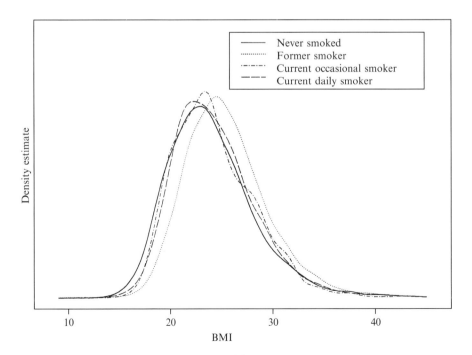

Figure 10.2 Distributions of BMI for smoking status.

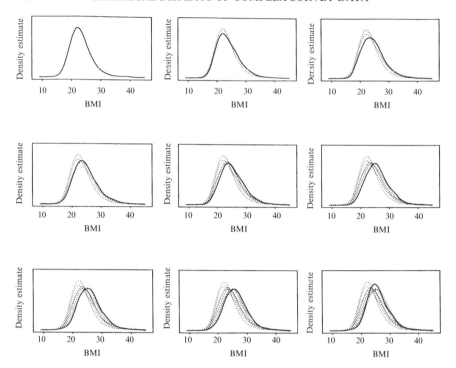

Figure 10.3 Change in the distributions of BMI for increasing age.

smoking status taking males and females together at all ages. In particular, we constructed density estimates $\hat{f}(y)$ for each of the domains defined by the groups 'never smoked', 'former smoker', 'current occasional smoker' and 'current daily smoker'. These superimposed distributions are shown in Figure 10.2. As superimposed histograms, the equivalent picture would have been quite cluttered. From Figure 10.2, however, it may be easily seen that three of the four distributions are very similar. Although the distribution for the group 'former smoker' has the same shape as the other three, it is shifted to the right of the others. Consequently, giving up smoking is associated with a mean shift in the distribution of the BMI. Weight gain associated with giving up smoking is well known and has been reported both more generally (US Department of Health and Human Services, 1990) and with particular reference to the Ontario Health Survey (Østbye *et al.*, 1995).

In comparisons of the BMI over domains defined by age groups, it has been found that BMI tends to increase with age except at the upper end of the range near age 65 (Østbye *et al.*, 1995). Figure 10.3 shows in a dynamic way how BMI changes with age, again with both males and females together. The graph at the top left is the distribution of BMI for the age group 20–24. As each new five-year age group is added (25–29, 30–39, ..., 60–64), the solid line shows the current density estimate and the dotted lines show all the density estimates at earlier ages. It may be seen from Figure 10.3 that BMI tends to increase with age until the upper end of the age range, and the variance also increases with age until about the age of 40.

As noted from Figure 10.1, BMI and DBMI may be related. This may be pursued further by examining the estimate of the joint density function. Figure 10.4 shows a 3-D plot and a contour plot for $\hat{f}(x, y)$ for females, where x is BMI and y is DBMI. The speculation that was made from Figure 10.1 is seen more clearly in the contour plots. In Figure 10.4, if a vertical line is drawn from the BMI axis then one can visually obtain an idea of the DBMI distribution conditional on the value of BMI where the vertical line crosses the BMI axis. When the vertical line is drawn from BMI = 20 then DBMI = 20 is in the upper tail of the distribution. At BMI = 28 or 29, the upper end of the contour plot falls below DBMI = 28. This illustrates that women generally want to weigh less for any given weight. It also shows that average desired weight loss increases with the value of BMI.

10.4. BIAS ADJUSTMENT TECHNIQUES

The density estimate $\hat{f}(y)$, as defined in (10.1), is biased for $f(y)$ with the bias given by $Bias\,[\hat{f}(y)] = E[\hat{f}(y)] - f(y)$. This is a model bias rather than a finite population bias since $f(y)$ is obtained from a limiting value for a nested sequence of increasing finite populations. If we can obtain an approximation to the bias then we can get a bias-adjusted estimate $\tilde{f}(y)$ from

$$\tilde{f}(y) = \hat{f}(y) - Bias\,[\hat{f}(y)]. \tag{10.5}$$

An approximation to the bias can be obtained in at least one of two ways: obtain either a mathematical or a bootstrap approximation to the bias at each y.

The mathematical approximation to the bias of the estimator in (10.1) is the same as that shown in Silverman (1986), namely

$$Bias[\hat{f}(y)] \cong \left\{ \frac{\sigma^2 h^2}{2} + \frac{b^2}{24} \right\} f''(y). \tag{10.6}$$

Since $f''(y)$ is unknown we substitute $\hat{f}''(y)$. Recall that h is the window width of the kernel smoother and b is the bin width of the histogram. When $K(y)$ is a standard normal kernel, then $\sigma^2 = 1$ and

$$\hat{f}''(y) = \frac{1}{h^3} \sum_{i=1}^{k} \hat{p}_i \left\{ \frac{(y - m_i)^2}{h^2} - 1 \right\} K\left(\frac{y - m_i}{h} \right). \tag{10.7}$$

The bias-corrected estimate is obtained from (10.5) and (10.6) after replacing $f''(y)$ by (10.7).

To develop the appropriate bootstrap algorithm, it is necessary to look at the parallel 'Real World' and then construct the appropriate 'Bootstrap World' as outlined in Efron and Tibshirani (1993, p. 87). For the 'Real World' assume that we have a density f. The 'Real World' algorithm is:

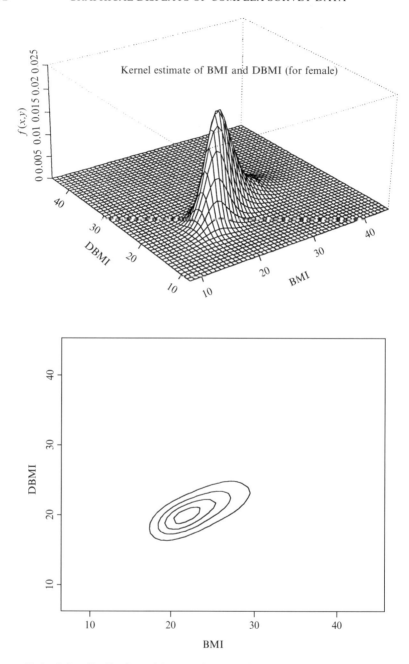

Figure 10.4   Joint distribution of BMI and DBMI for females.

(a)   Obtain a sample $\{y_1, y_2, \ldots, y_n\}$ from f.

(b)   Using the midpoints $m = (m_1, m_2, \ldots, m_K)$ bin the data to obtain $\hat{p} = (\hat{p}_1, \hat{p}_2, \ldots, \hat{p}_K)$.

(c) Smooth the binned data, or equivalently compute the kernel density estimate $\hat{f}_b$ on the binned data.

(d) Repeat steps (a)–(d) B times to get $\hat{f}_{b1}, \hat{f}_{b2}, \ldots, \hat{f}_{bB}$.

(e) Compute $\bar{f}_b = \sum \hat{f}_{bi}/B$. The bias $E(\hat{f}) - f$ is estimated by $\bar{f}_b - f$.

Here in the estimate of bias an expected value is replaced by a mean thus avoiding what might be a complicated analytical calculation. However, in practice this bias estimate cannot be computed because f is unknown and the simulation cannot be conducted. The solution is then to replace f with $\hat{f}$ in the first step and mimic the rest of the algorithm. This strategy is effective if the relationship between the f and the mean $\bar{f}_b$ is similar to the relationship between $\hat{f}$ and the corresponding bootstrap quantities given in the algorithm; that is, if the bias in the bootstrap mimics the bias in the 'Real World'. Assuming that we have only the binned data $\{m, \hat{p}\}$, the algorithm for the 'Bootstrap World' is:

(a) Smooth the binned data $\{m, \hat{p}\}$ to get $\hat{f}$ as in (10.1) and obtain a sample $\{y_1^*, y_2^*, \ldots, y_n^*\}$ from $\hat{f}$.

(b) Bin the data $\{y_1^*, y_2^*, \ldots, y_n^*\}$ to get $\{m, \hat{p}^*\}$.

(c) Obtain a kernel density estimate $\hat{f}_b^*$ on this binned data by smoothing $\{m, \hat{p}^*\}$.

(d) Repeat steps (b)–(d) B times to get $\hat{f}_{b1}^*, \hat{f}_{b2}^*, \ldots, \hat{f}_{bB}^*$.

(e) Compute $\bar{f}_b^* = \sum \hat{f}_{bi}^*/B$. The bias is $\bar{f}_b^* - \hat{f}$.

The sample is step (a) may be obtained by the rejection method in Rice (1995, p. 91) or by the smooth bootstrap technique in Efron and Tibshirani (1993). From (10.5) and step (e), the bias-corrected estimate is $2\hat{f}(y) - \bar{f}_b^*(y)$.

We first illustrate the effect of bias adjustment techniques with simulated data. We generated data from a standard normal distribution, binned the data and then smoothed the binned data. The results of this exercise are shown in Figure 10.5 where the standard normal density is shown as the solid line, the kernel density estimate based on the raw data is the dotted line and the smoothed histogram is the dashed line. The smoothed histogram is given for the ideal window size and then decreasing window sizes. The ideal window size h is obtained from the relationship $b = 1.25h$ given in both Jones (1989) for the standard case and Bellhouse and Stafford (1999) for complex surveys. It is evident from Figure 10.5 that the smoothed histogram has increased bias even though the ideal window size is used. This is due to iterating the smoothing process – smoothing, binning and smoothing again. The bias may be reduced to some extent by picking the window size to be smaller (top right plot). However, the bottom two plots in Figure 10.5 show that as the window size decreases the behaviour of the smoothed histogram becomes dominated by the bins and no further reduction in bias can be satisfactorily made. A more effective bias adjustment can be produced from the bootstrap scheme described above. The application of this scheme is shown in Figure 10.6, which shows bias-adjusted versions of the top two plots in Figure 10.5. In Figure 10.6 the solid line is the

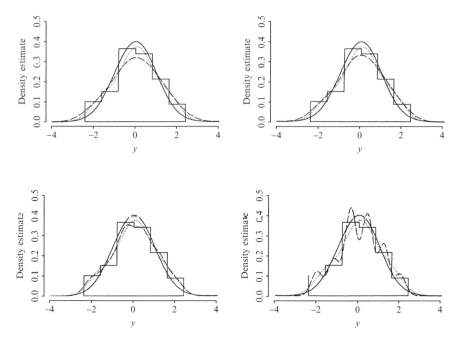

Figure 10.5   Density estimates of standard normal data.

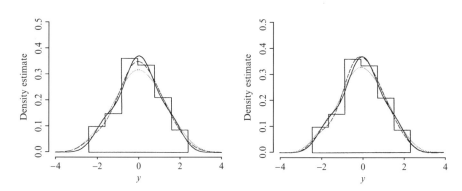

Figure 10.6   Bias-adjusted density estimates for standard normal data.

kernel density estimate based on the raw data, the dotted line is the smoothed histogram and the dashed line is the bias-adjusted smoothed histogram.

The plots in Figure 10.7 show the mathematical approximation approach (dotted line) and the bootstrap approach (dashed line) to bias adjustment applied to DBMI for the whole sample. We used a bin size larger than the precision of the data so that we could better illustrate the effect of bias adjustment. It appears that the bootstrap technique provides a better adjustment in the

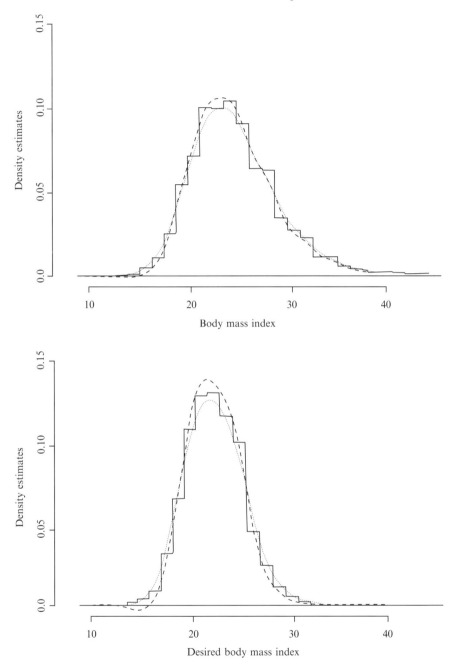

Figure 10.7 Bias-adjusted density estimates of BMI and DBMI.

centre of the distribution than the adjustment based on mathematical approximation. The reverse occurs in the tails of the distribution, especially in both left tails where the bootstrap provides a negative estimate of the density function.

10.5. LOCAL POLYNOMIAL REGRESSION

Local polynomial regression is a nonparametric technique used to discover the relationship between the variate of interest y and the covariate x. Suppose that x has I distinct values, or that it can be categorized into I bins. Let x_i be the value of x representing the ith distinct value or the ith bin, and assume that the values of x_i are equally spaced. The finite population mean for the variate of interest y at x_i is $\bar{y}_{U_i}$. On using the survey weights and the measurements on y from the survey data, we calculate the estimate $\hat{\bar{y}}_i$ of $\bar{y}_{U_i}$. For large surveys, a plot of $\hat{\bar{y}}_i$ against x_i may be more informative and less cluttered than a plot of the raw data. The survey estimates $\hat{\bar{y}}_i$ have variance–covariance matrix V. The estimate of V is $\hat{V}$.

As in density estimation we take an asymptotic approach to determine the function of interest. As before, we assume that we have a nested sequence of increasing finite populations. Now we assume that for a given range on x in the sequence of populations, $I \to \infty$ and the spacing between the x approaches zero. Further, the limiting population mean at x, denoted by $\bar{y}_x$ or $m(x)$, is assumed continuous in x. The limiting function $m(x)$ is the function of interest.

We can investigate the relationship between y and x through local polynomial regression of $\hat{\bar{y}}_i$ on x_i. The regression relationship is obtained on plotting the fit $\hat{m}(x) = \hat{\beta}_0$ against x for each particular choice of x. The term $\hat{\beta}_0$ is the estimated slope parameter in the weighted least squares objective function

$$\sum_{i=1}^{I} \hat{p}_i \{\hat{\bar{y}}_i - \beta_0 - \beta_1(x_i - x) - \cdots - \beta_q(x_i - x)^q\}^2 K((x_i - x)/h)/h, \quad (10.8)$$

where $K(x)$ is the kernel evaluated at a point x, and h is the window width. The weights in this procedure are $\hat{p}_i K((x_i - x)/h)/h$, where $\hat{p}_i$ is the estimate of the proportion p_i of observations with distinct value x_i. Korn and Graubard (1998b) have considered an objective function similar to (10.8) for the raw data with $\hat{p}_i$ replaced by the sampling weights, but provided no properties for their procedure.

The estimate $\hat{m}(x)$ can be written as

$$\hat{m}(x) = e^{\mathrm{T}} (X_x^{\mathrm{T}} \hat{W}_x X_x)^{-1} X_x^{\mathrm{T}} \hat{W}_x \hat{\bar{y}} \quad (10.9)$$

and its variance estimate as

$$e^{\mathrm{T}} (X_x^{\mathrm{T}} \hat{W}_x X_x)^{-1} X_x^{\mathrm{T}} \hat{W}_x \hat{V} \hat{W}_x X_x (X_x^{\mathrm{T}} \hat{W}_x X_x)^{-1} e, \quad (10.10)$$

where

$$\hat{\bar{y}} = (\hat{\bar{y}}_1, \ldots, \hat{\bar{y}}_I)^{\mathrm{T}}, \quad X_x = \begin{bmatrix} 1 & x_1 - x & \ldots & (x_1 - x)^q \\ 1 & x_2 - x & \ldots & (x_2 - x)^q \\ \vdots & \vdots & \vdots & \vdots \\ 1 & x_I - x & \ldots & (x_I - x)^q \end{bmatrix}$$

and

$$\hat{W}_x = \frac{1}{h}\text{diag}\{\hat{p}_1 K((x_1 - x)/h), \, \hat{p}_2 K((x_2 - x)/h), \ldots, \hat{p}_I K((x_I - x)/h)\}.$$

On assuming that $\hat{\bar{y}}$ is approximately multivariate normal, then confidence bands for $m(x)$ can be obtained from (10.9) and (10.10).

The asymptotic bias and variance of $\hat{m}(x)$, as defined by Wand and Jones (1995), turn out to be the same as those obtained in the standard case by Wand and Jones (1995) for the fixed design framework as opposed to the random design framework. Details are given in Bellhouse and Stafford (2001). In addition to the asymptotic framework assumed here, we also need to assume that $\hat{\bar{y}}_i$ is asymptotically unbiased for $\bar{y}_i$ in the sense of Särndal, Swensson and Wretman (1992, pp. 166–7).

10.6.   REGRESSION EXAMPLES FROM THE ONTARIO HEALTH SURVEY

In minimizing (10.10) to obtain local polynomial regression estimates, there are two possibilities for binning on x. The first is to bin to the accuracy of the data so that $\hat{\bar{y}}_x$ is calculated at each distinct outcome of x. In other situations it may be practical to pursue a binning on x that is rougher than the accuracy of the data.

When there are only a few distinct outcomes of x, binning on x is done in a natural way. For example, in investigating the relationship between BMI and age, the age of the respondent was reported only at integral values. The solid dots in Figure 10.8 are the survey estimates of the average BMI $(\hat{\bar{y}}_i)$ for women

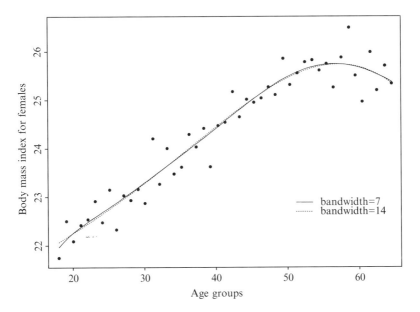

Figure 10.8   Age trend in BMI for females.

at each of the ages 18 through 65 (x_i). The solid and dotted lines show the plot of $\hat{m}(x)$ against x using bandwidths $h = 7$ and $h = 14$ respectively. It may be seen from Figure 10.8 that BMI increases approximately linearly with age until around age 50. The increase slows in the early fifties, peaks at age 55 or so, and then begins to decrease. On plotting the trend lines only for BMI and DBMI for females as shown in Figure 10.9, it may be seen that, on average, women desire to reduce their BMI at every age by approximately two units.

On examining Figure 10.8 it might be conjectured that the trend line follows a second-degree polynomial. Since the full set of data was available, a second-degree polynomial was fitted to the data using SUDAAN. The 95% confidence bands for the general trend line $m(x)$ were calculated according to (10.9) and (10.10). Figure 10.10 shows the second-degree polynomial line superimposed on the confidence bands for $m(x)$. Since the polynomial line falls at the upper limit of the band for women in their thirties and outside the band for women in their sixties, the plot indicates that a second-degree polynomial may not adequately describe the trend.

In other situations it is practical to construct bins on x wider than the precision of the data. To investigate the relationship between what women desire for their weight (DBMI $= \hat{\bar{y}}_i$) and what women actually weigh (BMI $= x_i$) the x_i were grouped. Since the data were very sparse for values of BMI below 15 and above 42, these data were removed from consideration. The remaining groups were 15.0 to 15.2, 15.3 to 15.4 and so on, with the value of x_i chosen as the middle value in each group. The binning was done in this way to obtain a wide range of equally spaced nonempty bins. For each group the survey estimate $\hat{\bar{y}}_i$ was calculated. The solid dots in Figure 10.11 shows the

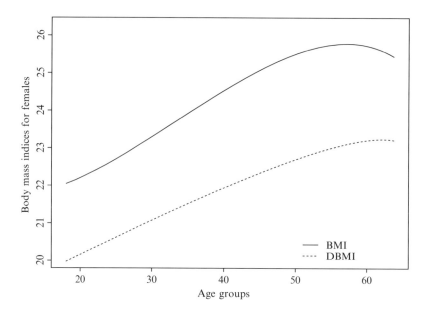

Figure 10.9   Age trends in BMI and DBMI for females.

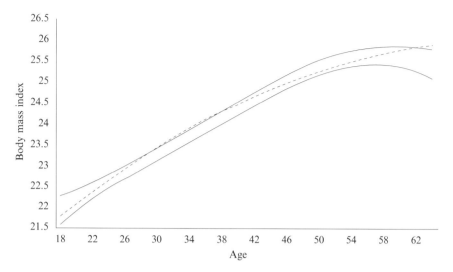

Figure 10.10   Confidence bands for the age trend in BMI for females.

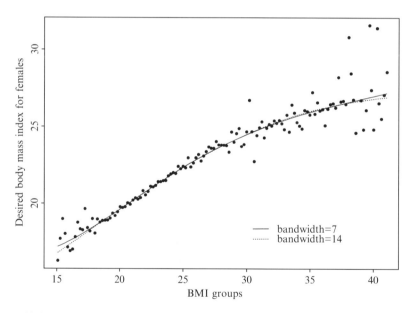

Figure 10.11   BMI trend in DBMI for females.

survey estimates of women's DBMI for each grouped value of their respective BMI. The scatter at either end of the line reflects that sampling variability due to low sample sizes. The plot shows a slight desire to gain weight when the BMI is at 15. This desire is reversed by the time the BMI reaches 20 and the gap between the desire (DBMI) and reality (BMI) widens as BMI increases.

ACKNOWLEDGEMENTS

Some of the work in this chapter was done while the second author was an M.Sc. student at the University of Western Ontario. She is now a Ph.D. student at the University of Toronto. J. E. Stafford is now at the University of Toronto. This research was supported by grants from the Natural Sciences and Engineering Research Council of Canada.

CHAPTER 11

Nonparametric Regression with Complex Survey Data

R. L. Chambers, A. H. Dorfman and
M. Yu. Sverchkov

11.1. INTRODUCTION

The problem considered here is one familiar to analysts carrying out exploratory data analysis (EDA) of data obtained via a complex sample survey design.
How does one adjust for the effects, if any, induced by the method of sampling
used in the survey when applying EDA methods to these data? In particular, are
adjustments to standard EDA methods necessary when the analyst's objective
is identification of 'interesting' population (rather than sample) structures?

A variety of methods for adjusting for complex sample design when carrying
out parametric inference have been suggested. See, for example, Skinner, Holt
and Smith (1989) (abbreviated to SHS), Pfeffermann (1993) and Breckling *et al.*
(1994). However, comparatively little work has been done to date on extending
these ideas to EDA, where a parametric formulation of the problem is typically
inappropriate.

We focus on a popular EDA technique, nonparametric regression or scatterplot smoothing. The literature contains a limited number of applications of this
type of analysis to survey data, usually based on some form of sample
weighting. The design-based theory set out in Chapter 10, with associated
references, provides an introduction to this work. See also Chesher (1997).

The approach taken here is somewhat different. In particular, it is model
based, building on the sample distribution concept discussed in Section 2.3.
Here we develop this idea further, using it to motivate a number of methods for
adjusting for the effect of a complex sample design when estimating a population regression function. The chapter itself consists of seven sections. In
Section 11.2 we describe the sample distribution-based approach to inference,
and the different types of survey data configurations for which we develop
estimation methods. In Section 11.3 we set out a number of key identities that

Analysis of Survey Data Edited by R. L. Chambers and C. J. Skinner

allow us to reexpress the population regression function of interest in terms of related sample regression quantities. In Section 11.4 we use these identities to suggest appropriate smoothers for the sample data configurations described in Section 11.2. The performances of these smoothers are compared in a small simulation study reported in Section 11.5. In Section 11.6 we digress to explore diagnostics for informative sampling. Section 11.7 provides a conclusion with a discussion of some extensions to the theory.

Before moving on, it should be noted that the development in this chapter is an extension of Smith (1988) and Skinner (1994), see also Pfeffermann, Krieger and Rinott (1998) and Pfeffermann and Sverchkov (1999). The notation we employ is largely based on Skinner (1994).

To keep the discussion focused, we assume throughout that nonsampling error, from whatever source (e.g. lack of coverage, nonresponse, interviewer bias, measurement error, processing error), is not a problem as far as the survey data are concerned. We are only interested in the impact of the uncertainty due to the sampling process on nonparametric smoothing of these data. We also assume a basic familiarity with nonparametric regression concepts, comparable to the level of discussion in Härdle (1990).

11.2. SETTING THE SCENE

Since we are interested in scatterplot smoothing we suppose that two (scalar) random variables Y and X can be defined for a target population U of size N and values of these variables are observed for a sample taken from U. We are interested in estimating the smooth function $g_U(x)$ equal to the expected value of Y given $X = x$ over the target population U. Sample selection is assumed to be probability based, with π denoting the value of the sample inclusion probability for a generic population unit. We assume that the sample selection process can be (at least partially) characterized in terms of the values of a multivariate sample design variable Z (not necessarily scalar and not necessarily continuous). For example, Z can contain measures of size, stratum indicators and cluster indicators. In the case of ignorable sampling, π is completely determined by the values in Z. In this chapter, however, we generalize this to allow π to depend on the population values of Y, X and Z. The value π is therefore itself a realization of a random variable, which we denote by Π. Define the sample inclusion indicator I, which, for every unit in U, takes the value 1 if that unit is in the sample and is zero otherwise. The distribution of I for any particular population unit is completely specified by the value of Π for that unit, and so

$$\Pr(I = 1 \,|\, Y = y, X = x, Z = z, \Pi = \pi) = \Pr(I = 1 | \Pi = \pi) = \pi.$$

11.2.1. A key assumption

In many cases it is possible to assume that the population values of the row vector (Y, X, Z) are jointly independent and identically distributed (*iid*).

Unfortunately, the same is usually not true for the sample values of these variables. However, since the methods developed in this chapter depend, to a greater or lesser extent, on some form of exchangeability for the sample data we make the following assumption:

Random indexing: The population values of the random row vector (Y, X, Z, I, Π) are *iid*.

That is, the values of Y, X and Z for any two distinct population units are generated independently, and, furthermore, the subsequent values of I and Π for a particular population unit only depend on that unit's values of Y, X and Z. Note that in general this assumption does not hold, e.g. where the population values of Y and X are clustered. In any case the joint distribution of the bivariate random variable (I, Π) will depend on the *population* values of Z (and sometimes on those of Y and X as well), so an *iid* assumption for (Y, X, Z, I, Π) fails. However, in large populations the level of dependence between values of (I, Π) for different population units will be small given their respective values of Y, X and Z, and so this assumption will be a reasonable one. A similar assumption underpins the parametric estimation methods described in Chapter 12, and is, to some extent, justified by asymptotics described in Pfeffermann, Krieger and Rinott (1998).

11.2.2. What are the data?

The words 'complex survey data' mask the huge variety of forms in which survey data appear. A basic problem with any form of survey data analysis therefore is identification of the relevant data for the analysis.

The method used to select the sample will have typically involved a combination of complex sample design procedures, including multi-way stratification, multi-stage clustering and unequal probability sampling. In general, the information available to the survey data analyst about the sampling method can vary considerably and hence we consider below a number of alternative data scenarios. In many cases we are secondary analysts, unconnected with the organization that actually carried out the survey, and therefore denied access to sample design information on confidentiality grounds. Even if we are primary analysts, however, it is often the case that this information is not easily accessible because of the time that has elapsed since the survey data were collected.

What *is* generally available, however, is the value of the sample weight associated with each sample unit. That is, the weight that is typically applied to the value of a sample variable before summing these values in order to 'unbiasedly' estimate the population total of the variable. For the sake of simplicity, we shall assume that this sample weight is either the inverse of the sample inclusion probability π of the sample unit, or a close proxy. Our dataset therefore includes the sample values of these inclusion probabilities. This leads us to:

Data scenario A: Sample values of Y, X and Π are known. No other information is available.

This scenario is our base scenario. We envisage that it represents the minimum information set where methods of data analysis which allow for complex sampling are possible. The methods described in Chapter 10 and Chapter 12 are essentially designed for sample data of this type.

The next situation we consider is where some extra information about how the sampled units were selected is also available. For example, if a stratified design was used, we know the strata to which the different sample units belong. Following standard practice, we characterize this information in terms of the values of a vector-valued design covariate Z known for all the sample units. Thus, in the case where only stratum membership is known, Z corresponds to a set of stratum indicators. In general Z will consist of a mix of such indicators and continuous size measures. This leads us to:

Data scenario B: Sample values of Y, X, Z and Π are known. No other information is available.

Note that Π will typically be related to Z. However, this probability need not be completely determined by Z.

We now turn to the situation where we not only have access to sample data, but also have information about the nonsampled units in the population. The extent of this information can vary considerably. The simplest case is where we have population summary information on Z, say the population average $\bar{z}_u$. Another type of summary information we may have relates to the sample inclusion probabilities Π. We may know that the method of sampling used corresponds to a fixed size design, in which case the population average of Π is n/N. Both these situations are combined in:

Data scenario C: Sample values of Y, X, Z and Π are known. The population average $\bar{z}_u$ of Z is known, as is the fact that the population average of Π is n/N.

Finally, we consider the situation where we have access to the values of both Z and Π for *all* units in the population, e.g. from a population frame. This leads to:

Data scenario D: Sample values of Y, X, Z and Π are known, as are the nonsample values of Z and Π.

11.2.3. Informative sampling and ignorable sample designs

A key concern of this chapter is where the sampling process somehow confounds standard methods for inference about the population characteristics of interest. It is a fundamental (and often unspoken) 'given' that such standard methods assume that the distribution of the sample data and the corresponding population distribution are the same, so inferential statements about the former apply to the latter. However, with data collected via complex sample designs this situation no longer applies.

A sample design where the distribution of the sample values and population values for a variable Y differ is said to be *informative* about Y. Thus, if an unequal probability sample is taken, with inclusion probabilities proportional to a positive-valued size variable Z, then, provided Y and Z are positively correlated, the sample distribution of Y will be skewed to the right of its corresponding population distribution. That is, this type of unequal probability sampling is informative.

An extreme type of informative sampling discussed in Chapter 8 by Scott and Wild is case–control sampling. In its simplest form this is where the variable Y takes two values, 0 (a control) and 1 (a case), and sampling is such that all cases in the population (of which there are $n \ll N$) are selected, with a corresponding random sample of n of the controls also selected. Obviously the population proportion of cases is n/N. However, the corresponding sample proportion (0.5) is very different.

In some cases an informative sampling design may become uninformative given additional information. For example, data collected via a stratified design with nonproportional allocation will typically be distributed differently from the corresponding population distribution. This difference is more marked the stronger the relationship between the variable(s) of interest and the stratum indicator variables. Within a stratum, however, there may be no difference between the population and sample data distributions, and so the overall difference between these distributions is completely explained by the difference in the sample and population distributions of the stratum indicator variable.

It is standard to characterize this type of situation by saying a sampling method is *ignorable* for inference about the population distribution of a variable Y given the population values of another variable Z if Y is independent of the sample indicator I given the population values of Z. Thus, if Z denotes the stratum indicator referred to in the previous paragraph, and if sampling is carried out at random within each stratum, then it is easy to see that I and Y are independent within a stratum and so this method of sampling is ignorable given Z.

In the rest of this chapter we explore methods for fitting the population regression function $g_U(x)$ in situations where an informative sampling method has been used. In doing so, we consider both ignorable and nonignorable sampling situations.

11.3. RE-EXPRESSING THE REGRESSION FUNCTION

In this section we develop identities which allow us to re-express $g_U(x)$ in terms of sample-based quantities as well as quantities which depend on Z. These identities underpin the estimation methods defined in Section 11.4.

We use $f_U(w)$ to denote the value of the population density of a variable W at the value w, and $f_s(w)$ to denote the corresponding value of the sample density of this variable. This sample density is defined as the density of the conditional variable $W|I = 1$. See also Chapter 12. We write this (conditional) density as $f_s(w) = f_U(w|I = 1)$. To reduce notational clutter, conditional densities $f(w|V = v)$ will be denoted $f(w|v)$. We also use $E_U(W)$ to denote the expectation of W over the population (i.e. with respect to f_U) and $E_s(W)$ to denote the expectation of W over the sample (i.e. with respect to f_s). Since development of expressions for the regression of Y on one or more variables will be our focus, we introduce special notation for this case. Thus, the population and sample regressions of Y on another (possibly vector-valued) variable W will be denoted $g_U(w) = E_U(Y|W = w) = E_U(Y|w)$ and $g_s(w) = E_s(Y|W = w) = E_s(Y|w)$ respectively below.

We now state two identities. Their proofs are straightforward given the definitions of I and Π and the random indexing assumption of Section 11.2.1:

$$f_s(w|\pi) = f_U(w|\pi) \tag{11.1}$$

and

$$f_U(\pi) = f_s(\pi)E_U(\Pi)/\pi = f_s(\pi)/(\pi E_s(1/\Pi)). \tag{11.2}$$

Consequently

$$f_U(w) = \frac{\int \pi^{-1} f_s(w|\pi) f_s(\pi) d\pi}{E_s[\Pi^{-1}]} \tag{11.3}$$

and so

$$E_U(W) = E_s[\Pi^{-1} E_s(W|\Pi)]/E_s[\Pi^{-1}].$$

Recollect that $g_U(x)$ is the regression of Y on X at $X = x$ in the population. Following an application of Bayes' theorem, one can then show

$$g_U(x) = \frac{E_s[\Pi^{-1} f_s(x|\Pi) g_s(x, \Pi)]}{E_s[\Pi^{-1} f_s(x|\Pi)]} \tag{11.4}.$$

From the right hand side of (11.4) we see that $g_U(x)$ can be expressed in terms of the ratio of two sample-based unconditional expectations. As we see later, these quantities can be estimated from the sample data, and a plug-in estimate of $g_U(x)$ obtained.

11.3.1. Incorporating a covariate

So far, no attempt has been made to incorporate information from the design covariate Z. However, since the development leading to (11.4) holds for arbitrary X, and in particular when X and Z are amalgamated, and since $g_U(x) = E_U(g_U(x, Z)|x)$, we can apply (11.4) twice to obtain

$$g_U(x) = \frac{E_s[\Pi^{-1} f_s(x|\Pi) E_s(g_U(x, Z)|x, \Pi)]}{E_s[\Pi^{-1} f_s(x|\Pi)]} \tag{11.5a}$$

where

$$g_U(x, z) = \frac{E_s\left[\Pi^{-1}f_s(x, z|\Pi)g_s(x, z, \Pi)\right]}{E_s\left[\Pi^{-1}f_s(x, z|\Pi)\right]} \tag{11.5b}$$

An important special case is where the method of sampling is *ignorable* given Z; that is, the random variables Y and X are independent of the sample indicator I (and hence Π) given Z. This implies that $g_U(x, z) = g_s(x, z)$ and hence

$$g_U(x) = \frac{E_s\left[\Pi^{-1}f_s(x|\Pi)E_s(g_s(x, Z)|x, \Pi)\right]}{E_s\left[\Pi^{-1}f_s(x|\Pi)\right]}. \tag{11.6}$$

Under ignorability given Z, it can be seen that $E_s(g_s(x, Z)|x, \pi) = g_s(x, \pi)$, and hence (11.6) reduces to (11.4). Further simplification of (11.4) using this ignorability then leads to

$$g_U(x) = E_s\left[\Pi^{-1}f_s(x|Z)g_s(x, Z)\right]/E_s\left[\Pi^{-1}f_s(x|Z)\right], \tag{11.7}$$

which can be compared to (11.4).

11.3.2. Incorporating population information

The identities (11.4), (11.5) and (11.7) all express $g_U(x)$ in terms of sample moments. However, there are situations where we have access to population information, typically about Z and Π. In such cases we can weave this information into estimation of $g_U(x)$ by expressing this function in terms of estimable population and sample moments.

To start, note that

$$1/E_s\left[\Pi^{-1}|x\right] = E[\Pi f_s(x|\Pi)]/E[f_s(x|\Pi)]$$

and so we can rewrite (11.4) as

$$g_U(x) = \frac{E_s\left[\Pi^{-1}f_s(x|\Pi)g_s(x, \Pi)\right]}{E_s[f_s(x|\Pi)]} \frac{E_U[\Pi f_s(x|\Pi)]}{E_U[f_s(x|\Pi)]}. \tag{11.8}$$

The usefulness of this reexpression of (11.4) depends on whether the ratio of population moments on the right hand side of (11.8) can be evaluated from the available data. For example, suppose all we know is that $E_U(\Pi) = n/N$, and that $f_s(x|\Pi) = f_s(x)$. Here n is the sample size. Then (11.8) reduces to

$$g_U(x) = \frac{n}{N} E_s\left[\Pi^{-1}g_s(x, \Pi)\right].$$

Similarly, when population information on both Π and Z is available, we can replace (11.5) by

$$g_U(x) = \frac{E_s\left[\Pi^{-1}f_s(x|\Pi)E_s(g_U(x, Z)|x, \Pi)\right]}{E_s[f_s(x|\Pi)]} \frac{E_U[\Pi f_s(x|\Pi)]}{E_U[f_s(x|\Pi)]} \tag{11.9a}$$

where

$$g_U(x,z) = \frac{E_s\left[\Pi^{-1}f_s(x,z|\Pi)g_s(x,z,\Pi)\right]}{E_s[f_s(x,z|\Pi)]} \frac{E_U[\Pi f_s(x,z|\Pi)]}{E_U[f_s(x,z|\Pi)]}. \tag{11.9b}$$

The expressions above are rather complicated. Simplification does occur, however, when the sampling method is ignorable given Z. As noted earlier, in this case $g_U(x,z) = g_s(x,z)$, so $g_U(x) = E_U(g_s(x,Z)|x)$. However, since $f_U(x|z) = f_s(x|z)$ it immediately follows

$$g_U(x) = E_U[f_s(x|Z)g_s(x,Z)]/E_U[f_s(x|Z)]. \tag{11.10}$$

A method of sampling where $f_U(y|x) = f_s(y|x)$, and so $g_U(x) = g_s(x)$, is *noninformative*. Observe that ignorability given Z is not the same as being noninformative since it does not generally lead to $g_U(x) = g_s(x)$. For this we also require that the population and sample distributions of Z are the same, i.e. $f_U(z) = f_s(z)$.

We now combine these results on $g_U(x)$ obtained in the previous section with the data scenarios earlier described to develop estimators that capitalize on the extent of the survey data that are available.

11.4.   DESIGN-ADJUSTED SMOOTHING

11.4.1.   Plug-in methods based on sample data only

The basis of the plug-in approach is simple. We replace sample-based quantities in an appropriately chosen representation of $g_U(x)$ by corresponding sample estimates. Effectively this is a method of moments estimation of $g_U(x)$. Thus, in scenario A in Section 11.2.2 we only have sample data on Y, X and Π. The identity (11.4) seems most appropriate here since it depends only on the sample values of Y, X and Π. Our plug-in estimator of $g_U(x)$ is

$$\hat{g}_U(x) = \sum_s \pi_t^{-1}\hat{f}_s(x|\pi_t)\hat{g}_s(x,\pi_t)\Big/\sum_s \pi_t^{-1}\hat{f}_s(x|\pi_t) \tag{11.11}$$

where $\hat{f}_s(x|\pi)$ denotes the value at x of a nonparametric estimate of the conditional density of the sample X-values given $\Pi = \pi$, and $\hat{g}_s(x,\pi)$ denotes the value at (x,π) of a nonparametric smooth of the sample Y-values against the sample X- and Π-values. Both these nonparametric estimates can be computed using standard kernel-based methods, see Silverman (1986) and Härdle (1990).

Under scenario B we have extra sample information, consisting of the sample values of Z. If these values explain a substantial part of the variability in Π, then it is reasonable to assume that the sampling method is ignorable given Z, and representation (11.7) applies. Our plug-in estimator of $g_U(x)$ is consequently

$$\hat{g}_U(x) = \sum_s \pi_t^{-1}f_s(x|z_t)\hat{g}_s(x,z_t)\Big/\sum_s \pi_t^{-1}f_s(x|z_t). \tag{11.12}$$

If the information in Z is not sufficient to allow one to assume ignorability then one can fall back on the two-level representation (11.5). That is, one first computes an estimate of the population regression of Y on X and Z,

$$\hat{g}_U(x, z) = \sum_s \pi_t^{-1} \hat{f}_s(x, z | \pi_t) \hat{g}_s(x, z, \pi_t) \Big/ \sum_s \pi_t^{-1} \hat{f}_s(x, z | \pi_t), \qquad (11.13a)$$

and then smooths this estimate further (as a function of Z) against X and Π to obtain

$$\hat{g}_U(x) = \sum_s \pi_t^{-1} \hat{f}_s(x | \pi_t) \hat{E}_s(\hat{g}_U(x, Z) | x, \pi_t) \Big/ \sum_s \pi_t^{-1} \hat{f}_s(x | \pi_t) \qquad (11.13b)$$

where $\hat{E}_s(\hat{g}_U(x, Z) | x, \pi_t)$ denotes the value at (x, π_t) of a sample smooth of the values $\hat{g}_U(x, z_t)$ against the sample X-and Π-values.

11.4.2. Examples

The precise form and properties of these estimators will depend on the nature of the relationship between Y, X, Z and Π. To illustrate, we consider two situations, corresponding to different sample designs.

Stratified sampling on Z
We assume a scenario B situation where Z is a mix of stratum indicators Z_1 and auxiliary covariates Z_2. We further suppose that sampling is ignorable within a stratum, so (11.12) applies. Let h index the overall stratification, with s_h denoting the sample units in stratum h. Then (11.12) leads to the estimator

$$\hat{g}_U(x) = \sum_h \sum_{t \in s_h} \pi_t^{-1} \hat{f}_{sh}(x | z_{2t}) \hat{g}_{sh}(x, z_{2t}) \Big/ \sum_h \sum_{t \in s_h} \pi_t^{-1} \hat{f}_{sh}(x | z_{2t}) \qquad (11.14)$$

where $\hat{f}_{sh}$ denotes a nonparametric density estimate based on the sample data from stratum h. In some circumstances, however, we will be unsure whether it is reasonable to assume ignorability given Z. For example, it could be the case that Π is actually a function of $Z = (Z_1, Z_2)$ and an unobserved third variable Z_3 that is correlated with Y and X. Here the two-stage estimator (11.13) is appropriate, leading to

$$\hat{g}_U(x, z_1 = h, z_2) = \hat{g}_h(x, z_2) = \frac{\sum_s \pi_t^{-1} \hat{f}_{sh}(x, z_2 | \pi_t) \hat{f}_{sh}(\pi_t) \hat{g}_{sh}(x, z_2, \pi_t)}{\sum_s \pi_t^{-1} \hat{f}_{sh}(x, z_2 | \pi_t) \hat{f}_{sh}(\pi_t)} \qquad (11.15a)$$

and hence

$$\hat{g}_U(x) = \sum_s \pi_t^{-1} \hat{f}_s(x | \pi_t) \hat{E}_s(\hat{g}_{z_{1t}}(x, z_{2t}) | x, \pi_t) \Big/ \sum_s \pi_t^{-1} \hat{f}_s(x | \pi_t) \qquad (11.15b)$$

where $\hat{f}_{sh}(\pi)$ denotes an estimate of the probability that a sample unit with $\Pi = \pi$ is in stratum h, and $\hat{E}_s(\hat{g}_{z_1}(x, z_2) | x, \pi)$ denotes the value at (x, π) of a nonparametric smooth of the sample $\hat{g}_{z_1}(x, z_2)$-values defined by (11.15a) against the sample (X, Π)-values.

  Calculation of (11.15) requires 'smoothing within smoothing' and so will be computer intensive. A further complication is that the sample $\hat{g}_{z_1}(x, z_2)$ values smoothed in (11.15b) will typically be discontinuous between strata, so that standard methods of smoothing may be inappropriate.

Array Sampling on X

Suppose the random variable X takes n distinct values $\{x_t; t = 1, 2, \ldots, n\}$ on the population U. Suppose furthermore that Z is univariate, taking m_t distinct (and strictly positive) values $\{z_{jt}; j = 1, 2, \ldots, m_t\}$ when $X = x_t$, and that we have $Y = X + Z$. The population values of Y and Z so formed can thus be thought of as defining an array, with each row corresponding to a distinct value of X. Finally suppose that the sampling method chooses one population unit for each value x_t of X (so the overall sample size is n) with probability

$$\pi_{jt} = z_{jt} \Big/ \sum_{j \in t} z_{jt} = z_{jt}/s_t.$$

Given this set-up, inspection of the sample data (which includes the values of Π) allows one to immediately observe that $g_g(x, z) = x + z$ and to calculate the realized value of s_t as z_{st}/π_{st} where z_{st} and π_{st} denote the sample values of Z and Π corresponding to $X = x_t$. Furthermore, the method of sampling is ignorable given Z, so $g_U(x, z) = g_s(x, z)$ and hence $g_U(x) = x + \mu$, where $\mu = E_U(Z)$. If the total population size N were known, an obvious (and efficient) estimator of μ would be

$$\hat{\mu} = \left[\sum_{t=1}^{n} m_t \right]^{-1} \left[\sum_{t=1}^{n} s_t \right] = N^{-1} \sum_{t=1}^{n} \pi_{st}^{-1} z_{st}$$

and so $g_U(x)$ could be estimated with considerable precision. However, we do not know N and so are in scenario B. The above method of sampling ensures that every distinct value of X in the sample is observed once and only once. Hence $\hat{f}_s(x|z) = 1/n$. Using (11.12), our estimator of $g_U(x)$ then becomes

$$\hat{g}_U(x) = \left[\sum_{t=1}^{n} \pi_{st}^{-1} \right]^{-1} \left[\sum_{t=1}^{n} \pi_{st}^{-1}(x + z_{st}) \right] = x + \left[\sum_{t=1}^{n} \pi_{st}^{-1} \right]^{-1} \cdot \left[\sum_{t=1}^{n} \pi_{st}^{-1} z_{st} \right].$$

$$(11.16)$$

This is an approximately unbiased estimator of $g_U(x)$. To see this we note that, by construction, each value of π_{st} represents an independent realization from a distribution defined on the values $\{\pi_{jt}\}$ with probabilities $\{\pi_{jt}\}$, and so $E_U(\pi_{st}^{-1}) = m_t$. Hence

$$E_U(\hat{g}_U(x)) \approx x + \left[\sum_{t=1}^{n} m_t \right]^{-1} \left[\mu \sum_{t=1}^{n} m_t \right] = x + \mu = g_U(x).$$

11.4.3. Plug-in methods which use population information

We now turn to the data scenarios where population information is available. To start, consider scenario C. This corresponds to having additional summary population information, typically population average values, available for Z and Π. More formally, we know the value of the population size N and one or both of the values of the population averages $\bar{\pi}$ and $\bar{z}$.

How this information can generally be used to improve upon direct sample-based estimation of $g_U(x)$ is not entirely clear. The fact that this information *can* be useful, however, is evident from the array sampling example described in the previous section. There we see that, given the sample data, this function can be estimated very precisely once either the population mean of Z is known, or, if the method of sampling is known, we know the population size N. This represents a considerable improvement over the estimator (11.16) which is only approximately unbiased for this value.

However, it is not always the case that such dramatic improvement is possible. For example, suppose that in the array sampling situation we are not told (i) the sample values of Z and (ii) that sampling is proportional to Z within each row of the array. The only population information is the value of N. This is precisely the situation described following (11.8), and so we could use the estimator

$$\hat{g}_U(x) = N^{-1} \sum_{t=1}^{n} \pi_{st}^{-1} \hat{g}_s(x, \pi_{st}) \tag{11.17}$$

where $\hat{g}_s(x, \pi)$ is the estimated sample regression of Y on X and Π. It is not immediately clear why (11.17) should in general represent an improvement over (11.16). Note that the regression function $\hat{g}_s(x, \pi)$ in (11.17) is not 'exact' (unlike $g_s(x, z) = x + z$) and so (11.17) will include an error arising from this approximation.

Finally, we consider scenario D. Here we know the population values of Z and Π. In this case we can use the two-level representation (11.9) to define a plug-in estimator of $g_U(x)$ if we have reason to believe that sampling is not ignorable given Z. However, as noted earlier, this approach seems overly complex. Equation (11.10) represents a more promising alternative provided that the sampling is ignorable given Z, as will often be the case for this type of scenario (detailed population information available). This leads to the estimator

$$\hat{g}_U(x) = \sum_{t=1}^{N} \hat{f}_s(x|z_t)\hat{g}_s(x, z_t) \Big/ \sum_{t=1}^{N} \hat{f}_s(x|z_t). \tag{11.18}$$

Clearly, under stratified sampling on Z, (11.18) takes the form

$$\hat{g}_U(x) = \sum_h \sum_{t \in h} \hat{f}_{sh}(x|z_{2t})\hat{g}_{sh}(x, z_{2t}) \Big/ \sum_h \sum_{t \in h} \hat{f}_{sh}(x|z_{2t}) \tag{11.19}$$

which can be compared to (11.14). In contrast, knowing the population values of Z under array sampling on X means we know the values m_t and s_t for each t and so we can compute a precise estimate of $g_U(x)$ from the sample data.

11.4.4.   The estimating equation approach

The starting point for this approach is Equation (11.4), which can be equivalently written

$$g_U(x) = E_s(\Pi^{-1}Y|X = x)/E_s(\Pi^{-1}|X = x) \qquad (11.20)$$

providing $E_s(\Pi^{-1}|X = x) > 0$. That is, $g_U(x)$ can always be represented as the solution of the equation

$$E_s[\Pi^{-1}(Y - g_U(x))|X = x] = 0. \qquad (11.21)$$

Replacing the left hand side of (11.21) by a kernel-based estimate of the regression function value leads to the estimating equation ($h_s(x)$ is the bandwidth)

$$\sum_s K\left(\frac{x - x_t}{h_s(x)}\right)\pi_t^{-1}(Y_t - \hat{g}_U(x)) = 0 \qquad (11.22)$$

which has the solution

$$\hat{g}_U(x) = \sum_s K\left(\frac{x - x_t}{h_s(x)}\right)\pi_t^{-1}y_t \bigg/ \sum_s K\left(\frac{x - x_t}{h_s(x)}\right)\pi_t^{-1}. \qquad (11.23)$$

This is a standard Nadaraya–Watson-type estimator of the sample regression of Y on X, but with kernel weights modified by multiplying them by the inverses of the sample inclusion probabilities.

The Nadaraya–Watson nonparametric regression estimator is known to be inefficient compared to local polynomial alternatives when the sample regression function is reasonably smooth (Fan, 1992). Since this will typically be the case, a popular alternative solution to (11.22) is one that parameterizes $\hat{g}_U(x)$ as being *locally linear* in x. That is, we write

$$\hat{g}_U(x) = \hat{a}(x) + \hat{b}(x)x \qquad (11.24)$$

and hence replace (11.22) by

$$\sum_s K\left(\frac{x - x_t}{h_s(x)}\right)\pi_t^{-1}(y_t - \hat{a}(x) - \hat{b}(x)x_t) = 0. \qquad (11.25)$$

The parameters $\hat{a}(x)$ and $\hat{b}(x)$ are obtained by weighted local linear least squares estimation. That is, they are the solutions to

$$\sum_s \pi_t^{-1}K\left(\frac{x - x_t}{h_s(x)}\right)\left(y_t - \hat{a}(x) - \hat{b}(x)x_t\right)\binom{1}{x_t} = \binom{0}{0}. \qquad (11.26)$$

Following arguments outlined in Jones (1991) it can be shown that, for either (11.23) or (11.24), a suitable bandwidth $h_s(x)$, in terms of minimizing the mean squared error $E_s(\hat{g}_U(x) - g_U(x))^2$, must be of order $n^{-1/5}$.

One potential drawback with this approach is that there seems no straightforward way to incorporate population auxiliary information. These estimators are essentially scenario A estimators. One ad hoc solution to this is to replace the inverse sample inclusion probabilities in (11.22) and (11.26) by sample weights which reflect this information (e.g. weights that are calibrated to known population totals). However, it is not obvious why this modification should lead to improved performance for either (11.23) or (11.24).

11.4.5. The bias calibration approach

Suppose $\hat{g}_s(x)$ is a standard (i.e. unweighted) nonparametric estimate of the sample regression of Y on X at x. The theory outlined in Section 11.3 shows that this estimate will generally be biased for the value of the population regression of Y on X at x if the sample and population regression functions differ. One way around this problem therefore is to nonparametrically bias-calibrate this estimate. That is, we compute the sample residuals, $r_t = y_t - \hat{g}_s(x_t)$, and re-smooth these against X using a methodology that gives a consistent estimator of their population regression on X. This smooth is then added to $\hat{g}_s(x)$. For example, if (11.11) is used to estimate this residual population regression, then our final estimate of $g_U(x)$ is

$$\hat{g}_U(x) = \hat{g}_s(x) + \frac{\sum_s \pi_t^{-1}\hat{f}_s(x|\pi_t)\hat{g}_{sR}(x,\pi_t)}{\sum_s \pi_t^{-1}\hat{f}_s(x|\pi_t)} \tag{11.27}$$

where $\hat{g}_{sR}(x,\pi)$ denotes the value at (x,π) of a sample smooth of the residuals r_t against the sample X and Π values. Other forms of (11.27) can be easily written down, based on alternative methods of consistently estimating the population regression of the residuals at x.

This approach is closely connected to the concept of 'twicing' or double smoothing for bias correction (Tukey, 1977; Chambers, Dorfman and Wehrly, 1993).

11.5. SIMULATION RESULTS

Some appreciation for the behaviour of the different nonparametric regression methods described in the previous section can be obtained by simulation. We therefore simulated two types of populations, both of size $N = 1000$, and for each we considered two methods of sampling, one noninformative and the other informative. Both methods had $n = 100$, and were based on application of the procedure described in Rao, Hartley and Cochran (1962), with inclusion probabilities as defined below.

The first population simulated was defined by the equations:

$$Y = 1 + X + XZ + \varepsilon_Y \tag{11.28a}$$

$$X = 4 + 0.5Z + \varepsilon_X \tag{11.28b}$$

$$Z = 4 + 2\varepsilon_Z \tag{11.28c}$$

where $\varepsilon_Y, \varepsilon_X$ and ε_Z are independent standard normal variates. It is straight-forward to see that $g_U(x) = E_U(Y|X = x) = 1 - x + x^2$. Figure 11.1 shows a realization of this population. The two sampling procedures we used for this population were based on inclusion probabilities that were approximately proportional to the population values of Z (*PPZ*: an informative sampling method) and X (*PPX*: a noninformative sampling method). These probabilities were defined by

$$PPZ: \pi_t = 100(z_t + \min_U(z) + 0.1) \Big/ \sum_{u=1}^{N} (z_u + \min_U(z) + 0.1)$$

and

$$PPX: \pi_t = 100(x_t + \min_U(X) + 0.1) \Big/ \sum_{u=1}^{N} (x_u + \min_U(X) + 0.1)$$

respectively.

The second population type we simulated reflected the heterogeneity that is typical of many real populations and was defined by the equations

$$Y = 30 + X + 0.0005ZX^2 + 3\sqrt{X\varepsilon_Y} \tag{11.29a}$$

$$X = 20 + 100\eta_X \tag{11.29b}$$

where ε_Y is a standard normal variate, η_X is distributed as Gamma(2) inde-pendently of ε_Y, and Z is a binary variable that takes the values 1 and 0 with

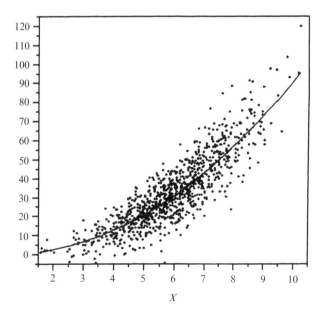

Figure 11.1   Scatterplot of Y vs. X for a population of type 1. Solid line is $g_U(x)$.

probabilities 0.4 and 0.6 respectively, independently of both ε_Y and η_X. Figure 11.2 shows a realization of this population. Again, straightforward calculation shows that for this case $g_U(x) = E_U(Y|X = x) = 30 + x + 0.0002x^2$.

As with the first population type (11.28), we used two sampling methods with this population, corresponding to inclusion probabilities proportional to Z (*PPZ*: an informative sampling method) and proportional to X (*PPX*: a noninformative sampling method). These probabilities were defined by

$$PPZ: \pi_t = 100(z_t + 0.5) \Big/ \sum_{u=1}^{N} (z_u + 0.5)$$

and

$$PPX: \pi_t = 100x_t \Big/ \sum_{u=1}^{N} x_u$$

respectively. Note that *PPZ* above corresponds to a form of stratified sampling, in that all population units with $Z = 1$ have a sample inclusion probability that is three times greater than that of population units with $Z = 0$.

Each population type was independently simulated 200 times and for each simulation two samples were independently drawn using *PPZ* and *PPX* inclusion probabilities. Data from these samples were then used to fit a selection of the nonparametric regression smooths described in the previous section. These methods are identified in Table 11.1. The naming convention used in this table

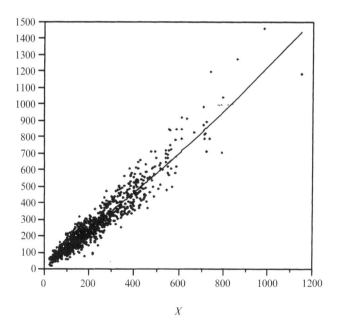

Figure 11.2   Scatterplot of Y vs. X for a population of type 2. Solid line is $g_U(x)$.

Table 11.1 Nonparametric smoothing methods evaluated in the simulation study.

Name	Estimator type	Definition
$M(\Pi_s)$	Scenario A plug-in	(11.11)
$M(Z_s\Pi_s)$	Scenario B plug-in	(11.12)
$M(Z_{strs}\Pi_{strs})$	Stratified scenario B plug-in	(11.14)
$M(Z)$	Scenario D plug-in	(11.18)
$M(Z_{str})$	Stratified scenario D plug-in	(11.19)
$Elin(\Pi_s)$	Weighted local linear smooth	(11.24)
$Elin + Elin(\Pi_s)$	Unweighted+weighted smooth	(11.27[b])
$Elin$	Unweighted local linear smooth	(11.24[a])
$Elin+Elin$	Unweighted+unweighted smooth	(11.27[b])

[a] Based on (11.25) with $\pi_i = 1$.
[b] These are bias-calibrated or 'twiced' methods, representing two variations on the form defined by (11.27). In each case the first component of the method's name is the name of the initial (potentially biased) smooth and the second component (after '+') is the name of the second smooth that is applied to the residuals from the first. Note that the 'twiced' standard smooth Elin+Elin was only investigated in the case of *PPZ* (i.e. informative) sampling.

denotes a plug-in (moment) estimator by 'M' and one based on solution of an estimating equation by 'E'. The amount of sample information required in order to compute an estimate is denoted by the symbols that appear within the parentheses associated with its name. Thus, $M(\Pi_s)$ denotes the plug-in estimator defined by (11.11) that requires one to know the sample values of Π and nothing more (the sample values of Y and X are assumed to be always available). In contrast, $M(Z)$ is the plug-in estimator defined by (11.25) that requires knowledge of the population values of Z before it can be computed. Note the twiced estimator Elin + Elin. This uses an unweighted locally linear smooth at each stage. In contrast, the twiced estimator Elin + Elin(Π_s) uses a weighted local linear smoother at the second stage. We included Elin + Elin in our study for the *PPZ* sampling scenarios only to see how much bias from informative sampling would be soaked up just by reapplication of the smoother. We also investigated the performance of the Nadaraya–Watson weighted smooth in our simulations. However, since its performance was uniformly worse than that of the locally linear weighted smooth Elin(Π_s), we have not included it in the results we discuss below.

Each smooth was evaluated at the 5th, 6th, ... , 95th percentiles of the population distribution of X and two measures of goodness of fit were calculated. These were the mean error, defined by the average difference between the smooth and the actual values of $g_U(x)$ at these percentiles, and the root mean squared error (RMSE), defined by the square root of the average of the squares of these differences. These summary measures were then averaged over the 200 simulations. Tables 11.2 to 11.5 show these average values.

Table 11.2 Simulation results for population type 1 (11.28) under *PPZ* sampling.

Method	0.5	1	2	3	4	5	6	7
				Bandwidth coefficient				
				Average mean error				
$M(\Pi_s)$	0.20	−0.09	−0.31	−0.35	−0.17	0.09	0.29	0.39
$M(Z_s\Pi_s)$	1.51	1.79	2.09	2.23	2.33	2.39	2.43	2.44
$M(Z)$	0.22	0.02	−0.36	−0.38	−0.18	0.08	0.26	0.35
$Elin(\Pi_s)$	0.28	0.24	0.40	0.64	0.84	0.93	0.95	0.91
$Elin + Elin(\Pi_s)$	0.27	0.17	0.13	0.17	0.25	0.36	0.45	0.52
$Elin$	1.85	1.91	2.13	2.38	2.56	2.64	2.64	2.59
$Elin+Elin$	1.82	1.83	1.86	1.91	1.99	2.09	2.18	2.24
				Average root mean squared error				
$M(\Pi_s)$	3.80	2.48	2.21	3.54	4.81	5.66	6.13	6.36
$M(Z_s\Pi_s)$	14.08	5.03	2.93	2.65	2.65	2.71	2.77	2.83
$M(Z)$	3.36	2.28	2.16	3.54	4.79	5.60	6.03	6.24
$Elin(\Pi_s)$	2.50	1.94	1.48	1.40	1.43	1.48	1.54	1.61
$Elin + Elin(\Pi_s)$	2.95	2.27	1.67	1.44	1.33	1.26	1.25	1.29
$Elin$	2.93	2.57	2.45	2.60	2.74	2.84	2.88	2.90
$Elin+Elin$	3.30	2.77	2.36	2.26	2.67	2.32	2.40	2.48

Table 11.3 Simulation results for population type 1 (11.28) under *PPX* sampling.

Method	0.5	1	2	3	4	5	6	7
				Bandwidth coefficient				
				Average mean error				
$M(\Pi_s)$	0.01	0.04	0.05	−2.25	0.05	0.10	0.18	0.25
$M(Z_s\Pi_s)$	−0.17	0.29	0.31	0.44	0.53	0.57	0.58	0.56
$M(Z)$	−1.13	−1.19	−1.24	−0.95	−0.52	−0.12	0.14	0.26
$Elin(\Pi_s)$	0.09	0.14	0.36	0.62	0.82	0.92	0.93	0.89
$Elin + Elin(\Pi_s)$	0.05	0.06	0.11	0.18	0.26	0.35	0.43	0.50
$Elin$	0.09	0.14	0.35	0.58	0.73	0.78	0.75	0.68
				Average root mean squared error				
$M(\Pi_s)$	3.79	2.93	1.92	4.34	1.41	1.37	1.44	1.55
$M(Z_s\Pi_s)$	12.71	5.47	2.26	1.69	1.60	1.61	1.66	1.74
$M(Z)$	4.05	2.49	2.59	3.81	4.98	5.77	6.21	6.42
$Elin(\Pi_s)$	2.46	1.92	1.47	1.38	1.42	1.49	1.57	1.64
$Elin + Elin(\Pi_s)$	2.96	2.29	1.70	1.45	1.33	1.27	1.27	1.32
$Elin$	2.46	1.93	1.48	1.37	1.39	1.45	1.53	1.63

Table 11.4   Simulation results for population type 2 (11.29) under *PPZ* sampling.

Method	Bandwidth coefficient							
	0.5	1	2	3	4	5	6	7
	Average mean error							
$M(\Pi_s)$	1.25	1.01	1.05	1.17	1.35	1.45	1.48	1.52
$M(Z_s\Pi_s)$	7.90	9.49	8.97	9.05	9.40	9.74	10.11	10.46
$M(Z_{strs}\Pi_{strs})$	1.72	1.11	1.10	1.28	1.38	1.42	1.39	1.36
$M(Z)$	1.12	0.95	0.99	1.11	1.29	1.39	1.42	1.45
$M(Z_{str})$	1.67	1.06	1.06	1.25	1.35	1.39	1.37	1.33
Elin(Π_s)	1.52	1.29	1.44	1.58	1.70	1.80	1.86	1.93
Elin + Elin(Π_s)	−5.28	−5.57	−5.58	−5.99	−1.07	−6.02	−5.93	−5.81
Elin	8.06	8.28	8.90	9.41	9.75	9.99	10.15	10.30
Elin+Elin	0.20	0.21	0.30	0.41	0.43	0.51	0.64	0.78
	Average root mean squared error							
$M(\Pi_s)$	28.19	17.11	10.40	8.98	8.18	7.68	7.44	7.38
$M(Z_s\Pi_s)$	75.21	41.54	23.44	16.65	15.03	14.92	15.21	15.79
$M(Z_{strs}\Pi_{strs})$	15.78	12.20	9.66	8.61	8.15	7.90	7.73	7.67
$M(Z)$	28.47	17.51	10.34	8.90	8.07	7.55	7.27	7.20
$M(Z_{str})$	15.75	12.12	9.56	8.54	8.07	7.79	7.61	7.52
Elin(Π_s)	15.41	12.77	10.58	9.49	8.87	8.54	8.39	8.41
Elin + Elin(Π_s)	17.60	14.86	12.38	11.43	10.82	10.40	10.13	9.90
Elin	17.15	15.48	14.50	14.36	14.46	14.72	15.05	15.47
Elin+Elin	16.05	12.94	9.91	8.72	8.00	7.58	7.33	7.21

All kernel-based methods used an Epanechnikov kernel. In order to examine the impact of bandwidth choice on the different methods, results are presented for a range of bandwidths. These values are defined by a bandwidth coefficient, corresponding to the value of b in the bandwidth formula

$$\text{bandwidth} = b \times \text{sample range of } X \times n^{-1/5}.$$

Examination of the results shown in Tables 11.2–11.5 shows a number of features:

1. RMSE-optimal bandwidths differ considerably between estimation methods, population types and methods of sampling. In particular, twicing-based estimators tend to perform better at longer bandwidths, while plug-in methods seem to be more sensitive to bandwidth choice than methods based on estimating equations.

2. For population type 1 with *PPZ* (informative) sampling we see a substantial bias for the unweighted methods Elin and Elin + Elin, and consequently large RMSE values. Somewhat surprisingly the plug-in method $M(Z_s\Pi_s)$ also displays a substantial bias even though its RMSE values are

Table 11.5 Simulation results for population type 2 (11.29) under PPX sampling.

Method	Bandwidth coefficient							
	0.5	1	2	3	4	5	6	7
	Average mean error							
$M(\Pi_s)$	0.09	0.07	0.03	−0.09	−0.11	−0.09	−0.13	−0.19
$M(Z_s\Pi_s)$	16.62	0.58	0.50	0.51	0.51	0.59	0.59	0.54
$M(Z_{strs}\Pi_{strs})$	0.36	0.47	0.72	0.80	0.82	0.75	0.60	0.34
$M(Z)$	−0.09	0.46	0.66	0.71	0.77	0.70	0.50	0.28
$M(Z_{str})$	0.31	0.42	0.66	0.74	0.76	0.71	0.56	0.29
Elin(Π_s)	0.27	0.49	0.83	1.06	1.19	1.27	1.32	1.37
Elin + Elin(Π_s)	−0.43	−0.37	−0.32	−0.18	−0.02	0.08	0.18	0.31
Elin	0.27	0.47	0.61	0.71	0.73	0.65	0.48	0.30
	Average root mean squared error							
$M(\Pi_s)$	20.38	14.44	10.03	8.26	7.37	7.01	6.97	7.10
$M(Z_s\Pi_s)$	217.7	29.72	12.04	10.13	9.43	8.94	8.74	8.79
$M(Z_{strs}\Pi_{strs})$	12.06	9.67	7.89	7.18	6.92	6.97	7.19	7.51
$M(Z)$	15.94	10.23	7.88	7.01	6.62	6.52	6.66	6.99
$M(Z_{str})$	11.92	9.47	7.70	6.97	6.71	6.75	6.96	7.26
Elin(Π_s)	12.79	9.90	7.82	7.08	6.69	6.49	6.42	6.44
Elin + Elin(Π_s)	15.76	12.49	9.52	8.40	7.61	7.08	6.75	6.58
Elin	12.79	9.94	8.03	7.42	7.16	7.10	7.23	7.51

not excessive. Another surprise is the comparatively poor RMSE performance of the plug-in method $M(Z)$ that incorporates population information. Clearly the bias-calibrated method Elin + Elin(Π_s) is the best overall performer in terms of RMSE, followed by Elin(Π_s).

3. For population type 1 with PPX (noninformative) sampling there is little to choose between any of the methods as far as bias is concerned. Again, we see that the plug-in method $M(Z)$ that incorporates population information is rather unstable and the overall best performer is bias-calibrated method Elin + Elin(Π_s). Not unsurprisingly, for this situation there is virtually no difference between Elin and Elin(Π_s).

4. For population type 2 and PPZ (informative) sampling there are further surprises. The plug-in method $M(Z_s\Pi_s)$, the standard method Elin and the bias-calibrated method Elin + Elin(Π_s) all display bias. In the case of $M(Z_s\Pi_s)$ this reflects behaviour already observed for population type 1. Similarly, it is not surprising that Elin is biased under informative sampling. However, the bias behaviour of Elin + Elin(Π_s) is hard to understand, especially when compared to the good performance of the unweighted twiced method Elin + Elin. In contrast, for this population all the plug-in methods (with the exception of $M(Z_s\Pi_s)$) work well. Finally

we see that in this population longer bandwidths are definitely better, with the optimal bandwidth coefficients for all methods except Elin and $M(Z_s\Pi_s)$ most likely greater than $b = 7$.

5. Finally, for population type 2 and *PPX* (noninformative) sampling, all methods (with the exception of $M(Z_s\Pi_s)$) have basically similar bias and RMSE performance. Although the bias performance of $M(Z_s\Pi_s)$ is unremarkable we see that in terms of RMSE, $M(Z_s\Pi_s)$ is again a relatively poor performer. Given the *PPX* sampling method is noninformative, it is a little surprising that the unweighted smoother Elin performs worse than the weighted smoother $Elin(\Pi_s)$. In part this may be due to the heteroskedasticity inherent in population type 2 (see Figure 11.2), with the latter smoother giving less weight to the more variable sample values.

6. The preceding analysis has attempted to present an overview of the performance of the different methods across a variety of bandwidths. In practice, however, users may well adopt a default bandwidth that seems to work well in a variety of situations. In the case of local linear smoothing (the underlying smoothing method for all the methods for which we present results), this corresponds to using a bandwidth coefficient of $b = 3$. Consequently it is also of some interest to just look at the performances of the different methods at this bandwidth. Here we see a much clearer picture. For population type 1, $Elin(\Pi_s)$ is the method of choice, while $M(Z_{str})$ is the method of choice for population type 2.

It is not possible to come up with a single recommendation for what design-adjusted smoothing method to use on the basis of these (very limited) simulation results. Certainly they indicate that the bias-calibrated method $Elin + Elin(\Pi_s)$, preferably with a larger bandwidth than would be natural, seems an acceptable general purpose method of nonparametrically estimating a population regression function. However, it can be inefficient in cases where plug-in estimators like $M(Z_{strs}\Pi_{strs})$ and $M(Z_{str})$ can take advantage of nonlinearities in population structure caused by stratum shifts in the relationship between Y and X. In contrast, the plug-in method $M(Z_s\Pi_s)$ that combines sample information on both Π and Z seems too unstable to be seriously considered, while both the basic plug-in method $M(\Pi_s)$ and the more complex population Z-based $M(Z)$ are rather patchy in their performance – both are reasonable with population type 2, but are rather unstable with population type 1. In the latter case this seems to indicate that it is not always beneficial to include auxiliary information into estimation of population regression functions, even when the sampling method (like *PPZ*) is ignorable given this auxiliary information.

11.6. TO WEIGHT OR NOT TO WEIGHT? (WITH APOLOGIES TO SMITH, 1988)

The results obtained in the previous section indicate that some form of design adjustment is advisable when the sampling method is informative. However, adopting a blanket rule to always use a design-adjusted method may lead to loss of efficiency when the sampling method is actually noninformative. On the other hand it is not always the case that inappropriate design adjustment leads to efficiency loss, compare Elin and Elin(Π_s) in Table 11.5.

Is there a way of deciding whether one should use design-adjusted (e.g. weighting) methods? Equivalently, can one, on the basis of the sample data, check whether the sample is likely to have been drawn via an informative sampling method?

A definitive answer to this question is unavailable at present. However, we describe two easily implementable procedures that can provide some guidance.

To start, we observe that if a sampling method is noninformative, i.e. if $f_U(y|x) = f_s(y|x)$, then design adjustment is unnecessary and we can estimate $g_U(x)$ by simply carrying out a standard smooth of the sample data. As was demonstrated in Pfeffermann, Krieger and Rinott (1998) and Pfeffermann and Sverchkov (1999), noninformativeness is equivalent to conditional independence of the sample inclusion indicator I and Y given X in the population, which in turn is equivalent to the identity

$$E_s(\Pi^{-1}|X = x, Y = y) = E_U(\Pi|X = x, Y = y)$$
$$= E_U(\Pi|X = x) = E_s(\Pi^{-1}|X = x).$$

This identity holds if the sample values of Π and Y are independent given X. Hence, one way of checking whether a sample design is noninformative is to test this conditional independence hypothesis using the sample data. If the sample size is very large, this can be accomplished by partitioning the sample distributions of Π, Y and X, and performing chi-square tests of independence based on the categorized values of Π and Y within each category of X. Unfortunately this approach depends on the overall sample size and the degree of categorization, and it did not perform well when we applied it to the sample data obtained in our simulation study.

Another approach, and one that reflects practice in this area, is to only use a design-adjusted method if it leads to significantly different results compared to a corresponding standard (e.g. unweighted) method. Thus, if we let $\hat{g}_s(x)$ denote the standard estimate of $g_U(x)$, with $\hat{g}_U(x)$ denoting the corresponding design-adjusted estimate, we can compute a jackknife estimate $v_J(x)$ of var($\hat{g}_s(x) - \hat{g}_U(x)$) and hence calculate the standardized difference

$$W(x) = [\hat{g}_s(x) - \hat{g}_U(x)]\big/\sqrt{v_J(x)}. \tag{11.30}$$

We would then only use $\hat{g}_U(x)$ if $W(x) > 2$. A multivariate version of this test, based on a Wald statistic version of (11.30), is easily defined. In this case we would be testing for a significant difference between these two estimates at k different x-values of interest. This would require calculation of a jackknife estimate of the variance–covariance matrix of the vector of differences between the g-values for the standard method and those for the design-adjusted method at these k different x-values. The Wald statistic value could then be compared to, say, the 95th percentile of a chi-squared distribution with k degrees of freedom. Table 11.6 shows how this approach performed with the sample data obtained in our simulations. We see that it works well at identifying the fact that the *PPX* schemes are noninformative. However, its ability to detect the informativeness of the *PPZ* sampling schemes is not as good, particularly in the case of population type 2. We hypothesize that this poor performance is due to the heteroskedasticity implicit in this population's generation mechanism.

The problem with the Wald statistic approach is that it does not test the hypothesis that is really of interest to us in this situation. This is the hypothesis that $g_U(x) = g_s(x)$. In addition, nonrejection of the noninformative sampling 'hypothesis' may result purely because the variance of the design-adjusted estimator is large compared to the bias of the standard estimator.

A testing approach which focuses directly on whether $g_U(x) = g_s(x)$ can be motivated by extending the work of Pfeffermann and Sverchkov (1999). In particular, from (11.21) we see that this equality is equivalent to

$$E_s[\Pi^{-1}(Y - g_s(x))|X = x] = 0. \tag{11.31}$$

The twicing approach of Section 11.4.5 estimates the left hand side of (11.31), replacing $g_s(x)$ by the standard sample smooth $\hat{g}_s(x)$. Consequently, testing $g_U(x) = g_s(x)$ is equivalent to testing whether the twicing adjustment is significantly different from zero.

A smooth function can be approximated by a polynomial, so we can always write

$$E_s(\Pi^{-1}(Y - g_s(x))|X = x) = \sum_{j\geq0} a_j x^j. \tag{11.32}$$

Table 11.6 Results for tests of informativeness. Proportion of samples where null hypothesis of noninformativeness is rejected at the (approximate) 5% level of significance.

Test method	*PPX* sampling (noninformative)		*PPZ* sampling (informative)	
	Population type 1	Population type 2	Population type 1	Population type 2
Wald statistic test[a]	0.015	0.050	1.000	0.525
Correlation test[b]	0.010	0.000	0.995	0.910

[a] The Wald statistic test was carried out by setting $\hat{g}_s = $ Elin and $\hat{g}_U = $ Elin(Π_s), both with bandwidth coefficient $b = 3$. The x-values where these two estimators were compared were the deciles of the sample distribution of X.
[b] The correlation test was carried out using $\hat{g}_s = $ Elin with bandwidth coefficient $b = 3$. A sample was rejected as being 'informative' if any *one* of the subhypotheses H_0, H_1 and H_2 was rejected at the 5% level.

We can test H: $g_U(x) = g_s(x)$ by testing whether the coefficients a_j of the regression model (11.32) are identically zero. This is equivalent to testing the sequence of subhypotheses

$$H_0: cor_s(\Pi^{-1}, (Y - g_s(X))) = 0$$
$$H_1: cor_s(\Pi^{-1}(Y - g_s(X)), X) = 0$$
$$H_2: cor_s(\Pi^{-1}(Y - g_s(X)), X^2) = 0$$

$$\vdots$$

In practice $g_s(x)$ in these subhypotheses is replaced by $\hat{g}_s(x)$. Assuming normality of all quantities, a suitable test statistic for each of these subhypotheses is then the corresponding t-value generated by the *cor.test* function within S-Plus (Statistical Sciences, 1990). We therefore propose that the hypothesis H be rejected at a '95% level' if any *one* of the absolute values of these t-statistics exceeds 2. Table 11.6 shows the results of applying this procedure to the sample data in our simulations, with the testing restricted to the subhypotheses H_0, H_1 and H_2. Clearly the test performs rather well across all sampling methods and population types considered in our simulations. However, even in this case we see that there are still problems with identifying the informativeness of *PPZ* sampling for population type 2.

11.7. DISCUSSION

The preceding development has outlined a framework for incorporating information about sample design, sample selection and auxiliary population information into nonparametric regression applied to sample survey data. Our results provide some guidance on choosing between the different estimators that are suggested by this framework.

An important conclusion that can be drawn from our simulations is that it pays to use some form of design-adjusted method when the sample data have been drawn using an informative sampling scheme, and there is little loss of efficiency when the sampling scheme is noninformative. However, we see that there is no single design-adjusted method that stands out. Local linear smoothing incorporating inverse selection probability weights seems to be an approach that provides reasonable efficiency at a wide range of bandwidths. But locally linear plug-in methods that capitalize on 'stratum structure' in the data can improve considerably on such a weighting approach. For nonparametric regression, as with any other type of analysis of complex survey data, it helps to find out as much as possible about the population structure the sample is supposed to represent. Unfortunately there appears to be no general advantage from incorporating information on an auxiliary variable into estimation, and in fact there can be a considerable loss of efficiency probably due to the use of higher dimensional smoothers with such data.

An important ancillary question we have also considered is identifying situations where in fact the sampling method is informative. Here an examination of the correlations between weighted residuals from a standard (unweighted) nonparametric smooth and powers of the X variable is a promising diagnostic tool.

Our results indicate there is a need for much more research in this area. The most obvious is development of algorithms for choosing an optimal bandwidth (or bandwidths) to use with the different design-adjusted methods described in this chapter. In this context Breunig (1999, 2001) has investigated optimal bandwidth choice for nonparametric density estimation from stratified and clustered sample survey data, and it should be possible to apply these ideas to the nonparametric regression methods investigated here. The extension to clustered sample data is an obvious next step.

CHAPTER 12

Fitting Generalized Linear Models under Informative Sampling

Danny Pfeffermann and M. Yu. Sverchkov

12.1. INTRODUCTION

Survey data are frequently used for analytic inference on population models. Familiar examples include the computation of income elasticities from household surveys, the analysis of labor market dynamics from labor force surveys and studies of the relationships between disease incidence and risk factors from health surveys.

The sampling designs underlying sample surveys induce a set of weights for the sample data that reflect unequal selection probabilities, differential non-response or poststratification adjustments. When the weights are related to the values of the outcome variable, even after conditioning on the independent variables in the model, the sampling process becomes informative and the model holding for the sample data is different from the model holding in the population (Rubin, 1976; Little, 1982; Sugden and Smith, 1984). Failure to account for the effects of informative sampling may yield large biases and erroneous conclusions. The books edited by Kasprzyk *et al.* (1989) and by Skinner, Holt and Smith (SHS hereafter) (1989) contain a large number of examples illustrating the effects of ignoring informative sampling processes. See also the review articles by Pfeffermann (1993, 1996).

In fairly recent articles by Krieger and Pfeffermann (1997), Pfeffermann, Krieger and Rinott (1998) and Pfeffermann and Sverchkov (1999), the authors propose a new approach for inference on population models from complex survey data. The approach consists of approximating the parametric distribution of the sample data (or moments of that distribution) as a function of the population distribution (moments of this distribution) and the sampling weights, and basing the inference on that distribution. The (conditional) sample

Analysis of Survey Data Edited by R. L. Chambers and C. J. Skinner

probability density function (*pdf*) $f_s(y_t|x_t)$ of an outcome variable Y, corresponding to sample unit t, is defined as $f(y_t|x_t; t \in s)$ where s denotes the sample and $x_t = (x_{t0}, x_{t1}, \ldots, x_{tk})'$ represents the values of auxiliary variables $X_0 \ldots X_k$ (usually $x_{t0} = 1$ for all t). Denoting the corresponding population *pdf* (before sampling) by $f_U(y_t|x_t)$, an application of Bayes' theorem yields the relationship

$$f_s(y_t|x_t) = f(y_t|x_t; t \in s) = \Pr(t \in s|y_t, x_t)f_U(y_t|x_t)/\Pr(t \in s|x_t). \qquad (12.1)$$

Note that unless $\Pr(t \in s|y_t, x_t) = \Pr(t \in s|x_t)$ for all possible values y_t, the sample and population *pdfs* differ, in which case the sampling process is informative. Empirical results contained in Krieger and Pfeffermann (1997) and Pfeffermann and Sverchkov (1999), based on simulated and real datasets, illustrate the potentially better performance of regression estimators and tests of distribution functions derived from use of the sample distribution as compared to the use of standard inverse probability weighting. For a brief discussion of the latter method (with references) and other more recent approaches that address the problem of informative sampling, see the articles by Pfeffermann (1993, 1996).

The main purpose of the present chapter is to extend the proposed methodology to likelihood-based inference, focusing in particular on situations where the models generating the population values belong to the family of generalized linear models (GLM; McCullagh and Nelder, 1989). The GLM consists of three components:

1. The *random component*; independent observations y_t, each drawn from a distribution belonging to the exponential family with mean μ_t.
2. The *systematic component*; covariates $x_{0t} \ldots x_{kt}$ that produce a linear predictor $\sum_{j=0}^{k} \beta_j x_{jt}$.
3. A *link function* $h(\mu_t) = \sum_{j=0}^{k} \beta_j x_{jt}$ between the random and systematic components.

The unknown parameters consist of the vector $\beta = (\beta_0, \beta_1, \ldots, \beta_k)'$ and possibly additional parameters defining the other moments of the distribution of y_t. A trivial example of the GLM is the classical regression model with the distribution of y_t being normal, in which case the link function is the identity function. Another familiar example considered in the empirical study of this chapter and in Chapters 6, 7, and 8 is logistic regression, in which case the link function is $h(\mu_t) = \log[\mu_t/(1 - \mu_t)]$. See the book by McCullagh and Nelder (1989) for many other models in common use. The GLM is widely used for modeling discrete data and for situations where the expectations μ_t are nonlinear functions of the covariates.

In Section 12.2 we outline the main features of the sample distribution. Section 12.3 considers three plausible sets of estimating equations for fitting the GLM to the sample data that are based on the sample distribution and discusses their advantages over the use of the randomization (design-based) distribution. A simple test statistic for testing the informativeness of the sampling process is developed. Section 12.4 proposes appropriate variance estimators. Section 12.5 contains simulation results used to assess the performance of the point estimators and the test statistic proposed in Sections 12.3 and 12.4. We close in Section 12.6 with a brief summary and possible extensions.

12.2. POPULATION AND SAMPLE DISTRIBUTIONS

12.2.1. Parametric distributions of sample data

Let $U = \{1 \ldots N\}$ define the finite population of size N. In what follows we consider single-stage sampling with selection probabilities $\pi_t = \Pr(t \in s)$, $t = 1 \ldots N$. In practice, π_t may depend on the population values $y_U = (y_1 \ldots y_N)'$ of the outcome variable, the values $x_U = [x_1 \ldots x_N]'$ of the auxiliary variables and values $z_U = [z_1 \ldots z_N]'$ of design variables used for the sample selection but not included in the working model under consideration. We express this by writing $\pi_t = g_t(y_U, x_U, z_U)$ for some function g_t.

The probabilities π_t are generally not the same as the probabilities $\Pr(t \in s | y_t, x_t)$ defining the sample distribution in (12.1) because the latter probabilities condition on (y_t, x_t) only. Nonetheless, by regarding π_t as a random variable, the following relationship holds:

$$\Pr(t \in s | y_t, x_t) = E_U(\pi_t | y_t, x_t) \tag{12.2}$$

where $E_U(.)$ defines the expectation operator under the corresponding population distribution. The relationship (12.2) follows by defining I_t to be the sample inclusion indicator and writing $\Pr(t \in s | y_t, x_t) = E(I_t | y_t, x_t) = E[E(I_t | y_u, x_u, z_u) | y_t, x_t] = E(\pi_t | y_t, x_t)$. Substituting (12.2) in (12.1) yields an alternative, and often more convenient, expression for the sample pdf,

$$f_s(y_t | x_t) = E_U(\pi_t | y_t, x_t) f_U(y_t | x_t) / E_U(\pi_t | x_t). \tag{12.3}$$

It follows from (12.3) that for a given population *pdf*, the marginal sample *pdf* is fully determined by the conditional expectation $\pi(y_t, x_t) = E_U(\pi_t | y_t, x_t)$. The paper by Pfeffermann, Krieger and Rinott (1998, hereafter PKR) contains many examples of pairs $[f_U(y, x), \pi(y, x)]$ for which the sample *pdf* is of the same family as the population *pdf*, although possibly with different parameters. In practice, the form of the expectations $\pi(y_t, x_t)$ is often unknown, but it can generally be identified and estimated from the sample data (see below).

For independent population measurements PKR establish an asymptotic independence of the sample measurements with respect to the sample distribution, under commonly used sampling schemes for selection with unequal probabilities. The asymptotic independence assumes that the population size increases holding, the sample size fixed. Thus, the use of the sample distribution permits in principle the application of standard statistical procedures like likelihood-based inference and residual analysis to the sample data.

The representation (12.3) enables also the establishment of relationships between moments of the population and the sample distributions. Denote the expectation operators under the two distributions by E_U and E_s and let

$w_t = 1/\pi_t$ define the sampling weights. Pfeffermann and Sverchkov (1999, hereafter PS) develop the following relationship for pairs of (vector) random variables (u_t, v_t):

$$E_U(u_t|v_t) = E_s(w_t u_t|v_t)/E_s(w_t|v_t). \qquad (12.4)$$

For $u_t = y_t$; $v_t = x_t$, the relationship (12.4) can be utilized for *semi-parametric* estimation of moments of the population distribution in situations where the parametric form of this distribution is unknown. The term semi-parametric estimation refers in this context to the estimation of population moments (for example, regression relationships) from the corresponding sample moments defined by (12.4), using least squares, the method of moments, or other estimation procedures that do not require full specification of the underlying distribution. See PS for examples. Notice, also, that by defining $u_t = \pi_t$ and $v_t = (y_t, x_t)$ or $v_t = x_t$ in (12.4), one obtains

$$E_U(\pi_t|y_t, x_t) = 1/E_s(w_t|y_t, x_t); E_U(\pi_t|x_t) = 1/E_s(w_t|x_t). \qquad (12.5)$$

Equation (12.5) shows how the conditional population expectations of π_t that define the sample distribution in (12.3) can be evaluated from the sample data. The relationship (12.4) can be used also for *nonparametric* estimation of regression models; see Chapter 11. The relationships between the population and sample distributions and between the moments of the two distributions are exploited in subsequent sections.

12.2.2. Distinction between the sample and the randomization distributions

The sample distribution defined by (12.3) is different from the randomization (design) distribution underlying classical survey sampling inference. The randomization distribution of a statistic is the distribution of the values for this statistic induced by all possible sample selections, with the finite population values held fixed. The sample distribution, on the other hand, accounts for the (superpopulation) distribution of the population values and the sample selection process, but the sample units are held fixed. Modeling of the sample distribution requires therefore the specification of the population distribution, but it permits the computation of the joint and marginal distributions of the sample measurements. This is not possible under the randomization distribution because the values of the outcome variable (and possibly also the auxiliary variables) are unknown for units outside the sample. Consequently, the use of this distribution for inference on models is restricted mostly to estimation problems and probabilistic conclusions generally require asymptotic normality assumptions.

Another important advantage of the use of the sample distribution is that it allows conditioning on the values of auxiliary variables measured for the sample units. In fact, the definition of the sample distribution already uses a conditional formulation. The use of the randomization distribution for conditional inference is very limited at present; see Rao (1999) for discussion and illustrations.

The sample distribution is different also from the familiar $p\xi$ distribution, defined as the combined distribution over all possible realizations of the finite population measurements (the population ξ distribution) and all possible sample values for a given population (the randomization p distribution). The $p\xi$ distribution is often used for comparing the performance of design-based estimators in situations where direct comparisons of randomization variances or mean square errors are not feasible. The obvious difference between the sample distribution and the $p\xi$ distribution is that the former conditions on the selected sample (and values of auxiliary variables measured for units in the sample), whereas the latter accounts for all possible sample selections.

Finally, rather than conditioning on the selected sample when constructing the sample distribution (and hence the sample likelihood), one could compute instead the joint distribution of the selected sample and the corresponding sample measurements. Denote by $y_s = \{y_t, t \in s\}$ the outcome variable values measured for the sample units and by $x_s = \{x_t, t \in s\}$ and $x_{\bar{s}} = \{x_t, t \notin s\}$ the values of the auxiliary variables corresponding to the sampled and nonsampled units. Assuming independence of the population measurements and independent sampling of the population units (Poisson sampling), the joint *pdf* of $(s, y_s)|(x_s, x_{\bar{s}})$ can be written as

$$f(s, y_s|x_s, x_{\bar{s}}) = \prod_{t \in s} [\pi(y_t, x_t) f_U(y_t|x_t)/\pi(x_t)] \prod_{t \in s} \pi(x_t) \prod_{t \notin s} [1 - \pi(x_t)] \quad (12.6)$$

where $\pi(y_t, x_t) = E_U(\pi_t|y_t, x_t)$ and $\pi(x_t) = E_U(\pi_t|x_t)$. Note that the product of the terms in the first set of square brackets on the right hand side of (12.6) is the joint sample *pdf*, $f_s(y_s|x_s, s)$, for units in the sample as obtained from (12.3). The use of (12.6) for likelihood-based inference has the theoretical advantage of employing the information on the sample selection probabilities for units outside the sample, but it requires knowledge of the expectations $\pi(x_t) = E_U(\pi_t|x_t)$ for all $t \in U$ and hence the values $x_{\bar{s}}$. This information is not needed when inference is based on the sample *pdf*, $f_s(y_s|x_s, s)$. When the values $x_{\bar{s}}$ are unknown, it is possible in theory to regard the values $\{x_t, t \notin s\}$ as random realizations from some *pdf* $g_{\bar{s}}(x_t)$ and replace the expectations $\pi(x_t)$ for units $t \notin s$ by the unconditional expectations $\pi(t) = \int \pi(x_t) g_{\bar{s}}(x_t) dx_t$. See, for example, Rotnitzky and Robins (1997) for a similar analysis in a different context. However, modeling the distribution of the auxiliary variables might be formidable and the resulting likelihood $f(s, y_s|x_s, x_{\bar{s}})$ could be very cumbersome.

12.3.   INFERENCE UNDER INFORMATIVE PROBABILITY SAMPLING

12.3.1.   Estimating equations with application to the GLM

In this section we consider four different approaches for defining estimating equations under informative probability sampling. We compare the various approaches empirically in Section 12.5.

Suppose that the population measurements $(y_U, x_U) = \{(y_t, x_t), t = 1 \ldots N\}$ can be regarded as N independent realizations from some *pdf* $f_{y,x}$. Denote by $f_U(y|x; \beta)$ the conditional *pdf* of y_t given x_t. The true value of the vector parameter $\beta = (\beta_0, \beta_1, \ldots, \beta_k)'$ is defined as the unique solution of the equations

$$W_U(\beta) = \sum_{t=1}^{N} E_U[d_{Ut}|x_t] = 0 \qquad (12.7)$$

where $d_{Ut} = (d_{Ut,0} d_{Ut,1} \ldots d_{Ut,k})' = \partial \log f_U(y_t|x_t; \beta)/\partial\beta$ is the tth score function. We refer to (12.7) as the 'parameter equations' since they define the vector parameter β. For the GLM defined in the introduction with the distribution of y_t belonging to the exponential family, $f_U(y; \theta, \phi) = \exp\{[y\theta - b(\theta)]/a(\phi) + c(y, \phi)\}$ where $a(.) > 0, b(.)$ and $c(.)$ are known functions, and ϕ is known. It follows that $\mu(\theta) = E(y) = \partial b(\theta)/\partial\theta$ so that if $\theta = h(x\beta)$ for some function $h(.)$ with derivative $g(.)$,

$$d_{Ut,j} = \{y_t - \mu(h(x_t'\beta))\}[g(x_t'\beta)]x_{t,j}. \qquad (12.8)$$

The 'census parameter' (Binder, 1983) corresponding to (12.7) is defined as the solution B_U of the equations

$$W_U^*(\beta) = \sum_{t=1}^{N} d_{Ut} = 0. \qquad (12.9)$$

Note that (12.9) defines the maximum likelihood estimating equations based on all the population values.

Let $s = \{1 \ldots n\}$ denote the sample, assumed to be selected by some probability sampling scheme with first-order selection probabilities $\pi_t = \Pr(t \in s)$. (The sample size n can be random.) The first approach that we consider for specifying the estimating equations involves redefining the parameter equations with respect to the sample distribution $f_s(y_t|x_t)$ (Equation (12.3)), rather than the population distribution as in (12.7). Assuming that the form of the conditional expectations $E_U(\pi_t|y_t, x_t)$ is known (see Section 12.3.2) and that the expectations $E_U(\pi_t|x_t) = \int_y E_U(\pi_t|y, x_t) f_U(y|x_t; \beta) dy$ are differentiable with respect to β, the parameter equations corresponding to the sample distribution are

$$W_{1s}(\beta) = \sum_s E_s\{[\partial \log f_s(y_t|x_t; \beta)/\partial\beta]|x_t\}$$
$$= \sum_s E_s\{[d_{Ut} + \partial \log E_s(w_t|x_t)/\partial\beta]|x_t\} = 0. \qquad (12.10)$$

(The second equation follows from (12.3) and (12.5).) The parameters β are estimated under this approach by solving the equations

$$W_{1s,e}(\beta) = \sum_s [d_{Ut} + \partial \log E_s(w_t|x_t)/\partial\beta] = 0. \qquad (12.11)$$

Note that (12.11) defines the sample likelihood equations.

The second approach uses the relationship (12.4) in order to convert the population expectations in (12.7) into sample expectations. Assuming a random sample of size n from the sample distribution, the parameter equations then have the form

$$W_{2s}(\beta) = \sum_s E_s(q_t d_{Ut}|x_t) = 0 \qquad (12.12)$$

where $q_t = w_t/E_s(w_t|x_t)$. The vector β is estimated under this approach by solving the equations

$$W_{2s,e}(\beta) = \sum_s q_t d_{Ut} = 0. \qquad (12.13)$$

The third approach is based on the property that if β solves Equations (12.7), then it solves also the equations

$$\tilde{W}_U(\beta) = \sum_{t=1}^{N} E_U(d_{Ut}) = E_x\left[\sum_{t=1}^{N} E_U(d_{Ut}|x_t)\right] = 0$$

where the expectation E_x is over the population distribution of the x_t. Application of (12.4) to each of the terms $E_U(d_{Ut})$ (without conditioning on x_t) yields the following parameter equations for a random sample of size n from the sample distribution:

$$W_{3s}(\beta) = \sum_s E_s(w_t d_{Ut})/E_s(w_t) = 0. \qquad (12.14)$$

The corresponding estimating equations are

$$W_{3s,e}(\beta) = \sum_s w_t d_{Ut} = 0. \qquad (12.15)$$

An interesting feature of the equations in (12.15) is that they coincide with the pseudo-likelihood equations as obtained when estimating the census equations (12.9) by the Horvitz–Thompson estimators. (For the concept and uses of pseudo-likelihood see Binder, 1983; SHS; Godambe and Thompson, 1986; Pfeffermann, 1993; and Chapter 2.) Comparing (12.13) with (12.15) shows that the former equations use the adjusted weights, $q_t = w_t/E_s(w_t|x_t)$ instead of the standard weights w_t used in (12.15). As discussed in PS, the weights q_t account for the net sampling effects on the target conditional distribution of $y_t|x_t$, whereas the weights w_t account also for the sampling effects on the marginal distribution of x_t. In particular, when w is a deterministic function of x so that the sampling process is noninformative, $q_t \equiv 1$ and Equations (12.13) reduce to the ordinary likelihood equations (see (12.16) below). The use of (12.15) on the other hand may yield highly variable estimators in such cases, depending on the variability of the x_t.

The three separate sets of estimating equations defined by (12.11), (12.13), and (12.15) all account for the sampling effects. On the other hand, ignoring the sampling process results in the use of the ordinary (face value) likelihood equations

$$W_{4s,e}(\beta) = \sum_s d_{Ut} = 0. \qquad (12.16)$$

We consider the solution to (12.16) as a benchmark for the assessment of the performance of the other estimators.

Comment The estimating equations proposed in this section employ the scores $d_{Ut} = \partial \log f_U(y_t|x_t;\beta)/\partial\beta$. However, similar equations can be obtained for other functions d_{Ut}; see Bickel *et al.* (1993) for examples of alternative definitions.

12.3.2. Estimation of $E_s(w_t|x_t)$

The estimating equations defined by (12.11) and (12.13) contain the expectations $E_s(w_t|x_t)$ that depend on the unknown parameters β. When the w_t are continuous as in probability proportional to size (PPS) sampling with a continuous size variable, the form of these expectations can be identified from the sample data by the following three-step procedure that utilizes (12.5):

1. Regress w_t against (y_t, x_t) to obtain an estimate of $E_s(w_t|y_t, x_t)$.
2. Integrate $\int_y E_U(\pi_t|y, x_t)f_U(y|x_t;\beta)dy = \int_y [1/E_s(w_t|y, x_t)]f_U(y|x_t;\beta)dy$ to obtain an estimate of $E_U(\pi_t|x_t)$ as a function of β.
3. Compute $E_s(w_t|x_t) = 1/E_U(\pi_t|x_t)$.

(The computations in steps 2 and 3 use the estimates obtained in the previous step.) The articles by PKR and PS contain several plausible models for $E_U(\pi_t|y_t, x_t)$ and examples for which the integral in step 2 can be carried out analytically. In practice, however, the specific form of the expectation $E_U(\pi_t|y_t, x_t)$ will usually be unknown but the expectation $E_s(w_t|y_t, x_t)$ can be identified and estimated in this case from the sample. (The expectation depends on unknown parameters that are estimated in step 1.)

Comment 1 Rather than estimating the coefficients indexing the expectation $E_s(w_t|y_t, x_t)$ from the sample (step 1), these coefficients can be considered as additional unknown parameters, with the estimating equations extended accordingly. This, however, may complicate the solution of the estimating equations and also result in identifiability problems under certain models. See PKR for examples and discussion.

Comment 2 For the estimating equations (12.13) that use the weights $q_t = w_t/E_s(w_t|x_t)$, the estimation of the expectation $E_s(w_t|x_t)$ can be carried out by simply regressing w_t against x_t, thus avoiding steps 2 and 3. This is so because in this case there is no need to express the expectation as a function of the parameters β indexing the population distribution.

The discussion so far focuses on the case where the sample selection probabilities are continuous. The evaluation of the expectation $E_s(w_t|x_t)$ in the case of discrete selection probabilities is simpler. For example, in the empirical study of this chapter we consider the case of logistic regression with a discrete independent variable x and three possible values for the dependent variable y. For this case the expectation $E_s(w_t|x_t = k)$ is estimated as

$$E_s(w_t|x_t = k) = 1/E_U(\pi_t|x_t = k),$$

$$E_U(\pi_t|x_t = k) = \sum_{a=1}^{3} \mathrm{Pr}_U(y_t = a|x_t = k)E_U(\pi_t|y_t = a, x_t = k)$$

$$= \sum_{a=1}^{3} \mathrm{Pr}_U(y_t = a|x_t = k)/E_s(w_t|y_t = a, x_t = k) \qquad (12.17)$$

$$\hat{=} \sum_{a=1}^{3} \mathrm{Pr}_U(y_t = a|x_t = k)/\overline{w}_{ak}$$

where $\overline{w}_{ak} = [\sum_s w_t I(y_t = a, x_t = k)]/[\sum_s I(y_t = a, x_t = k)]$. Here $I(A)$ is the indicator function for the event A. Substituting the logistic function for $\mathrm{Pr}(y_t = a|x_t = k)$ in the last expression of (12.17) yields the required specification. The estimators $\overline{w}_{ak}$ are considered as fixed numbers when solving the estimating equations.

For the estimating equations (12.13), the expectations $E_s(w_t|x_t = k)$ in (12.13) in this example are estimated by

$$E_s(w_t|x_t = k) \hat{=} \overline{w}_k = \left[\sum_s w_t I(x_t = k)\right] \bigg/ \left[\sum_s I(x_t = k)\right] \qquad (12.18)$$

rather than using (12.17) that depends on the unknown logistic coefficients (see Comment 2 above).

For an example of the evaluation of the expectation $E_s(w_t|x_t)$ with discrete selection probabilities but continuous outcome and explanatory variables, see PS (section 5.2).

12.3.3.   Testing the informativeness of the sampling process

The estimating equations developed in Section 12.3.1 for the case of informative sampling involve the use of the sampling weights in various degrees of complexity. It is clear therefore that when the sampling process is in fact noninformative, the use of these equations yields more variable estimators than the use of the ordinary score function defined by (12.16). See Tables 12.2 and 12.4 below for illustrations. For the complex sampling schemes in common use, the sample selection probabilities are often determined by the values of several design variables, in which case the informativeness of the selection process is not always apparent. This raises the need for test procedures as a further indication of whether the sampling process is ignorable or not. Several tests have been proposed in the past for this problem. The common feature of these tests is that they compare the probability-weighted estimators of the target parameters to the ordinary (unweighted) estimators that ignore the sampling process, see Pfeffermann (1993) for review and discussion. For the classical linear regression model, PS propose a set of test statistics that compare the moments of the sample distribution of the regression residuals to the corresponding moments of the population distribution. The use of these tests is equivalent to testing that the correlations under the sample distribution between powers of the regression residuals and the sampling weights are all zero. In Chapter 11 the tests developed by PS are extended to situations where the moments of the model residuals are functions of the regressor variables, as under many of the GLMs in common use.

A drawback of these test procedures is that they involve the use of a series of tests with dependent test statistics, such that the interpretation of the results of these tests is not always clear-cut. For this reason, we propose below a single alternative test that compares the estimating equations that ignore the sampling process to estimating equations that account for it. As mentioned before, the question arising in practice is whether to use the estimating equations (12.16) that ignore the sample selection or one of the estimating equations (12.11), (12.13), or (12.15) that account for it, so that basing the test on these equations is very natural.

In what follows we restrict attention to the comparison of the estimating equations (12.13) and (12.16) (see Comment below). From a theoretical point of view, the sampling process can be ignored for inference if the corresponding parameter equations are equivalent or $\sum_s E_s(d_{Ut}|x_t) = \sum_s E_s(q_t d_{Ut}|x_t)$. Denoting $R(x_t) = E_s(d_{Ut}|x_t) - E_s(q_t d_{Ut}|x_t)$, the null hypothesis is therefore

$$H_0: R_n = n^{-1} \sum_s R(x_t) = 0. \tag{12.19}$$

Note that $\dim(R_n) = k + 1 = \dim(\beta)$. If β were known, the hypothesis could be tested by use of the Hotelling test statistic,

$$H(R) = \frac{n - (k+1)}{k+1} \hat{R}'_n S_n^{-1} \hat{R}_n \sim^{H_0} F_{k+1, n-(k+1)} \tag{12.20}$$

where

$$\hat{R}_n = n^{-1} \sum_s \hat{R}(x_t); \quad \hat{R}(x_t) = (d_{Ut} - q_t d_{Ut}) \text{ and}$$

$$S_n = n^{-1} \sum_s (\hat{R}(x_t) - \hat{R}_n)(\hat{R}(x_t) - \hat{R}_n)'.$$

In practice, β is unknown and the score d_{Ut} in $\hat{R}(x_t)$ has to be evaluated at a sample estimate of β. In principle, any of the estimates defined in Section 12.3.1 could be used for this purpose since under H_0 all the estimators are consistent for β, but we find that the use of the solution of (12.16) that ignores the sampling process is the simplest and yields the best results.

Let $\hat{d}_{Ut}$ define the value of d_{Ut} evaluated at $\hat{\beta}$ – the solution of (12.16) – and let $\tilde{R}(x_t)$, $\tilde{R}_n$, and $\tilde{S}_n$ be the corresponding values of $\hat{R}(x_t)$, $\hat{R}_n$, and S_n obtained after substituting $\hat{\beta}$ for β in (12.20). The test statistic is therefore

$$\tilde{H}(R) = \frac{n - (k + 1)}{k + 1} \tilde{R}'_n \tilde{S}_n^{-1} \tilde{R}_n \approx^{H_0} F_{k+1,\, n-(k+1)}. \qquad (12.21)$$

Note that $\sum_s \hat{d}_{Ut} = 0$ by virtue of (12.16), so $\tilde{R}_n = -n^{-1} \sum_s q_t \hat{d}_{Ut}$. The random variables $q_t \hat{d}_{Ut}$ are no longer independent since $\sum_s \hat{d}_{Ut} = 0$, but utilizing the property that $E_s(q_t | x_t) = 1$ implies that under the null hypothesis $\mathrm{var}_s[\sum_s q_t \hat{d}_{Ut}] = \mathrm{var}_s[\sum_s (q_t \hat{d}_{Ut} - \hat{d}_{Ut})] = \sum_s \mathrm{var}_s(\hat{d}_{Ut} - q_t \hat{d}_{Ut})$, thus justifying the use of $\tilde{S}_n/(n-1)$ as an estimator of $\mathrm{var}(\tilde{R}_n)$ in the construction of the test statistic in (12.21).

Comment   The Hotelling test statistic uses the estimating equations (12.13) for the comparison with (12.16) and here again, one could use instead the equations defined by (12.11) or (12.15): that is, replace $q_t d_{Ut}$ in the definition of $\hat{R}(x_t)$ by $d_{Ut} + \partial \log [E_s(w_t | x_t)]/\partial\beta$, or by $w_t d_{Ut}$ respectively. The use of (12.11) is more complicated since it requires evaluation of the expectation $E_s(w_t | x_t)$ as a function of β (see Section 12.3.2). The use of (12.15) is the simplest but it yields inferior results to the use of (12.13) in our simulation study.

12.4. VARIANCE ESTIMATION

Having estimated the model parameters by any of the solutions of the estimating equations in Section 12.3.1, the question arising is how to estimate the variances of these estimators. Unless stated otherwise, the true (estimated) variances are with respect to the sample distribution for a given sample of units, that is, the variance under the *pdf* obtained by the product of the sample *pdfs* (12.3). Note also that since the estimating equations are only for the β-parameters, with the coefficients indexing the expectations $E_s(w_t | y_t, x_t)$ held fixed at their estimators of these values, the first four variance estimators below do not account for the variability of the estimated coefficients.

For the estimator $\hat{\beta}_{1s}$ defined by the solution to the estimating equations (12.11), that is, the maximum likelihood estimator under the sample distribution, a variance estimator can be obtained from the inverse of the information matrix evaluated at this estimator. Thus,

$$\hat{V}(\hat{\beta}_{1s}) = \{ - E_s[\partial W_{1s,\,e}(\beta)/\partial\beta']_{\beta=\hat{\beta}_{1s}} \}^{-1}. \qquad (12.22)$$

For the estimators $\hat{\beta}_{2s}$ solving (12.13), we use a result from Bickel *et al.* (1993). By this result, if for the true vector parameter β_0, the left hand side of an estimating equation $W_n(\beta) = 0$ can be approximated as $W_n(\beta_0) = n^{-1} \sum_s \varphi(y_t, x_t; \beta_0) + O_p(n^{-1/2})$ for some function φ satisfying $E(\varphi) = 0$ and $E(\varphi^2) < \infty$, then under some additional regularity conditions on the order of convergence of certain functions,

$$n^{1/2}(\hat{\beta}_n - \beta_0) = n^{-1/2} \sum_s [\dot{W}(\beta_0)]^{-1} \varphi(y_t, x_t; \beta_0) + o_p(1) \qquad (12.23)$$

where $\hat{\beta}_n$ is the solution of $W_n(\beta) = 0$, $\dot{W}(\beta_0) = [\partial W(\beta)/\partial \beta']_{\beta=\beta_0}$ and $W(\beta) = 0$ is the parameter equation with $\dot{W}(\beta_0)$ assumed to be nonsingular.

For the estimating equations (12.13), $\varphi(y_t, x_t; \beta) = q_t d_{Ut}$, implying that the variance of $\hat{\beta}_{2s}$ solving (12.13) can be estimated as

$$\hat{V}_s(\hat{\beta}_{2s}) = [\dot{W}_{2s,e}(\hat{\beta}_{2s})]^{-1} \left\{ \sum_s [q_t d_{Ut}(\hat{\beta}_{2s})]^2 \right\} [\dot{W}_{2s,e}(\hat{\beta}_{2s})]^{-1} \qquad (12.24)$$

where $\dot{W}_{2s,e}(\hat{\beta}_{2s}) = [\partial W_{2s,e}(\beta)/\beta']_{\beta=\hat{\beta}_{2s}}$ and $d_{Ut}(\hat{\beta}_{2s})$ is the value of d_{Ut} evaluated at $\hat{\beta}_{2s}$. Note that since $E_s(q_t d_{Ut}|x_t) = 0$ (and also $\sum_s q_t d_{Ut}(\hat{\beta}_{2s}) = 0$), the estimator (12.24) estimates the conditional variance $V_s(\hat{\beta}_{2s}|\{x_t, t \in s\})$; that is, the variance with respect to the conditional sample distribution of the outcome y.

The estimating equations (12.15) have been derived in Section 12.3.1 by two different approaches, implying therefore two separate variance estimators. Under the first approach, these equations estimate the parameter equations (12.14), which are defined in terms of the unconditional sample expectation (the expectation under the sample distribution of $\{(y_t, x_t), t \in s\}$). Application of the result from Bickel *et al.* (1993) mentioned before yields the following variance estimator (compare with (12.24)):

$$\hat{V}_s(\hat{\beta}_{3s}) = [\dot{W}_{3s,e}(\hat{\beta}_{3s})]^{-1} \left\{ \sum_s [w_t d_{Ut}(\hat{\beta}_{3s})]^2 \right\} [\dot{W}_{3s,e}(\hat{\beta}_{3s})]^{-1} \qquad (12.25)$$

where $\dot{W}_{3s,e}(\hat{\beta}_{3s}) = [\partial W_{3s,e}(\beta)/\partial \beta']_{\beta=\hat{\beta}_{3s}}$. For this case $E_s(w_t d_{Ut}) = 0$ (and also $\sum_s w_t d_{Ut}(\hat{\beta}_{3s}) = 0$) so that (12.25) estimates the unconditional variance over the joint sample distribution of $\{(y_t, x_t), t \in s\}$.

Under the second approach, the estimating equations (12.15) are the randomization unbiased estimators of the census equations (12.9). As such, the variance of $\hat{\beta}_{3s}$ can be evaluated with respect to the randomization distribution over all possible sample selections, with the population values held fixed. Following Binder (1983), the randomization variance is estimated as

$$\hat{V}_R(\hat{\beta}_{3s}) = [\hat{\dot{W}}_U^*(\hat{\beta}_{3s})]^{-1} \hat{V}_R \left[\sum_s w_t d_{Ut}(\hat{\beta}_{3s}) \right] [\hat{\dot{W}}_U^*(\hat{\beta}_{3s})]^{-1} \qquad (12.26)$$

where $\hat{\dot{W}}_U^*(\hat{\beta}_{3s})$ is design (randomization) consistent for $[\partial W_U^*/\partial \beta']_{\beta=\beta_0}$ and $\hat{V}_R[\sum_s w_t d_{Ut}(\hat{\beta}_{3s})]$ is an estimator of the randomization variance of $\sum_s w_t d_{Ut}(\beta)$, evaluated at $\hat{\beta}_{3s}$.

In order to illustrate the difference between the variance estimators (12.25) and (12.26), consider the case where the sample is drawn by Poisson sampling such that units are selected into the sample independently by Bernoulli trials with probabilities of success $\pi_t = \Pr(t \in s)$. Simple calculations imply that for this case the randomization variance estimator (12.26) has the form

$$\hat{V}_R(\hat{\beta}_{3s}) = [\hat{\dot{W}}^*_U(\hat{\beta}_{3s})]^{-1} \left\{ \sum_s (1 - \pi_t)[w_t d_{Ut}(\hat{\beta}_{3s})]^2 \right\} [\hat{\dot{W}}^*_U(\hat{\beta}_{3s})]^{-1} \quad (12.27)$$

where $\hat{\dot{W}}^*_U(\hat{\beta}_{3s}) = \dot{W}_{3s,e}(\hat{\beta}_{3s})$. Thus, the difference between the estimator defined by (12.25) and the randomization variance estimator (12.26) is in this case in the weighting of the products $w_t d_{Ut}(\hat{\beta}_{3s})$ by the weights $(1 - \pi_t)$ in the latter estimator. Since $0 < (1 - \pi_t) < 1$, the randomization variance estimators are smaller than the variance estimators obtained under the sample distribution. This is expected since the randomization variances measure the variation around the (fixed) population values and if some of the selection probabilities are large, a correspondingly large portion of the population is included in the sample (in high probability), thus reducing the variance.

Another plausible variance estimation procedure is the use of bootstrap samples. As mentioned before, under general conditions on the sample selection scheme listed in PKR, the sample measurements are asymptotically independent with respect to the sample distribution, implying that the use of the (classical) bootstrap method for variance estimation is well founded. In contrast, the use of the bootstrap method for variance estimation under the randomization distribution is limited, and often requires extra modifications; see Sitter (1992) for an overview of bootstrap methods for sample surveys. Let $\hat{\beta}_s$ stand for any of the preceding estimators and denote by $\hat{\beta}^b_s$ the estimator computed from bootstrap sample b ($b = 1 \ldots B$), drawn by simple random sampling with replacement from the original sample (with the same sample size). The bootstrap variance estimator of $\hat{\beta}_s$ is defined as

$$\hat{V}_{boot}(\hat{\beta}_s) = B^{-1} \sum_{b=1}^{B} (\hat{\beta}^b_s - \bar{\beta}_{boot})(\hat{\beta}^b_s - \bar{\beta}_{boot})' \quad (12.28)$$

where

$$\bar{\beta}_{boot} = B^{-1} \sum_{b=1}^{B} \hat{\beta}^b_s.$$

It follows from the construction of the bootstrap samples that the estimator (12.28) estimates the unconditional variance of $\hat{\beta}_s$. A possible advantage of the use of the bootstrap variance estimator in the present context is that it accounts in principle for all the sources of variation. This includes the identification of the form of the expectations $E_s(w_t|y_t, x_t)$ when unknown, and the estimation of the vector coefficient λ indexing that expectation, which is carried out for each of the bootstrap samples but not accounted for by the other variance estimation methods unless the coefficients λ are considered as part of the unknown model parameters (see Section 12.3.2).

12.5. SIMULATION RESULTS

12.5.1. Generation of population and sample selection

In order to assess and compare the performance of the parameter estimators, variance estimators, and the test statistic proposed in Sections 12.3 and 12.4, we designed a Monte Carlo study that consists of the following stages:

A. Generate a univariate population of x-values of size $N = 3000$, drawn independently from the discrete $U[1, 5]$ probability function, $\Pr(X = j) = 0.2, j = 1 \ldots 5$.

B. Generate corresponding y-values from the logistic probability function,

$$\Pr(y_t = 1|x_t) = [\exp(\beta_{10} + \beta_{11}x_t)]/C$$
$$\Pr(y_t = 2|x_t) = [\exp(\beta_{20} + \beta_{21}x_t)]/C \qquad (12.29)$$
$$\Pr(y_t = 3|x_t) = 1 - \Pr(y_t = 1|x_t) - \Pr(y_t = 2|x_t)$$

where $C = 1 + \exp(\beta_{10} + \beta_{11}x_t) + \exp(\beta_{20} + \beta_{21}x_t)$.
Stages **A** and **B** were repeated independently $R = 1000$ times.

C. From every population generated in stages **A** and **B**, draw a single sample using the following sampling schemes (one sample under each scheme):

    Ca. Poisson sampling: units are selected independently with probabilities $\pi_t = nz_t/\sum_{u=1}^{N} z_u$, where $n = 300$ is the expected sample size and the values z_t are computed in two separate ways:

$$\textbf{Ca(1):}\ z_t(1) = Int[(5/9)y_t^2 u_t + 2x_t];$$
$$\textbf{Ca(2):}\ z_t(2) = Int[5u_t + 2x_t]. \qquad (12.30)$$

    The notation $Int[\cdot]$ defines the integer value and $u_t \sim U(0,1)$.

    Cb. Stratified sampling: the population units are stratified based on either the values $z_t(1)$ (scheme **Cb(1)**) or the values $z_t(2)$ (scheme **Cb(2)**), yielding a total of 13 strata in each case. Denote by $S_{(h)}(j)$ the strata defined by the values $z_t(j)$ such that for units $t \in S_h(j), z_t(j) \equiv z_{(h)}(j), j = 1, 2$. Let $N_{(h)}(j)$ represent the corresponding strata sizes. The selection of units within the strata was carried out by simple random sampling without replacement (SRSWR), with the sample sizes $n_{(h)}(j)$ fixed in advance. The sample sizes were determined so that the selection probabilities are similar to the corresponding selection probabilities under the Poisson sampling scheme and $\sum_h n_h(j) = 300, j = 1, 2$.

The following points are worth noting:

1. The sampling schemes that use the values $z_t(1)$ are *informative*, as the selection probabilities depend on the y-values. The sampling schemes that

use the values $z_t(2)$ are *noninformative* since the selection probabilities depend only on the x-values and the inference is targeted at the population model of the conditional probabilities of $y_t|x_t$ defined by (12.29).

2. For the Poisson sampling schemes, $E_U[\pi_t|\{(y_u, x_u), u = 1 \ldots N\}]$ depends only on the values (y_t, x_t) when $z_t = z_t(1)$, and only on the value x_t when $z_t = z_t(2)$. (With large populations, the totals $\sum_{t=1}^{N} z_t(j)$ can be regarded as fixed.) For the stratified sampling schemes, however, the selection probabilities depend on the strata sizes $N_{(h)}(j)$ that are random (they vary between populations), so that they depend in principle on all the population values $\{(y_t, x_t), t = 1 \ldots N\}$, although with large populations the variation of the strata sizes between populations will generally be minor.

3. The stratified sampling scheme **Cb** corresponds to a *case–control study* whereby the strata are defined based on the values of the outcome variable (y) and possibly some of the covariate variables. See Chapter 8. (Such sampling schemes are known as *choice-based sampling* in the econometric literature.) For the case where the strata are defined based only on y-values generated by a logistic model that contains intercept terms and the sampling fractions within the strata are fixed in advance, it is shown in PKR that the standard MLE of the *slope* coefficients (that is, the MLE that ignores the sampling scheme) coincides with the MLE under the sample distribution. As illustrated in the empirical results below, this is no longer true when the stratification depends also on the x-values. (As pointed out above, under the design **Cb** the sampling fractions within the strata have some small variation. We considered also the case of a stratified sample with fixed sampling fractions within the strata and obtained almost identical results as for the scheme **Cb**.) In order to assess the performance of the estimators derived in the previous sections in situations where the stratification depends only on the y-values, we consider also a third stratified sampling scheme:

Cc. Stratified sampling: The population units are stratified based on the values y_t (three strata); select n_h units from stratum h by SRSWR with the sample sizes fixed in advance, $\sum_h n_h = 300$.

12.5.2. Computations and results

The estimators of the logistic model coefficients $\beta_k = \{(\beta_{k0}, \beta_{k1}), k = 1, 2\}$ in (12.29), obtained by solving the estimating equations (12.11), (12.13), (12.15), and (12.16), have been computed for each of the samples drawn by the sampling methods described in Section 12.5.1, yielding four separate sets of estimators. The expectations $E_s(w_t|x_t)$ have been estimated using the procedures described in Section 12.3.2. For each point estimator we computed the corresponding variance estimator as defined by (12.22), (12.24), and (12.25) for the first three estimators and by use of the inverse information matrix (ignoring the sampling process) for the ordinary MLE. In addition, we computed for each of the point estimators the bootstrap variance estimator defined by (12.27). Due to computation time constraints, the bootstrap variance estimators are based on only 100 samples for each of 100 parent samples. Finally, we computed for each sample the Hotelling test statistic (12.21) for sample informativeness developed in Section 12.3.3.

The results of the simulation study are summarized in Tables 12.1–12.5. These tables show for each of the five sampling schemes and each point estimator the mean estimate, the empirical standard deviation (Std) and the mean of the Std estimates (denoted Mean Std est. in the tables) over the 1000 samples; and the mean of the bootstrap Std estimates (denoted Mean Std est. Boot in the tables) over the 100 samples. Table 12.6 compares the theoretical and empirical distribution of the Hotelling test statistic under H_0 for the two noninformative sampling schemes defined by **Ca(2)** and **Cb(2)**.

The main conclusions from the results set out in Tables 12.1–12.5 are as follows:

Table 12.1 Means, standard deviations (Std) and mean Std estimates of logistic regression coefficients. Poisson sampling, informative scheme **Ca (1)**.

Coefficients	β_{10}	β_{11}	β_{20}	β_{21}
MLE (sample *pdf*, Equation (12.11)):				
True values	1.00	0.30	0.50	0.50
Mean estimate	1.03	0.30	0.54	0.50
Empirical Std	0.62	0.18	0.61	0.18
Mean Std est.	0.58	0.18	0.58	0.18
Mean Std est. Boot	0.56	0.18	0.56	0.18
Q-weighting (Equation (12.13)):				
True values	1.00	0.30	0.50	0.50
Mean estimate	0.99	0.31	0.51	0.51
Empirical Std	0.63	0.19	0.63	0.19
Mean Std est.	0.59	0.18	0.59	0.18
Mean Std est. Boot	0.68	0.21	0.66	0.20
W-weighting (Equation (12.15)):				
True values	1.00	0.30	0.50	0.50
Mean estimate	1.00	0.31	0.53	0.51
Empirical Std	0.67	0.21	0.67	0.20
Mean Std est.	0.63	0.19	0.62	0.19
Mean Std est. Boot	0.65	0.21	0.64	0.20
Ordinary MLE (Equation (12.16)):				
True values	1.00	0.30	0.50	0.50
Mean estimate	0.29	0.43	0.06	0.59
Empirical Std	0.60	0.19	0.60	0.18
Mean Std est.	0.59	0.18	0.58	0.18
Mean Std est. Boot	0.57	0.18	0.58	0.18

Table 12.2 Means, standard deviations (Std) and mean Std estimates of logistic regression coefficients. Poisson sampling, noninformative scheme **Ca (2)**.

Coefficients	β_{10}	β_{11}	β_{20}	β_{21}
MLE (sample *pdf*, Equation (12.11)):				
True values	1.00	0.30	0.50	0.50
Mean estimate	1.01	0.31	0.53	0.50
Empirical Std	0.66	0.20	0.67	0.20
Mean Std est.	0.62	0.20	0.62	0.20
Mean Std est. Boot	0.62	0.20	0.62	0.20
Q-weighting (Equation (12.13)):				
True values	1.00	0.30	0.50	0.50
Mean estimate	0.99	0.31	0.51	0.51
Empirical Std	0.67	0.20	0.68	0.20
Mean Std est.	0.63	0.19	0.63	0.20
Mean Std est. Boot	0.66	0.22	0.66	0.22
W-weighting (Equation (12.15)):				
True values	1.00	0.30	0.50	0.50
Mean estimate	1.00	0.31	0.52	0.51
Empirical Std	0.71	0.22	0.72	0.22
Mean Std est.	0.65	0.21	0.66	0.21
Mean Std est. Boot	0.68	0.22	0.69	0.22
Ordinary MLE (Equation (12.16)):				
True values	1.00	0.30	0.50	0.50
Mean estimate	0.99	0.31	0.50	0.51
Empirical Std	0.63	0.19	0.63	0.19
Mean Std est.	0.60	0.19	0.61	0.19
Mean Std est. Boot	0.62	0.20	0.63	0.20

1. The three sets of estimating equations defined by (12.11), (12.13), and (12.15) perform well in eliminating the sampling effects under informative sampling. On the other hand, ignoring the sampling process and using the ordinary MLE (Equation (12.16)) yields highly biased estimators.
2. The use of Equations (12.11) and (12.13) produces very similar results. This outcome, found also in PS for ordinary regression analysis, is important since the use of (12.13) is much simpler and requires fewer assumptions than the use of the full equations defined by (12.11); see also Section 12.3.2. The use of simple weighting (Equation (12.15)) that corresponds to the application of the pseudo-likelihood approach again performs well in eliminating the bias, but except for Table 12.5 the variances under this approach are consistently larger than under the first two approaches, illustrating the discussion in Section 12.3.1. On the other hand, the ordinary MLE that does not involve any weighting has in most cases the smallest standard deviation. The last outcome is known also from other studies.

Table 12.3   Means, standard deviations (Std) and mean Std estimates of logistic regression coefficients. Stratified sampling, informative scheme **Cb (1)**.

Coefficients	β_{10}	β_{11}	β_{20}	β_{21}
MLE (sample *pdf*, Equation (12.11)):				
True values	1.00	0.30	0.50	0.50
Mean estimate	1.05	0.30	0.53	0.51
Empirical Std	0.56	0.17	0.59	0.17
Mean Std est.	0.56	0.17	0.56	0.17
Mean Std est. Boot	0.50	0.17	0.50	0.17
Q-weighting (Equation (12.13)):				
True values	1.00	0.30	0.50	0.50
Mean estimate	1.05	0.30	0.53	0.50
Empirical Std	0.55	0.16	0.58	0.17
Mean Std est.	0.58	0.18	0.58	0.18
Mean Std est. Boot	0.66	0.21	0.64	0.20
W-weighting (Equation (12.15)):				
True values	1.00	0.30	0.50	0.50
Mean estimate	1.07	0.29	0.55	0.50
Empirical Std	0.58	0.18	0.61	0.18
Mean Std est.	0.61	0.19	0.61	0.19
Mean Std est. Boot	0.63	0.21	0.63	0.21
Ordinary MLE (Equation (12.16)):				
True values	1.00	0.30	0.50	0.50
Mean estimate	0.37	0.41	0.16	0.56
Empirical Std	0.52	0.15	0.55	0.16
Mean Std est.	0.57	0.18	0.57	0.18
Mean Std est. Boot	0.52	0.18	0.52	0.18

3. The MLE variance estimator (12.22) and the semi-parametric estimators (12.24) and (12.25) underestimate in most cases the true (empirical) variance with an underestimation of less than 8 %. As anticipated in Section 12.4, the use of the bootstrap variance estimators corrects for this underestimation by better accounting for all the sources of variation, but this only occurs with the estimating equations (12.13) and (12.15). We have no clear explanation for why the bootstrap variance estimators perform less satisfactory for the MLE equations defined by (12.11) and (12.16) but we emphasize again that we have only used 100 bootstrap samples for the variance estimation, which in view of the complexity of the estimating equations is clearly not sufficient. It should be mentioned also in this respect that the standard deviation estimators (12.22), (12.24), and (12.25) are more stable (in terms of their standard deviation) than the bootstrap standard deviation estimators, which again can possibly be attributed to the relatively small number of bootstrap samples. (The standard deviations of the standard deviation estimators are not shown.)

Table 12.4 Means, standard deviations (Std) and mean Std estimates of logistic regression coefficients. Stratified sampling, noninformative scheme **Cb(2)**.

Coefficients	β_{10}	β_{11}	β_{20}	β_{21}
MLE (sample *pdf*, Equation (12.11)):				
True values	1.00	0.30	0.50	0.50
Mean estimate	1.06	0.29	0.53	0.50
Empirical Std	0.63	0.19	0.64	0.20
Mean Std est.	0.60	0.19	0.61	0.19
Mean Std est. Boot	0.56	0.20	0.57	0.20
Q-weighting (Equation (12.13)):				
True values	1.00	0.30	0.50	0.50
Mean estimate	1.04	0.30	0.51	0.51
Empirical Std	0.63	0.20	0.64	0.20
Mean Std est.	0.60	0.19	0.61	0.19
Mean Std est. Boot	0.62	0.22	0.63	0.22
W-weighting (Equation (12.15)):				
True values	1.00	0.30	0.50	0.50
Mean estimate	1.05	0.30	0.52	0.51
Empirical Std	0.66	0.21	0.67	0.21
Mean Std est.	0.62	0.20	0.63	0.20
Mean Std est. Boot	0.63	0.22	0.65	0.23
Ordinary MLE (Equation (12.16)):				
True values	1.00	0.30	0.50	0.50
Mean estimate	1.03	0.30	0.49	0.52
Empirical Std	0.60	0.19	0.61	0.20
Mean Std est.	0.58	0.19	0.59	0.19
Mean Std est. Boot	0.56	0.20	0.57	0.20

4. Our last comment refers to Table 12.5 that relates to the sampling scheme **Cc** by which the stratification is based only on the *y*-values. For this case the first three sets of parameter estimators and the two semi-parametric variance estimators perform equally well. Perhaps the most notable outcome from this table is that ignoring the sampling process in this case and using Equations (12.16) yields similar mean estimates (with smaller standard deviations) for the two slope coefficients as the use of the other equations. Note, however, the very large biases of the intercept estimators. As mentioned before, PKR show that the use of (12.16) yields the correct MLE for the slope coefficients under Poisson sampling, which is close to the stratified sampling scheme underlying this table.

Table 12.6 compares the empirical distribution of the Hotelling test statistic (12.21) over the 1000 samples with the corresponding nominal levels of the theoretical distribution for the two noninformative sampling schemes **Ca(2)** and **Cb(2)**. As can be seen, the empirical distribution matches almost perfectly the theoretical distribution. We computed the test statistic also under the

Table 12.5 Means, standard deviations (Std) and mean Std estimates of logistic regression coefficients. Stratified sampling, informative scheme **Cc**.

Coefficients	β_{10}	β_{11}	β_{20}	β_{21}
MLE (sample *pdf*, Equation (12.11)):				
True values	1.00	0.30	0.50	0.50
Mean estimate	1.00	0.32	0.48	0.53
Empirical Std	0.42	0.16	0.42	0.15
Mean Std est.	0.40	0.14	0.38	0.14
Mean Std est. Boot	0.39	0.15	0.37	0.14
Q-weighting (Equation (12.13)):				
True values	1.00	0.30	0.50	0.50
Mean estimate	0.99	0.31	0.48	0.52
Empirical Std	0.42	0.16	0.42	0.15
Mean Std est.	0.43	0.15	0.42	0.15
Mean Std est. Boot	0.44	0.16	0.43	0.16
W-weighting (Equation (12.15)):				
True values	1.00	0.30	0.50	0.50
Mean estimate	1.00	0.31	0.48	0.52
Empirical Std	0.42	0.16	0.42	0.15
Mean Std est.	0.43	0.15	0.42	0.15
Mean Std est. Boot	0.44	0.16	0.43	0.16
Ordinary MLE (Equation (12.16)):				
True values	1.00	0.30	0.50	0.50
Mean estimate	0.01	0.31	−0.04	0.52
Empirical Std	0.36	0.14	0.36	0.13
Mean Std est.	0.40	0.14	0.38	0.14
Mean Std est. Boot	0.39	0.15	0.37	0.14

Table 12.6 Nominal levels and empirical distribution of Hotelling test statistic under H_0 for noninformative sampling schemes **Ca(2)** and **Cb(2)**.

Nominal levels	0.01	0.025	0.05	0.10	0.90	0.95	0.975	0.99
Emp. dist (**Ca(2)**)	0.01	0.02	0.04	0.09	0.89	0.95	0.98	0.99
Emp. dist (**Cb(2)**)	0.01	0.03	0.05	0.08	0.91	0.96	0.98	0.99

three informative sampling schemes and in all the 3 × 1000 samples, the null-hypothesis of noninformativeness of the sampling scheme had been rejected at the 1 % significance level, indicating very high power.

12.6. SUMMARY AND EXTENSIONS

This chapter considers three alternative approaches for the fitting of GLM under informative sampling. All three approaches utilize the relationships between the population distribution and the distribution of the sample observations as defined by (12.1), (12.3), and (12.4) and they are shown to perform well in eliminating the bias of point estimators that ignore the sampling process. The use of the pseudo-likelihood approach, derived here under the framework of the sample distribution, is the simplest, but it is shown to be somewhat inferior to the other two approaches. These two approaches require the modeling and estimation of the expectation of the sampling weights, either as a function of the outcome and the explanatory variables, or as a function of only the explanatory variables. This additional modeling is not always trivial but general guidelines are given in Section 12.3.2. It is important to emphasize in this respect that the use of the sample distribution as the basis for inference permits the application of standard model diagnostic tools so that the goodness of fit of the model to the sample data can be tested.

The estimating equations developed under the three approaches allow the construction of variance estimators based on these equations. These estimators have a small negative bias since they fail to account for the estimation of the expectations of the sampling weights. The use of the bootstrap method that is well founded under the sample distribution overcomes this problem for two of the three approaches, but the bootstrap estimators seem to be less stable. Finally, a new test statistic for the informativeness of the sampling process that compares the estimating equations that account for the sampling process with estimating equations that ignore it is developed and shown to perform extremely well in the simulation study.

An important use of the sample distribution not considered in this chapter is for prediction problems. Notice first that if the sampling process is informative, the model holding for the outcome variable for units outside the sample is again different from the population model. This implies that even if the population model is known with all its parameters, it cannot be used directly for the prediction of outcome values corresponding to nonsampled units. We mention in this respect that the familiar 'model-dependent estimators' of finite population totals assume noninformative sampling. On the other hand, it is possible to derive the distribution of the outcome variable for units outside the sample, similarly to the derivation of the sample distribution, and then obtain the optimal predictors under this distribution. See Sverchkov and Pfeffermann (2000) for application of this approach.

Another important extension is to two-level models with application to small-area estimation. Here again, if the second-level units (schools, geographic areas) are selected with probabilities that are related to the outcome values, the model holding for the second-level random effects might be different from the model holding in the population. Failure to account for the informativeness of the sampling process may yield biased estimates for the model parameters and biased predictions for the small-area means. Appropriate weighting may eliminate the first bias but not the second. Work on the use of the sample distribution for small-area estimation is in progress.

Longitudinal Data

CHAPTER 13

Introduction to Part D
C. J. Skinner

13.1. INTRODUCTION

The next three chapters bring the time dimension into survey analysis. Whereas Parts B and C have been primarily concerned with analysing variation in a response variable Y across a finite population fixed in time, the next three chapters consider methods for modelling variation in Y not only across the population but also across time.

In this part of the book, we shall use t to denote time, since this use is so natural and widespread in the literature on longitudinal data, and use i to denote a population unit, usually an individual person. As a result, the basic values of the response variable of interest will now be denoted y_{it} with two subscripts, i for unit and t for time. This represents a departure from the notation used so far, where t has denoted unit.

Models will be considered in which time may be either *discrete* or *continuous*. In the former case, t takes a sequence of possible values $t_1, t_2, t_3, \ldots$ usually equally spaced. In the latter case, t may take any value in a given interval. The response variable Y may also be either discrete or continuous, corresponding to the distinction between Parts B and C of this book.

The case of continuous Y will be discussed first, in Section 13.2 and in Chapter 14 (Skinner and Holmes). Only discrete time will be considered in this case with the basic values to be modelled consisting of $\{y_{it}; i = 1, \ldots, N; t = 1, \ldots, T\}$ for the N units in the population and T equally spaced time points. The case of discrete Y will then be discussed in Section 13.3 and in Chapters 15 (Lawless) and 16 (Mealli and Pudney). The emphasis in these chapters will be on continuous time.

A basic longitudinal survey design involves a series of waves of data collection, often equally spaced, for all units in a fixed sample, s. At a given wave, either the current value of a variable or the value for a recent reference period may be measured. In this case, the sample data are recorded in discrete time and will take the form $\{y_{it}; i \in s, t = 1, \ldots, T\}$ for variables that are measured at all waves, in the absence of nonresponse. In the simplest case, for discrete time

Analysis of Survey Data Edited by R. L. Chambers and C. J. Skinner
© 2003 John Wiley & Sons, Ltd

models of the kind considered by Skinner and Holmes (Chapter 14), the times $1, \ldots, T$ at which data are collected correspond to the time points for which the model of interest is specified. In this case, longitudinal analysis may be considered as a form of multivariate analysis of the vectors of T responses $(y_{i1}, \ldots, y_{iT})$ and the question of how to handle complex sampling schemes in the selection of s is a natural generalisation of this question for cross-section surveys.

More generally, however, sampling needs to be considered in the broader context of the 'observation plan' of the longitudinal survey. Thus, even if the same sample of units is retained for all time points, it is necessary to consider not only how the sample was selected, but also for what values of t Y will be measured for sample units. Particular care may be required when models are specified in continuous time, as by Lawless (Chapter 15) and by Mealli and Pudney (Chapter 16). Sometimes, continuous time data are collected retrospectively, over some time 'window'. Sometimes, only the current value of Y may be collected at each wave in a longitudinal survey. Methods of making inference about continuous time models will need to take account of these observation plans. For example, if y_{it} is binary, indicating whether unit i has experienced an event (first marriage, for example) by time t, then Lawless (1982, section 7.3) shows how data on y_{it} at a finite set of time points t can still be used to fit continuous time models, such as a survival model of age at which the event first occurs.

There are several further ways in which the sampling and observation plan may be more complicated (Kalton and Citro, 1993). There will usually be attrition and other forms of nonresponse as discussed by Skinner and Holmes (Chapter 14). The sample may be subject to panel rotation and may be updated between waves to allow for new entrants to the population. Problems with tracking and tracing may lead to additional sources of incomplete data. The time windows used in data collection may also create complex censoring and truncation patterns for model fitting, as discussed further in Section 15.3.

In addition to incomplete data issues, there are potentially new problems of measurement error in longitudinal surveys. Recall error is common in retrospective measurement and there may be difficulties in 'piecing together' different fragments of an event history from multiple waves of a longitudinal survey (Skinner, 2000). Lawless (Chapter 15) refers to some of these problems.

13.2. CONTINUOUS RESPONSE DATA

In this section we discuss the basic discrete time continuous response case, where values y_{it} and x_{it} are measured on a continuous response variable Y and a vector of covariates X respectively, at equally spaced time points $t = 1, \ldots, T$ for units i in a sample s. A basic regression model relating Y to X is then given by

$$y_{it} = x_{it}\beta + u_i + v_{it} \tag{13.1}$$

where β is a vector of regression coefficients and u_i and v_{it} are disturbance terms. The term u_i allows for a 'permanent' effect of the ith unit, above or below that predicted by the covariates X.

The term v_{it} will usually be assumed to be the outcome of a random variable, which has mean 0 (conditional on the covariates). This is analogous to the standard disturbance term in a regression model.

The term u_i may sometimes be assumed to be a fixed unknown parameter, a *fixed effect*, or sometimes, like v_{it}, the outcome of a random variable, a *random effect*. The two approaches may be linked by viewing the fixed effects specification as a conditional version of the random effects specification, i.e. with inference being conditional on the effects in the sample (Hsiao, 1986, p. 42). From a survey sampling perspective, this fixed effects approach seems somewhat unnatural. The actual units included in the sample will be arbitrary and a function of the sampling scheme – it seems preferable that the interpretation of the parameters should not be dependent upon the actual sample drawn. It is more natural therefore to specify the model in terms of random effects, where the distribution of the random effects may be interpreted in terms of a super-population from which the finite population has been drawn. This is the approach adopted by Skinner and Holmes (Chapter 14).

The potential disadvantage of the random effects specification is that it involves stronger modelling assumptions than the fixed effects model; in particular it requires that the u_i have mean 0, conditional on the x_{it}, whereas the fixed effects specification allows for dependence between the u_i and the x_{it}. Thus, if such dependence exists, estimation of β, based upon the random effects model, may be biased. See Hsiao (1986, section 3.4) and Baltagi (2001, section 2.3.1) for further discussion.

There are many possible ways of extending the basic random effects model in (13.1) (Goldstein, 1995, Ch. 6). The coefficients β may vary randomly across units, giving a random coefficient model in contrast to (13.1), which may be viewed as a random intercept model. The v_{it} terms may be serially correlated, as considered by Skinner and Holmes (Chapter 14) or the model may be dynamic in the sense that lagged values of the dependent variable, y_{it-1}, y_{it-2}, . . . may be included amongst the covariates. There are also alternative approaches to the analysis of continuous response discrete time longitudinal data, such as the use of 'marginal' regression models (Diggle *et al.*, 2002), which avoid specifying random effects.

A traditional approach to inference about random effects models is through maximum likelihood. The likelihood is determined by making parametric assumptions about the distribution of random effects, for example by assuming that the u_i and v_{it} in (13.1) are normally distributed. It will usually not be possible to maximise the likelihood analytically and, instead, iterative numerical methods may be used. One approach is to use extensions of estimation methods for regression models, such as iterative generalised least squares (Goldstein, 1995). Another approach is to view the model as a 'covariance structure model' for the covariance matrix of the y_{it}, treated as a multivariate outcome for the different time points. Both these approaches are described by

Skinner and Holmes (Chapter 14) together with their extensions to allow for complex sampling schemes.

13.3. DISCRETE RESPONSE DATA

We now turn to the case when the response variable Y is discrete. One approach to the analysis of such data is to use methods based upon discrete response extensions of the models discussed in the previous section. These extensions may be obtained in the same way that continuous response regression models are commonly generalised to discrete response models. For example, logistic or probit regression models might be used for the case of binary Y (Hsiao, 1986, Ch. 7; Baltagi, 2001, Ch. 11).

An alternative approach is to use the methods of event history analysis which focus on the 'spells' of time which units spend in the different categories or 'states' of Y. The transitions between these states define certain 'events'. For example, if Y is marital status then the transition from the state 'single' to the state 'married' arises at the event of marriage. Event history models provide a framework for representing spell durations and transitions and their dependence over time and on covariates.

The choice between the two approaches is not a matter of which model is correct. Both kinds of model may be valid at the same time – they may simply be alternative representations of the distribution of the observed outcomes. The choice depends rather on the objectives of the analysis. In simplified terms, the first approach is appropriate when the aim is, as in Section 13.2, to model the dependence of the response variable Y on the covariates and when the temporal nature of the variables is usually not central to interpretation. The event history approach is appropriate when the aim is to model the durations in states and/or model the transitions between states or equivalently the timing of events.

The first approach is usually based upon discrete time data, as in Section 13.2, with the observations usually corresponding to the waves of the longitudinal survey. We shall not consider this approach further in this chapter, but refer the reader to discussions of discrete response longitudinal data in Hsiao (1986, Ch. 7) and Diggle *et al.* (2002).

The alternative event history approach will be considered by Lawless (Chapter 15) and Mealli and Pudney (Chapter 15) and is introduced further in this section. This approach focuses on temporal social phenomena (events, transitions, durations), which are usually not tied to the timing of the waves of the longitudinal survey. For example, it is unlikely that the timing of births, marriages or job starts will coincide with the waves of a survey. Since such timing is central to the aims of event history analysis, it is natural to define event history models in a continuous time framework, unconstrained by the timing of the survey data collection, and this is the framework adopted in Chapters 15 and 16 and assumed henceforth in this section. Discrete time analogues are, nevertheless, sometimes useful – see some comments by Lawless (Chapter 15) and Allison (1982) and Yamaguchi(1991).

The period of time during which a unit is in a given state is referred to as an *episode* or *spell*. The transition between two states defines an *event*. For example, an individual may move between three states: employed, unemployed and not in the labour force. The individual may thus experience a spell of employment or unemployment. The event of finding employment corresponds to the transition between the state of unemployment (or not in the labour force) and the state of employment. While changes in states define events, events may conversely be taken to define states. For example, the event of marriage defines the two states 'never-married' and 'ever-married', which an individual may move between. A discrete response process in continuous time may thus be represented either as a sequence of states or as a sequence of events. This is formalised by Lawless (Chapter 15), who distinguishes two mathematically equivalent frameworks, which may be adopted for event history analysis. The *multi-state framework* defines the event history in terms of $Y_i(t)$, the state occupied by unit i at time t. In our discussion above, $Y_i(t)$ is just the same as y_{it}, the discrete response of unit i at time t. This is also the basic framework adopted by Mealli and Pudney (Chapter 16). The alternative framework considered by Lawless is the *event occurrence framework* in which the event history is defined in terms of the number of occurrences of events of different types at different times. The latter framework may be more convenient when, for example, there is interest in the frequency of certain repeated events, such as consumer purchases. The remaining discussion in this section focuses on the multi-state framework.

In order to understand the general models and methods of event history analysis considered in this part of the book, it is sensible first to understand the models and methods of survival analysis (e.g. Lawless,1982; Cox and Oakes, 1984). As discussed by Lawless (Chapter 15), survival analysis deals with the special case when there are just two states and there is only one allowable transition, from state 1 to state 2. The focus then is on the time spent in state 1 before the transition to state 2 occurs. This time is called the survival time or duration. Basic survival analysis models treat this survival time as a random variable. The probability distribution of this random variable may be represented in terms of the *hazard function* $\lambda(t)$, which specifies the instantaneous rate of transition to state 2 given that the unit is still in state 1 at time t (see Equation (15.7)). This hazard function is also called the *transition intensity*. It may depend on covariates x and otherwise on the unit.

The basic survival analysis model is extended to general event history models, by defining a series of transition intensity functions $\lambda_{kl}(t)$ for each pair of distinct states k and l. The function is given in Equations (15.4) and (16.2) and represents the instantaneous probability of a transition from state k into state l at time t. This function may depend on the past history of the $Y_i(t)$ process, in particular on the duration in the current state k, it may depend on covariates x and it may depend otherwise on the unit i.

In the simple survival analysis case, knowledge of the hazard function is equivalent to knowledge of the distribution of the duration (Equations (15.8) and (15.9) show how the latter distribution may be derived from the hazard function). In the more general event history model, the intensity functions capture two separate features of the process, which may each be of interest. First, the intensity functions imply the relative probabilities of transition to alternative states, given that a transition occurs. Second, as for survival analysis, the intensity functions imply the distribution of the duration in any given state (Equation (15.6)).

The corresponding distinction in the data between states and durations is emphasised in Mealli and Pudney's notation (Chapter 16). Consider the data $\{Y_i(t); t_1 \leq t \leq t_2\}$ generated by observing a unit i over a finite interval of time $[t_1, t_2]$. Mealli and Pudney summarise these data in terms of $\delta_1, \delta_2, \ldots, \delta_m$ and $r_0, r_1, \ldots, r_{m-1}$, where m is the number of observed episodes, δ_j is the duration of the jth episode and r_j is the destination state of this episode. The initial state is $r_0 = Y_i(t_1)$. The joint distribution of the δ_j and the r_j may then be represented in terms of the intensity functions in Equation (16.3). In Lawless' notation (Chapter 15) the times of the events for unit i are denoted $t_{i1} \leq t_{i2} \leq t_{i3} \leq \ldots$ so the durations δ_j are given by $t_{ij} - t_{ij-1}$.

Event history models may be represented parametrically, semi-parametrically or non-parametrically. A simple example of a parametric model is the Weibull model for the duration of breast feeding in Section 15.8. A more complex example is given by Mealli and Pudney (Chapter 16), who discuss a four-state application using a parametric model, which is primarily based upon the Burr form for the transition intensities in Equation (16.8). The most well-known example of a semi-parametric model is the proportional hazards model, defined in Equation (15.10) and illustrated also on the breast feeding data. Some discussion of non-parametric methods such as the Kaplan–Meier estimation of the survivor function is given in Section 15.5.

Right-censoring is a well-known issue in survival analysis, arising as the result of incomplete observation of survival times. A wide range of such observation issues may arise in sample surveys and these are discussed in Section 15.3.

Beyond the issues of inference about models in the presence of sample selection, incomplete data and measurement error, there is the further question of how to use the fitted models to address policy questions. Mealli and Pudney (Chapter 16) consider the question of how to measure the impact of a youth training programme and describe a simulation approach based upon a fitted event history model.

CHAPTER 14

Random Effects Models for Longitudinal Survey Data

C. J. Skinner and D. J. Holmes

14.1. INTRODUCTION

Random effects models have a number of important uses in the analysis of longitudinal survey data. The main use, which we shall focus on in this chapter, is in the study of individual-level dynamics. Random effects models enable variation in individual responses to be decomposed into variation between the 'permanent' characteristics of individuals and temporal 'transitory' variation within individuals.

Another important use of random effects models in the analysis of longitudinal data is in allowing for the effects of time-constant unobserved covariates in regression models (e.g. Solon, 1989; Hsiao, 1986; Baltagi, 2001). Failure to allow for these unobserved covariates in regression analysis of cross-sectional survey data may lead to inconsistent estimation of regression coefficients. Consistent estimation may, however, be achievable with the use of random effects models and longitudinal data.

A 'typical' random effects model may be conceived of as follows. It is supposed that a response variable Y is measured at each of a number of successive waves of the survey. The measurement for individual i at wave t is denoted y_{it} and this value is assumed to be generated in two stages. First, 'permanent' random effects θ_i are generated from some distribution for each individual i. Then, at each wave, y_{it} is generated from θ_i. In the simplest case this generation follows the same process independently at each wave. For example, we may have

$$\theta_i \sim N\left(\theta, \sigma_1^2\right), \quad y_{it} \mid \theta_i \sim N\left(\theta_i, \sigma_2^2\right). \tag{14.1}$$

Under this model, longitudinal data enable the 'cross-sectional' variance $\sigma_1^2 + \sigma_2^2$ of y_{it} to be decomposed into the variance σ_1^2 of the 'permanent' component θ_i and the variance σ_2^2 of the 'transitory' component at each wave.

Analysis of Survey Data Edited by R. L. Chambers and C. J. Skinner

This may aid understanding of the mobility of individuals over time in terms of their place in the distribution of the response variable. An example, which we shall focus on, is where the response variable is earnings, subject to a log transformation, and a model of the form (14.1) enables us to study the degree of mobility of an individual's place in the earnings distribution (e.g. Lillard and Willis, 1978).

Rich classes of random effects models for longitudinal data have been developed for the above purposes. A number of different terms have been used to describe these models including *variance component models, error component models, mixed effects models, multilevel models* and *hierarchical models* (Baltagi, 2001; Hsiao, 1986; Diggle, Liang and Zeger, 1994; Goldstein, 1995).

The general aim of this chapter is to consider how to take account of complex sampling designs in the fitting of such random effects models. We shall suppose that there is a known probability sampling scheme employed to select the sample of individuals followed over the waves of the survey. Two additional complications will be that there may be wave nonresponse, so that not all sampled individuals will respond at each wave, and that the target population of individuals may not be fixed over time.

To provide a specific focus, we will consider data on earnings of male employees over the first five waves of the British Household Panel Survey (BHPS), that is over the period 1991–5. As a basic model for the log earnings y_{it} of individual i at wave $t(= 1, \ldots, T)$ we shall suppose that

$$y_{it} = \beta_t + u_i + v_{it}, \qquad t = 1, \ldots, T \qquad (14.2)$$

where the random effect u_i is the 'permanent' random effect, referred to earlier, and the v_{it} are transitory random effects, whose effects on the response variable may last beyond the current wave t via the first-order autoregressive (AR(1)) model:

$$v_{it} = \rho v_{it-1} + \varepsilon_{it}, \qquad t = 1, \ldots, T. \qquad (14.3)$$

Both u_i and v_{it} may include the effects of measurement errors (Abowd and Card, 1989). The random variables u_i and ε_{it} are assumed to be mutually independent with

$$E(u_i) = E(\varepsilon_{it}) = 0, \qquad \text{var}(u_i) = \sigma_u^2, \qquad \text{var}(\varepsilon_{it}) = \sigma_\varepsilon^2.$$

The unknown fixed parameters $\beta_t (t = 1, \ldots, T)$ represent annual (inflation) effects. Lillard and Willis (1978) considered this model (amongst others) for log-earnings for seven years (1967–73) of data from the US Panel Study of Income Dynamics. Letting $\sigma_v^2 = \text{var}(v_{it})$ and assuming the ε_{it} and v_{it} are mutually independent and stationary, we obtain

$$\sigma_v^2 = \sigma_\varepsilon^2 / (1 - \rho^2). \qquad (14.4)$$

We refer to the above model as Model B and to the more restricted 'variance components' model in which $\rho = 0$ as Model A. See Goldstein, Healy and Rasbash (1994) for further discussion of such models.

We shall consider two broad approaches to fitting these models under a complex sample design. The first is a covariance structure approach, following Chamberlain (1982) and Skinner, Holt and Smith (1989, section 3.4.5, henceforth referred to SHS), in which the observations on the T waves are treated as a multivariate outcome with individuals as 'single-level' units. This approach is set out in Section 14.2. The second approach treats the data as two-level (Goldstein, 1995) with the level 1 units as the waves $t = 1, \ldots, T$ and the level 2 units as the individuals i. The aim is to apply the methods developed by Pfeffermann *et al.* (1998). This approach is set out in Section 14.3. A related approach is developed by Feder, Nathan and Pfeffermann (2000) for a model with time-varying random effects. The application of both our approaches to earnings data from the British Household Panel Survey will be considered in Section 14.4.

14.2. A COVARIANCE STRUCTURE APPROACH

Following the notation in Section 14.1, let $y_i = (y_{i1}, \ldots, y_{iT})'$ be the $T \times 1$ vector representing the profile of values of individual i over the T waves of the survey. Under the model defined by (14.2)–(14.4), these multivariate outcomes are independent with mean vector and covariance matrix given respectively by

$$E(y_i) = \beta = (\beta_1, \ldots, \beta_T)', \tag{14.5}$$

$$\mathrm{var}(y_i) = \sigma_u^2 J_T + \sigma_v^2 V_T(\rho), \tag{14.6}$$

where J_T is the $T \times T$ matrix of ones and $V_T(\rho)$ is the $T \times T$ matrix with the (tt')th element given by $\rho^{(t'-t)}$ $(1 \le t \le t' \le T)$.

These equations define a 'covariance structure' model in which the mean vector is unconstrained but the $k = T(T+1)/2$ distinct elements of the covariance matrix are constrained to be functions of the parameter vector $\theta = (\sigma_u^2, \sigma_e^2, \rho)'$. Inference about these parameters may follow the approach outlined in SHS (section 3.4.5).

Assuming first no nonresponse, let the data consist of the values y_i for units i in a sample s. The usual survey estimator of the finite population covariance matrix S is given by

$$\hat{S} = \sum_s w_i (y_i - \bar{y})(y_i - \bar{y})' / \sum_s w_i, \tag{14.7}$$

where

$$\bar{y} = \sum_s w_i y_i / \sum_s w_i,$$

and where w_i is the survey weight for individual i. Let $\hat{A} = \mathrm{vech}(\hat{S})$ denote the $k \times 1$ vector of distinct elements of $\hat{S}$ (the 'vector half' of $\hat{S}$: see Fuller, 1987, p. 382) and let $A(\theta) = \mathrm{vech}[\mathrm{var}(y_i)]$ denote the corresponding vector of elements of $\mathrm{var}(y_i)$ from (14.6).

Following Chamberlain (1982), Fuller (1984) and SHS (section 3.4.5), a general class of estimators of θ is obtained by minimising

$$\left[\hat{A} - A(\theta)\right]' V^{-1} \left[\hat{A} - A(\theta)\right] \tag{14.8}$$

where V is a given $k \times k$ non-singular matrix. A *generalised least squares* (GLS) estimator $\hat{\theta}_{GLS}$ is obtained by taking V to be a consistent estimator of the covariance matrix of $\hat{A}$. One choice, V_c, is obtained from the linearisation method (Wolter, 1985) by approximating the covariance matrix of the elements of $\hat{S}$ by the covariance matrix of the corresponding elements of the linear statistic

$$\sum_s z_i, \tag{14.9}$$

where $z_i = w_i[(y_i - \bar{y})(y_i - \bar{y})' - \hat{S}]/\sum_s w_i$ is treated as a fixed variable. The estimator V_c may allow in the usual way for the complex design (Wolter, 1985).

Since $A(\theta)$ is a non-linear function of θ, iterative minimisation of (14.8) is required. It may be noted that, for a given value of ρ under Model B, $A(\theta)$ is linear in (σ_u^2, σ_v^2) and so closed form expressions may be determined for the values $\hat{\sigma}_u^2(\rho)$ and $\hat{\sigma}_v^2(\rho)$, which minimise (14.8) for given ρ. The iterative minimisation may thus be reduced to a scalar problem. A consistent estimator of the covariance matrix of $\hat{\theta}_{GLS}$ is given by (Fuller, 1984)

$$V_L\left(\hat{\theta}_{GLS}\right) = \left[\dot{A}\left(\hat{\theta}_{GLS}\right)' V_c^{-1} \dot{A}\left(\hat{\theta}_{GLS}\right)\right]^{-1}, \tag{14.10}$$

where $\dot{A}(\theta) = \partial A(\theta)/\partial\theta$.

An advantage of the GLS approach is that it provides a ready-made goodness-of-fit test as the minimised value of the criterion in (14.8), namely the Wald statistic:

$$X_W^2 = \left[\hat{A} - A\left(\hat{\theta}_{GLS}\right)\right]' V_c^{-1} \left[\hat{A} - A\left(\hat{\theta}_{GLS}\right)\right]. \tag{14.11}$$

If the model is correct and if the sample is large enough for V_c to be a good approximation to the covariance matrix of $\hat{A}$, then X_W^2 should be distributed approximately as chi-squared with $k - q$ degrees of freedom, where $q = 2$ and 3 for Models A and B respectively.

One potential problem with the GLS estimator is that the covariance matrix estimator may be unstable if it is based on a relatively small number of degrees of freedom. This may lead to departures from the null distribution of X_W^2 assumed above. In this case, it may be preferable to consider alternative choices of V. One approach is to let V be an estimator of the covariance matrix of $\hat{A}$ based upon the (false) assumption that observations are independent and identically distributed. Thus, if we write

$$\hat{A} = \sum_s a_i, \tag{14.12}$$

where $a_i = \text{vech}(z_i)$ denotes the $k \times 1$ vector of distinct elements of z_i, then we may set V equal to

$$V_{iid} = n \sum_s (a_i - \bar{a})(a_i - \bar{a})'/(n-1), \qquad (14.13)$$

where $\bar{a} = \sum_s a_i/n$ and n denotes the sample size. Although V_{iid} may be more stable than a variance estimator which allows for the complex design, this choice of V is still correlated with $\hat{A}$ and, as discussed by Altonji and Segal (1996), may lead to serious bias in the estimation of θ. To avoid this problem, an even simpler approach is to set V equal to the identity matrix, when the estimator of θ obtained by minimising (14.8) may be viewed as an *ordinary least squares* (OLS) estimator. In both the cases when $V = V_{iid}$ and when V is the identity matrix, the resulting estimator $\hat{\theta}$ will still be consistent for θ but the Wald statistic X_W^2 will no longer follow a chi-squared distribution if the model is true. The large-sample distribution will instead be a mixture of chi-squared distributions and this may be approximated by a chi-squared distribution using one or two moment Rao–Scott approximations (SHS, Ch. 4). It is also no longer appropriate to use expression (14.10) to obtain standard errors for the elements of $\hat{\theta}$. Instead, as noted in SHS (Ch. 3), a consistent estimator of the covariance matrix of $\hat{\theta}$ is

$$V(\hat{\theta}; V = V_0) = [\dot{A}(\hat{\theta})' V_0^{-1} \dot{A}(\hat{\theta})]^{-1} [\dot{A}(\hat{\theta})' V_0^{-1} V_c V_0^{-1} \dot{A}(\hat{\theta})][\dot{A}(\hat{\theta})' V_0^{-1} \dot{A}(\hat{\theta})]^{-1},$$

where V_0 is the specified choice of V (V_{iid} or the identity matrix) used to determine $\hat{\theta}$ and V_c is a consistent estimator of the covariance matrix of $\hat{A}$ under the complex design. Note that this expression reduces to (14.10) when $V_0 = V_c$.

The approach considered so far in this section is based on the estimated covariance matrix $\hat{S}$ in (14.7) and assumes no nonresponse. This is an unrealistic assumption. The simplest way of handling nonresponse is to consider only those individuals who respond on all T waves, the so-called 'attrition sample', s_T, at wave T. For longitudinal surveys, designed for multipurpose longitudinal analyses, it is common to construct longitudinal weights w_{it} at each wave t, which are appropriate for longitudinal analysis based upon data for the attrition sample s_t of individuals who respond up to wave t (Lepkowski, 1989). Thus, the simplest approach is to use only data from attrition sample s_T and to replace the weights w_i, e.g. in (14.7), by the weights w_{iT}.

A more sophisticated approach, aimed at producing more efficient estimates, uses data from all attrition samples $s_1, \ldots, s_T$. A recursive approach to the estimation of the covariance matrix of y_i may then be developed. Let $y_i^{(t)} = (y_{i1}, \ldots, y_{it})'$ and let $\hat{S}^{(t)}$ denote the estimated $t \times t$ covariance matrix of $y_i^{(t)}$. Begin the recursion by setting

$$\hat{S}^{(1)} = \sum_{s_1} w_{i1}(y_{i1} - \bar{y}_1)^2 / \sum_{s_1} w_{i1},$$

where

$$\bar{y}_1 = \sum_{s_1} w_{i1} y_{i1} / \sum_{s_1} w_{i1}$$

as in (14.7). At the tth step of the recursion $(t = 2, \ldots, T)$ set the $(t-1) \times (t-1)$ submatrix of $\hat{S}^{(t)}$ corresponding to $y_i^{(t-1)}$ equal to $\hat{S}^{(t-1)}$. Let $b^{(t)}$ be the vector of weighted regression coefficients of y_{it} on $y_i^{(t-1)}$ given by

$$b^{(t)} = \left[\sum_{s_t} w_{it} (y_i^{(t-1)} - \bar{y}^{(t-1)})(y_i^{(t-1)} - \bar{y}^{(t-1)})' \right]^{-1} \sum_{s_t} w_{it} (y_i^{(t-1)} - \bar{y}^{(t-1)}) y_{it}$$

where

$$\bar{y}^{(t-1)} - \sum_{s_t} w_{it} y_i^{(t-1)} / \sum_{s_t} w_{it}$$

Then set the (tt)th element of $\hat{S}^{(t)}$, corresponding to the variance of y_{it}, equal to

$$\hat{\sigma}_{et}^2 + b^{(t)'} \hat{S}^{(t-1)} b^{(t)},$$

where

$$\hat{\sigma}_{et}^2 = \sum_{s_t} w_{it} (e_{it} - \bar{e}_t)^2 / \sum_{s_t} w_{it}, \quad e_{it} = y_{it} - y_i^{(t-1)'} b^{(t)}, \quad \bar{e}_t = \sum_{s_t} w_{it} e_{it} / \sum_{s_t} w_{it}.$$

Finally, let $\hat{S}_{t,\,t-1}^{(t)}$ denote the $1 \times (t-1)$ vector of remaining elements of $\hat{S}^{(t)}$ corresponding to the covariances between y_{it} and $y_i^{(t-1)}$ and let

$$\hat{S}_{t,\,t-1}^{(t)} = b^{(t)'} \hat{S}^{(t-1)}.$$

The recursive process is repeated for $t = 2, \ldots, T$. If y_i is multivariate normal and there are no weights the resulting $\hat{S}^{(t)}$ is a maximum likelihood estimator (Holt, Smith and Winter, 1980) for data from the set of attrition samples. In general, the estimator may be viewed as a form of pseudo-likelihood estimator (see Chapter 2). If the weights do not vary greatly, if y_i is approximately multivariate normal and the observations for most individuals fall into one of the attrition samples, the estimator $\hat{S}^{(t)}$ may be expected to be fairly efficient.

Weighting can become unwieldy if it is attempted to adjust for all possible wave nonresponse patterns in addition to the attrition samples. See, for example, Lepkowski (1989) for further discussion. For a more general discussion of inference in the presence of nonresponse see Chapter 18. We return in Section 14.4 to the application of the methods discussed in this section.

14.3. A MULTILEVEL MODELLING APPROACH

A second approach to handling complex survey designs in the fitting of the models defined in Section 14.1 is by adapting standard approaches, such as iterative generalised least squares (IGLS), used for fitting random effects models (Goldstein, 1995). Pfeffermann *et al.* (1998) have considered modifying

IGLS estimation using an approach analogous to the pseudo-likelihood method (see Chapter 2) for a model of the form (14.2), where the v_{it} are not serially correlated. Here we consider the extension of their approach to a longitudinal context, allowing for serial correlation. A potential advantage of this approach is that covariates may be handled more directly in the model. A potential disadvantage is that goodness-of-fit tests are not generated so directly.

In multilevel modelling terminology (Goldstein, 1995), the individuals are the level 2 units and the repeated measurements at the different waves represent level 1 units. Pfeffermann *et al.* (1998) allow for a two-stage sampling scheme, whereby the level 2 units i are selected with inclusion probabilities π_i and the level 1 units t with inclusion probabilities $\pi_{t|i}$ conditional on level 2 unit i being selected. Weights w_i and $w_{t|i}$ are then constructed equal to the reciprocals of these respective probabilities, which are assumed known. To adapt this approach to our context of longitudinal surveys subject to wave nonresponse, it seems natural to let π_i denote the probability that individual i is sampled and $\pi_{t|i}$ the probability that this individual responds at wave t. While we may reasonably suppose that the π_i are known, it is not straightforward to estimate the $\pi_{t|i}$ for general patterns of wave nonresponse (as noted in the covariance structure approach of Section 14.2). We therefore restrict attention to estimation using only the data derived from the attrition samples s_t. As noted in Section 14.2, it is common for longitudinal weights w_{it} to be available for use with these attrition samples and we shall suppose here that these approximate $(\pi_i \pi_{t|i})^{-1}$. We may then set w_i equal to the design weight π_i^{-1} and $w_{t|i}$ equal to w_{it}/w_i. Alternatively, given $w_{i1}, \ldots, w_{iT}$, we may set $w_i = w_{i1}$ and $w_{t|i} = w_{it}/w_{i1}$ ($t = 1 \ldots T$). Note, in particular, that in this case $w_{1|i} = 1$ for all i. This approach treats the sample selection and the response process at the first wave as a common selection process. In the approach of Pfeffermann *et al.* (1998), correction for bias by weighting tends to be more difficult at level 1 than at level 2, because there tends to be more non-linearity in the IGLS estimator as a function of level 1 sums than of level 2 sums. Hence setting $w_i = w_{i1}$ may be preferable to setting $w_i = \pi_i^{-1}$ because the resulting $w_{t|i}$ may be less variable and closer to one.

Having then constructed the weights w_i and $w_{t|i}$, the approach of Pfeffermann *et al.* (1998) may be applied to fit a model of form (14.2) where the v_{it} are not serially correlated. This is Model A. The basic approach is to modify the IGLS estimation procedure by weighting all sums over i by the weights w_i and weighting all sums over t by the weights $w_{t|i}$.

Often survey weights are only available in a scaled form; for example, so that they sum to the sample size. For inference about many regression-type models, as in Parts B and C of this book, estimation procedures for the model parameters are invariant to such scaling. Although this is also true for multilevel modelling if the w_i are scaled, it is not true if the weights $w_{t|i}$ are scaled. Pfeffermann *et al.* (1998) took advantage of this fact to choose a scaling to minimise small-sample estimation bias. In our context we consider scaling the weights $w_{t|i}$ to construct the scaled weights $w_{t|i}^*$ as

$$w_{t|i}^* = t^*(i)w_{t|i}\bigg/\left[\sum_{t=1}^{t^*(i)} w_{t|i}\right]$$

where $t^*(i)$ is the last wave at which individual i responds ($1 \le t^*(i) \le T$). Hence the average weight $w_{t|i}^*$ for individual i across waves $1, \ldots, t^*(i)$ is equal to one.

We now consider the question of how to adapt the approach of Pfeffermann *et al.* (1998) to allow for possible serial correlation of the v_{it} in Model B. We follow an approach similar to that in Hsiao (1986, section 3.7), which is based on observing that if we know ρ then Model B may be transformed to the form of Model A by

$$y_{it} - \rho y_{it-1} = (\beta_t - \rho\beta_{t-1}) + (1 - \rho)u_i + \varepsilon_{it}. \tag{14.14}$$

The estimation procedure involves two steps:

Step 1. Eliminate the random effect u_i by differencing the responses y_{it}

$$\Delta_{it} = y_{it} - y_{it-1}, \quad i \in s_t, \ t = 2, \ldots, T$$

and estimate the linear regression model

$$\Delta_{it} = \delta_t + \gamma\Delta_{it-1} + \eta_{it}$$

by OLS weighted by the weights w_{it} for observations i in the attrition samples $s_t(t = 2, \ldots, T)$, where the parameters δ_t are unconstrained. Under Model B, the least squares estimator $\hat{\gamma}$ of γ is consistent for

$$\gamma = \text{cov}(\Delta_{it}, \Delta_{it-1})/\text{var}(\Delta_{it-1}) \quad t = 2, \ldots, T$$
$$= [-(1 - \rho)\sigma_\varepsilon^2/(1 + \rho)]/[2\sigma_\varepsilon^2/(1 - \rho)]$$
$$= -(1 - \rho)/2.$$

Set $\hat{\rho} = 1 + 2\hat{\gamma}$.

Step 2. Let $\tilde{y}_{it} = y_{it} - \hat{\rho}y_{it-1}$ and fit the model obtained from (14.14) for the transformed data:

$$\tilde{y}_{it} = \tilde{\beta}_t + \tilde{u}_i + \tilde{\varepsilon}_{it} \tag{14.15}$$

using the approach of Pfeffermann *et al.* (1998) with the assumptions of Model A applying to the model in (14.15). The estimated variance of $\tilde{u}_i$ is then divided by $(1 - \hat{\rho})^2$ to obtain the estimate $\hat{\sigma}_u^2$.

This two-step approach produces consistent estimators of the parameters of Model B but the resulting standard errors of $\hat{\sigma}_u^2$ and $\hat{\sigma}_\varepsilon^2$ will not allow for uncertainty in the estimation of ρ.

Finally, we note that Pfeffermann *et al.* (1998) only allowed for the sample to be clustered into level 2 units. In the application in Section 14.4 the sampling design will also lead to geographical clustering of the sample individuals into primary sampling units. The procedure for standard error estimation proposed by Pfeffermann *et al.* (1998) therefore needs to be extended to handle this case. We shall not, however, consider this extension here, presenting only point estimates for the multilevel modelling approach in the next section.

14.4. AN APPLICATION: EARNINGS OF MALE EMPLOYEES IN GREAT BRITAIN

In this section we apply the approaches set out earlier to fit random effects models to longitudinal data on the monthly earnings of male full-time employees in Great Britain for the period 1991–5, using data from the British Household Panel Study (BHPS). The BHPS is a household panel survey, based on a sample of around 10 000 individuals. Data were first collected in 1991 and successive waves have taken place annually (Berthoud and Gershuny, 2000).

We base our analysis on the work of Ramos (1999). Like him, we consider only men over the first five waves of the BHPS and divide the men into four age cohorts in order to control for life cycle effects. These cohorts consist of men (i) born before 1941, (ii) born between 1941 and 1950, (iii) born between 1951 and 1960 and (iv) born after 1960.

The variable y is taken as the logarithm of earnings, with earnings being defined as the usual monthly earnings or salary payment before tax, for a reference period determined in the survey. We avoid the problem of zero earnings by defining the target population at wave t to consist of those men in the age cohorts who have positive earnings. It is thus possible for individuals to move in and out of the target population between waves. It is clearly plausible that the earnings behaviour of those moving in and out of the target population will differ systematically from those remaining in the target population. For simplicity, we shall, however, assume that the models defined in Section 14.1 apply to all individuals when they have positive earnings.

The panel sample was selected by stratified multistage sampling, with postal sectors as primary sampling units (PSUs). We use the standard linearisation approach to variance estimation for stratified multistage samples (e.g. SHS, p. 50). The BHPS involves a stratified sample of 250 PSUs. For the purpose of variance estimation, we approximate this stratified design as being defined by 75 strata, obtained by first breaking down each of 18 regional strata into 2 or 3 'major strata', defined according to proportion of 'head of households' in professional/managerial positions, and then by breaking down each of these major strata into 2 'minor strata', defined according to the proportion of the population of pensionable age.

We first assess the fit of Models A and B (defined in Section 14.1) for each of the four cohorts. The results are presented in Table 14.1. We use goodness-of-fit tests based on the covariance structure approach of Section 14.2, with three choices of the matrix V in (14.8):

Table 14.1   Goodness-of-fit test statistics for Models A and B for four cohorts and three estimation methods.

	Model A			Model B		
Cohort (when born)	OLS	GLS (iid)	GLS (complex)	OLS	GLS (iid)	GLS (complex)
Before 1941	11.3	13.0	15.1	9.2	8.7	10.0
1941–50	41.2^b	39.0^b	39.9^b	28.4^b	27.0^b	29.5^b
1951–60	17.2	39.0^b	43.3^b	6.5	15.5	16.5
1960	29.1^b	37.4^b	35.5^b	15.8	16.7	17.7

Notes: 1. Test statistics are weighted and are referred to the chi-squared distribution with 13 df for Model A and 12 df for Model B.
2. ^{a} significant at 5% level; ^{b} significant at 1% level.
3. OLS and GLS (iid) test statistics involve Rao–Scott first-order correction.

OLS . $V = I$, the identity matrix;
GLS (iid): $V = V_{iid}$, as defined in (14.13);
GLS (complex): $V = V_c$, the linearisation estimator of the covariance matrix of $\hat{A}$, based upon (14.9), allowing for the complex design.

For $V = V_c$, the test statistic is given by X_W^2 in (14.11) with the null distribution indicated in Section 14.2. For $V = I$ or V_{iid}, the values of $\hat{\theta}_{GLS}$ and V_c in (14.11) are replaced by the corresponding values of $\hat{\theta}$ and V and a first-order Rao–Scott adjustment is applied to the test statistic (SHS, Ch. 4). The same null distributions as for V_c are used. Test statistics based upon second-order Rao–Scott approximations were also calculated and led to similar results. All of the test statistics are based on data from the attrition sample s_5 at wave 5, for individuals who gave full interviews at each of the five waves. Longitudinal weights w_{i5} were used, which allow both for unequal sampling probabilities and for differential attrition from nonresponse over the five waves. To allow for the changing population, the expression for the estimated covariance matrix in (14.7) was modified by including only those who reported positive earnings at each wave in the estimation of the covariance between the log earnings at two waves.

The values of the test statistics in Table 14.1 are referred to a chi-squared null distribution with 13 degrees of freedom in the case of Model A and with 12 degrees of freedom in the case of Model B. The results suggest that Model A provides an adequate fit for the cohort born before 1941 but not for the other cohorts and that Model B provides an adequate fit for all cohorts, except the one consisting of those born between 1941 and 1950.

The values of the test statistics vary according to the three choices of V. The differences between the values of the test statistics for the GLS (iid) and GLS (complex) choices of V are not large, reflecting the fact that there is a large number of degrees of freedom for estimating the covariance matrix of $\hat{A}$ (relative to the dimension of the matrix) and that the pairs of V matrices tend not to be dramatically disproportionate. The value of the test statistic with V as the identity matrix suggests a much better fit of both Models A and B for the 1951–60 cohort and a somewhat better fit for the cohort born after 1960. This may be because this test statistic tends to be sensitive to different deviations from the null hypothesis than the GLS test statistics. The 1951–60 cohort is distinctive in having less variation among the estimated variances of log earnings over the five waves and, more generally, displays the least evidence of non-stationarity. Because of the high positive correlation between the elements of $\hat{A}$, the test statistic with V as the identity matrix may be expected to attach greater 'weight' to such departures from Model A than the GLS test statistics and this may lead to the noticeable difference in values for the 1951–60 cohort. Strong graphical evidence against Model A for this cohort is provided by Figure 14.1. This figure plots the elements $\hat{S}_{tt'}$ of $\hat{S}$ in (14.3) against $|t - t'|$ and there is a clear tendency for the covariances to decline as the number of years between waves increases. This suggests that the insignificant value of the test statistic for Model A, with V as the identity matrix, reflects lack of power.

Estimates of the parameters in Model B are presented in Table 14.2 for the three cohorts for which Model B shows no significant lack of fit in Table 14.1. Estimates are presented for the same three choices of V matrix as in Table 14.1. While the estimates based on the two GLS choices of V are fairly similar, the OLS estimates, with V as the identity matrix, can be noticeably different, especially for the 1951–60 cohort. The effect of the differences for the cohort born after 1960 is illustrated in Figure 14.2, in which the estimated variances and covariances from (14.7) are presented together with fitted lines, joining the variances and covariances under Model B, implied by the parameter estimates in Table 14.2. The lines for the GLS choices of V are surprisingly low, unlike the OLS line, which passes through the middle of the points. Similar underfitting of the variances and covariances occurs for the other cohorts and this finding may reflect downward bias in such estimates employing

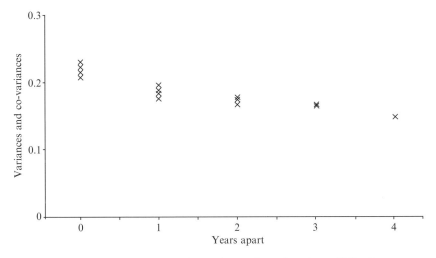

Figure 14.1   Estimated variances and covariances for cohort born 1951–60.

Table 14.2 Parameter estimates for Model B for three cohorts using covariance structure approach.

Cohort (when born)	Estimator	Parameter		
		ρ	σ_u^2	σ_ε^2
Before 1941	OLS	0.37 (0.16)	0.165 (0.028)	0.049 (0.018)
	GLS (iid)	0.35 (0.16)	0.150 (0.024)	0.034 (0.011)
	GLS (complex)	0.32 (0.13)	0.143 (0.022)	0.034 (0.009)
1951–60	OLS	0.56 (0.11)	0.146 (0.021)	0.048 (0.015)
	GLS (iid)	0.85 (0.09)	0.109 (0.047)	0.026 (0.047)
	GLS (complex)	0.85 (0.09)	0.106 (0.044)	0.026 (0.045)
After 1960	OLS	0.49 (0.08)	0.155 (0.018)	0.071 (0.014)
	GLS (iid)	0.41 (0.07)	0.154 (0.016)	0.063 (0.010)
	GLS (complex)	0.40 (0.07)	0.150 (0.016)	0.061 (0.009)

Notes: 1. Standard errors in parentheses.
2. Estimates are weighted and based only on data for attrition sample at wave 5.
3. 1941–50 cohort is excluded because of lack of fit of Model B in Table 14.1.

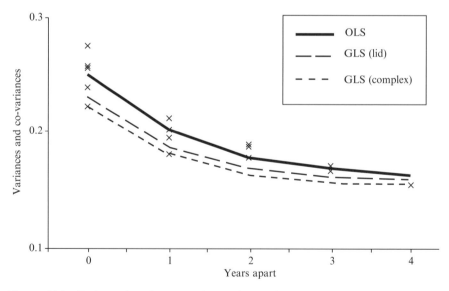

Figure 14.2 Estimated variances and covariances for cohort born after 1960 with values fitted under Model B.

sample-based V matrices, as discussed, for example, by Altonji and Segal (1996) and Browne (1984). The inversion of V implies that the lowest variances tend to receive most 'weight', leading to the fitted line following more the lower envelope of the points than the centre of them. The potential presence of non-negligible bias suggests that choosing V as the identity matrix may be preferable here for the purpose of parameter estimation, as concluded by Altonji and Segal (1996).

Table 14.3 shows for one cohort the effects of weighting, of the use of data from all attrition samples and of the use of the multilevel modelling approach of Section 14.3.

For the covariance structure approach, the impact of weighting is similar for all three choices of the matrix V. The fairly modest impact of weighting is expected here, since the BHPS weights do not vary greatly and are not strongly related to earnings.

The impact of using data from all attrition samples $s_1, \ldots, s_5$, not just from s_5, appears to be a little more marked than the impact of weighting. This may reflect the fact that the earnings behaviour of those men who leave the sample before 1995 may be different from those who remain in the sample for all five waves. In particular, this behaviour may be less stable leading to a reduction in the estimated correlation ρ. Control for possible informative attrition might be attempted by including covariates in the model.

Table 14.3 Parameter estimates for Model B for cohort born after 1960.

Estimator	Parameter		
	ρ	σ_u^2	σ_ε^2
Covariance structure approach			
Using attrition sample at wave 5 only			
Weighted			
OLS	0.49	0.155	0.071
GLS (iid)	0.41	0.154	0.063
GLS (complex)	0.40	0.150	0.061
Unweighted			
OLS	0.45	0.166	0.078
GLS (iid)	0.37	0.161	0.068
GLS (complex)	0.35	0.156	0.066
Using all five attrition samples (weighted)			
OLS	0.36	0.169	0.052
GLS (iid)	0.38	0.158	0.048
GLS (complex)	0.30	0.155	0.047
Multilevel modelling approach			
Using attrition sample at wave 5 only			
Weighted unscaled	0.41	0.169	0.041
Weighted scaled	0.41	0.167	0.042
Unweighted	0.41	0.170	0.045
Using all five attrition samples			
Weighted unscaled	0.43	0.167	0.043
Weighted scaled	0.43	0.163	0.045
Unweighted	0.43	0.165	0.047

The results for the multilevel modelling approach in Table 14.3 are based upon the two-step method described in Section 14.3. The estimated value of ρ is first determined and then estimates of $\hat{\sigma}_u^2$ and $\hat{\sigma}_\varepsilon^2$ are obtained by the method of Pfeffermann *et al.* (1998) either with or without weights and, in the former case, the weights may be scaled or not.

The impact of weighting on the multilevel approach is again modest, indeed somewhat more modest than for the covariance structure approach. This may be because a common estimate of ρ is used here. Scaling the weights also has little effect. This may be because all the weights $w_{t|i}$ are fairly close to one in this application and thus scaling has less of an impact than in the two-stage sampling application in Pfeffermann *et al.* (1998).

The differences between the estimates from the covariance structure approach and the corresponding multilevel modelling approaches are not especially large in Table 14.3 relative to the standard errors in Table 14.2. Nevertheless, across all four cohorts and both models, the main differences in the estimates between methods were between the three choices of V matrix for the covariance structure approach and between the covariance structure and the multilevel approaches. The impact of weighting and the scaling of weights tended to be less important.

14.5. CONCLUDING REMARKS

It is often useful to include random effects in the specification of models for longitudinal survey data. In this chapter we have considered two approaches to allowing for complex survey designs and sample attrition when fitting such models. The covariance structure approach is particularly natural with survey data. The complex survey design and attrition are allowed for when making inference about the covariance matrix of the longitudinal responses. Modelling of the structure of this matrix may then proceed in a standard way. The second approach is to adapt standard multilevel modelling procedures, extending the approach of Pfeffermann *et al.* (1998).

The two approaches may be compared in a number of ways:

• The multilevel approach incorporates the different attrition samples more directly, although the possible creation of bias with unequal $w_{t|i}$ (for given i) with small numbers of level 1 units (i.e. small T), as discussed by Pfeffermann *et al.* (1998), may be a problem.
• The multilevel approach incorporates covariates more naturally, although the extension of the covariance structure approach to include covariates using LISREL models is well established.
• The covariance structure approach handles serial correlation more easily.
• The covariance structure approach generates goodness-of-fit tests and residuals at the level of variances and covariances. The multilevel approach generates unit level residuals.

Finally, our application of the covariance structure approach to the BHPS data showed evidence of bias in the estimation of the variance components when using GLS with a covariance matrix V estimated from the data. This accords with the findings of Altonji and Segal (1996). This evidence suggests that it is safer to specify V as the identity matrix and use Rao – Scott adjustments for testing.

CHAPTER 15

Event History Analysis and Longitudinal Surveys

J. F. Lawless

15.1. INTRODUCTION

Event history analysis as discussed here deals with events that occur over the lifetimes of individuals in some population. For example, it can be used to examine educational attainment, employment, entry into marriage or parenthood, and other matters. In epidemiology and public health it can be used to study the relationship between the incidence of diseases and environmental, dietary, or economic factors. The main objectives of analysis are to model and understand event history processes of individuals. The timing, frequency and pattern of events are of interest, along with factors associated with them. 'Time' is often the age of an individual or the elapsed time from some event other than birth: for example, the time since a person married or the time since a disease was diagnosed. Occasionally, 'time' may refer to some other scale than calendar time.

Two closely related frameworks are used to describe and analyze event histories: the multi-state and event occurrence frameworks. In the former a finite set of states $\{1, 2, \ldots, K\}$ is defined such that at any time an individual occupies a unique state, for example employed, unemployed, or not in the labour force. In the latter the occurrences of specific types of events are emphasized. The two frameworks are equivalent since changes of state can be considered as types of events, and vice versa. This allows a unified statistical treatment but for description and interpretation we usually select one point of view or the other. Event history analysis includes as a special case the area of survival analysis. In particular, times to the occurrence of specific events (from a well-defined time origin), or the durations of sojourns in specific states, are often referred to as survival or duration times. This area is well developed (e.g. Kalbfleisch and Prentice, 2002; Lawless, 2002; Cox and Oakes, 1984).

Analysis of Survey Data Edited by R. L. Chambers and C. J. Skinner
© 2003 John Wiley & Sons, Ltd

Event history data typically consist of information about events and covariates over some time period, for a group of individuals. Ideally the individuals are randomly selected from a population and followed over time. Methods of modelling and analysis for such cohorts of closely monitored individuals are also well developed (e.g. Andersen *et al.*, 1993; Blossfeld, Hamerle and Mayer, 1989). However, in the case of longitudinal surveys there may be substantial departures from this ideal situation.

Large-scale longitudinal surveys collect data on matters such as health, fertility, educational attainment, employment, and economic status at successive interview or follow-up times, often spread over several years. For example, Statistics Canada's Survey of Labour and Income Dynamics (SLID) selects panels of individuals and interviews them once a year for six years, and its National Longitudinal Survey of Children and Youth (NLSCY) follows a sample of children aged 0–11 selected in 1994 with interviews every second year. The fact that individuals are followed longitudinally affords the possibility of studying individual event history processes. Problems of analysis can arise, however, because of the complexity of the populations and processes being studied, the use of complex sampling designs, and limitations in the frequency and length of follow-up. Missing data and measurement error may also occur, for example in obtaining information about individuals prior to their time of enrolment in the study or between widely spaced interviews. Attrition or losses to follow-up may be nonignorable if they are associated with the process under study.

This chapter reviews event history analysis and considers issues associated with longitudinal survey data. The emphasis is on individual-level explanatory analysis so the conceptual framework is the process that generates individuals and their life histories in the populations on which surveys are based. Section 15.2 reviews event history models, and section 15.3 discusses longitudinal observational schemes and conventional event history analysis. Section 15.4 discusses analytic inference from survey data. Sections 15.5, 15.6, and 15.7 deal with survival analysis, the analysis of event occurrences, and the analysis of transitions. Section 15.8 considers survival data from a survey and Section 15.9 concludes with a summary and list of areas needing further development.

15.2. EVENT HISTORY MODELS

The event occurrence and multi-state frameworks are mathematically equivalent, but for descriptive or explanatory purposes we usually adopt one framework or the other. For the former, we suppose J types of events are defined and for individual i let

$$Y_{ij}(t) = \text{number of occurrences of event type } j \text{ up to time } t. \qquad (15.1)$$

Covariates may be fixed or vary over time and so we let $x_i(t)$ denote the vector of all (fixed or time-varying) covariates associated with individual i at time t.

In the multi-state framework we define

$$Y_i(t) = \text{state occupied by individual } i \text{ at time } t, \tag{15.2}$$

where $Y_i(t)$ takes on values in $\{1, 2, \ldots, K\}$. Both (15.1) and (15.2) keep track of the occurrence and timing of events; in practice the data for an individual would include the times at which events occur, say $t_{i1} \le t_{i2} \le t_{i3} \le \ldots$, and the type of each event, say $A_{i1}, A_{i2}, A_{i3}, \ldots$. The multi-state framework is useful when transitions between states or the duration of spells in a state are of interest. For example, models for studying labour force dynamics often use states defined as: 1 – Employed, 2 – Unemployed but in the labour force, 3 – Out of the labour force. The event framework is convenient when patterns or numbers of events over a period of time are of interest. For example, in health-related surveys we may consider occurrences such as the use of hospital emergency or outpatient facilities, incidents of disease, and days of work missed due to illness.

Stochastic models for either setting may be specified in terms of event intensity functions (e.g. Andersen *et al.*, 1993). Let $H_i(t)$ denote the history of all events and covariates relevant to individual i, up to but not including time t. We shall treat time as a continuous variable, but discrete versions of the results below can also be given. The intensity function for a type j event ($j = 1, \ldots, J$) is then defined as

$$\lambda_{ij}(t|x_i(t), H_i(t)) = \lim_{\Delta t \to 0} \frac{Pr\{Y_{ij}[t, t + \Delta t) = 1|x_i(t), H_i(t)\}}{\Delta t}, \tag{15.3}$$

where $Y_{ij}[s, t) = Y_{ij}(t^-) - Y_{ij}(s^-)$ is the number of type j events in the interval $[s, t)$. That is, the conditional probability of a type j event occurring in $[t, t + \Delta t)$, given covariates and the prior event history, is approximately $\lambda_{ij}(t|x_i(t), H_i(t))\Delta t$ for small Δt.

For multi-state models there are correspondingly transition intensity functions,

$$\lambda_{ikl}(t|x_i(t), H_i(t)) = \lim_{\Delta t \to 0} \frac{Pr\{Y_i(t + \Delta t) = \ell|Y_i(t^-) = k, x_i(t), H_i(t)\}}{\Delta t}, \tag{15.4}$$

where $k \neq \ell$ and both k and ℓ range over $\{1, \ldots, K\}$.

If covariates are 'external' and it is assumed that no two events can occur simultaneously then the intensities specify the full event history process, conditional on the covariate historics. External covariates are ones whose values are determined independently from the event processes under study (Kalbfleisch and Prentice, 2002, Ch. 6). Fixed covariates are automatically external. 'Internal' covariates are more difficult to handle and are not considered in this chapter; a joint model for event occurrence and covariate evolution is generally required to study them.

Characteristics of the event history processes can be obtained from the intensity functions. In particular, for models based on (15.3) we have (e.g. Andersen *et al.*, 1993) that

$Pr\{$No events over $[t, t+s)|H_i(t), x_i(u)$ for $t \leq u \leq t+s\}$

$$= \exp\left\{ -\int_t^{t+s} \sum_{j=1}^J \lambda_{ij}(u|H_i(u), x_i(u))du \right\}. \quad (15.5)$$

Similarly, for multi-state models based on (15.4) we have

$Pr\{$No exit from state k by $t+s|Y_i(t) = k, H_i(t), x_i(u)$ for $t \leq u \leq t+s\}$

$$= \exp\left\{ -\int_t^{t+s} \sum_{l \neq k} \lambda_{ikl}(u|H_i(u), x_i(u))du \right\}. \quad (15.6)$$

The intensity function formulation is very flexible. For example, we may specify that the intensities depend on features such as the times since previous events or previous numbers of events, as well as covariates. However, it is obvious from (15.5) and (15.6) that even the calculation of simple features such as state sojourn probabilities may be complicated. In practice, we often restrict attention to simple models, in particular, Markov models, for which (15.3) or (15.4) depend only on $x(t)$ and t, and semi-Markov models, for which (15.3) or (15.4) depend on $H(t)$ only through the elapsed time since the most recent event or transition, and on $x(t)$.

Survival models are important in their own right and as building blocks for more detailed analysis. They deal with the time T from some starting point to the occurrence of a specific event, for example an individual's length of life, the duration of their first marriage, or the age at which they first enter the labour force. The terms failure time, duration, and lifetime are common synonyms for survival time. A survival model can be considered as a transitional model with two states, where the only allowable transition is from state 1 to state 2. The transition intensity (15.4) from state 1 to state 2 can then be written as

$$\lambda_i(t|x_i(t)) = \lim_{\Delta t \to 0} \frac{Pr\{T_i < t + \Delta t|T_i \geq t, x_i(t)\}}{\Delta t}, \quad (15.7)$$

where T_i represents the duration of individual i's sojourn in state 1. For survival models, (15.7) is called the hazard function. From (15.6),

$$Pr(T_i > t|x_i(u) \text{ for } 0 \leq u \leq t) = \exp\left\{ -\int_0^t \lambda_i(u|x_i(u))du \right\}. \quad (15.8)$$

When covariates x_i are all fixed, (15.7) becomes $\lambda_i(t|x_i)$ and (15.8) is the survivor function

$$S_i(t|x_i) = \exp\left\{ -\int_0^t \lambda_i(u|x_i)du \right\}. \quad (15.9)$$

Multiplicative regression models based on the hazard function are often used, following Cox (1972):

$$\lambda_i(t|x_i) = \lambda_0(t) \exp(\beta' x_i) \quad (15.10)$$

is common, where $\lambda_0(t)$ is a positive function and β is a vector of regression coefficients of the same length as x.

Models for repeated occurrences of the same event are also important; they correspond to (15.3) with $J = 1$. Poisson (Markov) and renewal (semi-Markov) processes are often useful. Models for which the event intensity function is of the form

$$\lambda_i(t|H_i(t), x_i(t)) = \lambda_0(t)g(x_i(t)) \qquad (15.11)$$

are called modulated Poisson processes. Models for which

$$\lambda_i(t|H_i(t), x_i(t)) = \lambda_0(u_i(t))g(x_i(t)), \qquad (15.12)$$

where $u_i(t)$ is the elapsed time since the last event (or since $t = 0$ if no event has yet occurred), are called modulated renewal processes.

Detailed treatments of the models above are given in books on event history analysis (e.g. Andersen et al., 1993; Blossfeld, Hamerle and Mayer, 1989), survival analysis (e.g. Kalbfleisch and Prentice, 2002; Lawless, 2002; Cox and Oakes, 1984), and stochastic processes (e.g. Cox and Isham, 1980; Ross 1983). Sections 15.5 to 15.8 outline a few basic methods of analysis.

The intensity functions fully specify a process and allow, for example, prediction of future events or the simulation of individual processes. If the data collected are not sufficient to identify or fit such models, we may consider a partial specification of the process. For example, for recurrent events the mean function is $M(t) = E\{Y(t)\}$; this can be considered without specifying a full model (Lawless and Nadeau, 1995).

In many populations the event processes for individuals in a certain group or cluster may not be mutually independent. For example, members of the same household or individuals living in a specific region may exhibit association, even after conditioning on covariates. The literature on multivariate models or association between processes is rather limited, except for the case of multivariate survival distributions (e.g. Joe, 1997). A common approach is to base specification of covariate effects and estimation on separate working models for different components of a process, but to allow for association in the computation of confidence regions or tests (e.g. Lee, Wei and Amato, 1992; Lin, 1994; Ng and Cook, 1999). This approach is discussed in Sections 15.5 and 15.6.

15.3. GENERAL OBSERVATIONAL ISSUES

The analysis of event history data is dependent on two key points: How were individuals selected for the study? What information was collected about individuals, and how was this done? In longitudinal surveys panels are usually selected according to a complex survey design; we discuss this and its implications in Section 15.4. In this section we consider observational issues associated with a generic individual, whose life history we wish to follow.

We consider studies which follow a group or panel of individuals longitudin-ally over time, recording events and covariates of interest; this is referred to as *prospective follow-up*. Limitations on data collection are generally imposed by time, cost, and other factors. Individuals are often observed over a time period which is shorter than needed to obtain a complete picture of the process in question, and they may be seen or interviewed only sporadically, for example annually. We assume for now that event history variables $Y(t)$ and covariates $x(t)$ for an individual over the time interval $[\tau_0, \tau_1]$ can be determined from the available data. The time scale could be calendar time or something specific to the individual, such as age. In any case, τ_0 will not in general correspond to the natural or physical origin of the process $\{Y(t)\}$, and we denote relevant history about events and covariates up to time τ_0 by $H(\tau_0)$. (Here, 'relevant' will depend on what is needed to model or analyze the event history process over the time interval $[\tau_0, \tau_1]$; see (15.13) below.) The times τ_0 or τ_1 may be random. For example, an individual may be *lost to follow-up* during a study, say if they move and cannot be traced, or if they refuse to participate further. We some-times say that the individual's event history $\{Y(t)\}$ is *(right-)censored* at time τ_1 and refer to τ_1 as a *censoring time*. The time τ_0 is often random as well; for example, we may wish to focus on a person's history following the random occurrence of some event such as entry to parenthood.

The distribution of $\{Y(t):\tau_0 \leq t \leq \tau_1\}$, conditional on $H(\tau_0)$ and relevant covariate information $X = \{x(t), t \leq \tau_1\}$, gives a likelihood function on which inferences can be based. If τ_0 and τ_1 are fixed by the study design (i.e. are non-random) then for an event history process specified by (15.3), we have (e.g. Andersen *et al.*, 1993, Ch. 2)

$Pr\{r$ events in $[\tau_0, \tau_1]$ at times $t_1 < \cdots < t_r$, of types $j_1, \ldots, j_r | H(\tau_0)\}$

$$= \prod_{\ell=1}^{r} \lambda_{j\ell}(t_\ell | H(t_\ell)) \exp\left\{-\int_{\tau_0}^{\tau_1} \sum_{j=1}^{J} \lambda_j(u | H(u)) du\right\}, \quad (15.13)$$

where 'Pr' denotes the probability density. For simplicity we have omitted covariates in (15.13); their inclusion merely involves replacing $H(u)$ with $H(u)$, $x(u)$.

If τ_0 or τ_1 is random then under certain conditions (15.13) is still valid for inference purposes; in particular, this allows τ_0 or τ_1 to depend upon past but not future events. In such cases (15.13) is not necessarily the probability density of $\{Y(t):\tau_0 \leq t \leq \tau_1\}$ conditional on τ_0, τ_1, and $H(\tau_0)$, but it is a *partial likeli-hood*. Andersen *et al.* (1993, Ch. 2) give a rigorous discussion.

Example 1.   Survival times

Suppose that $T \geq 0$ represents a survival time and that an individual is ran-domly selected at time $\tau_0 \geq 0$ and followed until time $\tau_1 > \tau_0$, where τ_0 and τ_1 are measured from the same time origin as T. An illustration concerning the duration of breast feeding of first-born children is discussed in Section 15.8, and duration of marital unions is considered later in this section. Assuming that

$T \geq \tau_0$, we observe $T = t$ if $t \leq \tau_1$, but otherwise it is right-censored at τ_1. Let $y = \min(t, \tau_1)$ and $\delta = I(y = t)$ indicate whether t was observed. If $\lambda(t)$ denotes the hazard function (15.7) for T (for simplicity we assume no covariates are present) then the right hand side of (15.13) with $J = 1$ reduces to

$$L = \lambda(y)^\delta \exp\left\{ -\int_{\tau_0}^{y} \lambda(u)du \right\}. \tag{15.14}$$

The likelihood (15.14) is often written in terms of the density and survival functions for T:

$$L = \left[\frac{f(y)}{S(\tau_0)}\right]^\delta \left[\frac{S(y)}{S(\tau_0)}\right]^{1-\delta} \tag{15.15}$$

where $S(t) = \exp\{-\int_0^t \lambda(u)du\}$ as in (15.9), and $f(t) = \lambda(t)S(t)$. When $\tau_0 = 0$ we have $S(\tau_0) = 1$ and (15.15) is the familiar censored data likelihood (see e.g. Lawless, 2002, section 2.2). If $\tau_0 > 0$ then (15.15) indicates that the relevant distribution is that of T, given that $T \geq \tau_0$; this is referred to as *left-truncation*. This is a consequence of the implicit fact that we are following an individual for whom 'failure' has not occurred before the time of selection τ_0. Failure to recognize this can severely bias results.

Example 2. A state duration problem

Many life history processes can be studied as a sequence of durations in specified states. As a concrete example we consider the entry of a person into their first marital union (event E_1) and the dissolution of that union by divorce or death (event E_2). In practice we would usually want to separate dissolutions by divorce or death but for simplicity we ignore this; see Trussell, Rodriguez and Vaughan (1992) and Hoem and Hoem (1992) for more detailed treatments. Figure 15.1 portrays the process.

We might wish to examine the occurrence of marriage and the length of marriage. We consider just the duration S of marriage, for which important covariates might include the calendar time of the marriage, ages of the partners at marriage, and time-varying factors such as the births of children. Suppose that the transition intensity from state 2 to 3 as defined in (15.4) is of the form

$$\lambda_{23}(t|H(t), x(t)) = \lambda(t - t_1|x(t)), \tag{15.16}$$

where t_1 is the time (age) of marriage and $x(t)$ represents fixed and time-varying covariates. The function $\lambda(s|x)$ is thus the hazard function for S.

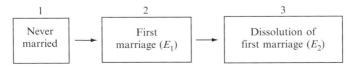

Figure 15.1   A model for first marriage.

Suppose that individuals are randomly selected and that an individual is followed prospectively over the time interval $[t_S, t_F]$. Figure 15.2 shows four different possibilities according to whether each of the events E_1 and E_2 occurs within $[t_S, t_F]$ or not. There may also be individuals for whom both E_1 and E_2 occurred before t_S and ones for whom E_1 does not occur by time t_F, but they contribute no information on the duration of marriage. By (15.13), the portion of the event history likelihood depending on (15.16) for any of cases A to D is

$$\lambda(y - t_1 | x(t_2))^\delta \exp\left\{ - \int_{\tau_0}^y \lambda(u - t_1 | x(u)) du \right\}, \tag{15.17}$$

where t_j is the time of event E_j ($j = 1, 2$), $\delta = I$ (event E_2 is observed), $\tau_0 = \max(t_1, t_S)$, and $y = \min(t_2, t_F)$. For all cases (15.17) reduces to the censored data likelihood (15.14) if we write $s = t_2 - t_1$ as the marriage duration and let $\lambda(u)$ depend on covariates. For cases C and D, we need to know the time $t_1 < t_S$ at which E_1 occurred. In some applications (but not usually in the case of marriage) the time t_1 might be unknown. If so an alternative to (15.17) must be sought, for example by considering $Pr\{E_2 \text{ occurs at } t_2 | E_1 \text{ occurs before } t_S\}$ instead of $Pr\{E_2 \text{ occurs at } t_2 | H(t_S)\}$, upon which (15.17) is based. This requires information about the intensity for events E_1, in addition to $\lambda(s|x)$. An alternative is to discard data for cases of type C and D. This is permissible and does not bias estimation for the model (15.16) (e.g. Aalen and Husebye, 1991; Guo, 1993) but often reduces the amount of information greatly.

Finally, we note that individuals could be selected differentially according to what state they are in at time t_S; this does not pose any problem as long as the probability of selection depends only on information contained in $H(t_S)$. For example, one might select only persons who are married, giving only data types C and D.

The density (15.13) and thus the likelihood function factors into a product over $j = 1, \ldots, J$ and so if intensity functions do not share common parameters,

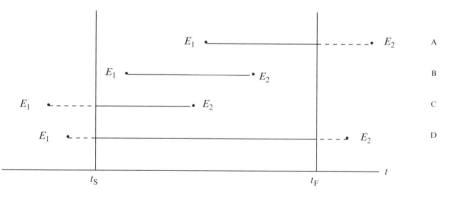

Figure 15.2 Observation of an individual re E_1 and E_2.

the events can be analyzed one type at a time. In the analysis of both multiple event data and survival data it has become customary to use the notation $(\tau_{i0}, y_i, \delta_i)$ introduced in Example 1. Therneau and Grambsch (2000) describe its use in connection with S-Plus and SAS procedures. The notation indicates that an individual is observed at risk for some specific event over the period $[\tau_{i0}, y_i]$; δ_i indicates whether the event occurred at y_i ($\delta_i = 1$) or whether no event was observed ($\delta_i = 0$).

Frequently individuals are seen only at periodic interviews or follow-up visits which are as much as one or two years apart. If it is possible to identify accurately the times of events and values of covariates through records or recall, the likelihoods (15.13) and (15.14) can be used. If information about the timing of events is unknown, however, then (15.13) or (15.14) must be replaced with expressions giving the joint probability of outcomes $Y(t)$ at the discrete time points at which the individual was seen; for certain models this is difficult (e.g. Kalbfleisch and Lawless, 1989). An important intermediate situation which has received little study is when information about events or covariates between follow-up visits is available, but subject to measurement error (e.g. Holt, McDonald and Skinner, 1991).

Right-censoring of event histories (at τ_1) is not a problem provided that the censoring process depends only on observable covariates or events in the past. However, if censoring depends on the current or future event history then observation is response selective and (15.13) is no longer the correct distribution of the observed data. For example, suppose that individuals are interviewed every year, at which time events over the past year are recorded. If an individual's nonresponse, refusal to be interviewed, or loss to follow-up is related to events during that year, then censoring of the event history at the previous year would depend on future events and thus violate the requirements for (15.13).

More generally, event or covariate information may be missing at certain follow-up times because of nonresponse. If nonresponse at a time point is independent of current and future events, given the past events and covariates, then standard missing data methods (e.g. Little and Rubin, 1987) may in principle be used. However, computation may be complicated, and modelling assumptions regarding covariates may be needed (e.g. Lipsitz and Ibrahim, 1996). Little (1992, 1995) and Carroll, Ruppert and Stefanski (1995) discuss general methodology, but this is an area where further work is needed.

We conclude this section with a remark about the retrospective ascertainment of information. There may in some studies be a desire to utilize portions of an individual's life history prior to their time of inclusion in the study (e.g. prior to t_S in Example 2) as responses, rather than simply as conditioning events, as in (15.13) or (15.17). This is especially tempting in settings where the typical duration of a state sojourn is long compared to the length of follow-up for individuals in the study. Treating past events as responses can generate selection effects, and care is needed to avoid bias; see Hoem (1985, 1989).

15.4. ANALYTIC INFERENCE FROM LONGITUDINAL SURVEY DATA

Panels in many longitudinal surveys are selected via a sample design that involves stratification and clustering. In addition, the surveys have numerous objectives, many of which are descriptive (e.g. Kalton and Citro, 1993; Binder, 1998). Because of their generality they may yield limited information about explanatory or causal mechanisms, but analytic inference about the life history processes of individuals is nevertheless an important goal.

Some aspects of analytic inference are controversial in survey sampling, and in particular, the use of weights (see e.g. Chapters 6 and 9); we consider this briefly. Let us drop for now the dependence upon time and write Y_i and x_i for response variables and covariates, respectively. It is assumed that there is a "superpopulation" model or process that generates individuals and their (y_i, x_i) values. At any given time there is a finite population of individuals from which a sample could be drawn, but in a process which evolves over time the numbers and make-up of the individuals and covariates in the population are constantly changing. Marginal or individual-specific models for the superpopulation process consider the distribution $f(y_i|x_i)$ of responses given covariates. Responses for individuals may not be (conditionally) independent, but for a complex population the specification of a joint model for different individuals is daunting, so association between individuals is often not modelled explicitly. For the survey, we assume for simplicity that a sample s is selected at a single time point at which there are N individuals in the finite population, and let $I_i = I(i \in s)$ indicate whether individual i is included in the sample. Let the vector z_i denote design-related factors such as stratum or cluster information, and assume that the sample inclusion probabilities

$$\pi_i = Pr(I_i = 1 | y_i, x_i, z_i), \quad i = 1, \ldots, N \tag{15.18}$$

depend only on the z_i.

The objective is inference about the marginal distributions $f(y_i|x_i)$ or joint distributions $f(y_1, y_2, \ldots | x_1, x_2, \ldots)$ based on the sample data $(x_i, y_i, i \in s; s)$. For convenience we use f to denote various density functions, with the distribution represented being clear from the arguments of the function. As discussed by Hoem (1985, 1989) and others the key issue is whether sampling is response selective or not. Suppose first that Y_i and z_i are independent, given x_i:

$$f(y_i|x_i, z_i) = f(y_i|x_i). \tag{15.19}$$

Then

$$Pr(y_i|x_i, I_i = 1) = f(y_i|x_i) \tag{15.20}$$

and if we are also willing to assume independence of the Y_i given s and the $x_i (i \in s)$, inference about $f(y_i|x_i)$ can be based on the likelihood

$$L = \prod_{i \in s} f(y_i|x_i), \tag{15.21}$$

for either parametric or semi-parametric model specifications. Independence may be a viable assumption when x_i includes sufficient information, but if it is not then an alternative to (15.21) must be sought. One option is to develop multivariate models that specify dependence. This is often difficult, and another approach is to base estimation of the marginal individual-level models on (15.21), with an adjustment made for variance estimation to recognize the possibility of dependence; we discuss this for survival analysis in Section 15.5.

If (15.20) does not hold then (15.21) is incorrect and leads to biased estimation of $f(y_i|x_i)$. Sometimes (e.g. see papers in Kasprzyk *et al.*, 1989) this is referred to as model misspecification, since (15.19) is violated, but that is not really the issue. The distribution $f(y_i|x_i)$ is well defined and, for example, if we use a non-parametric approach no strong assumptions about specification are made. The key issue is as identified by Hoem (1985, 1989): when (15.20) does not hold, sampling is response selective, and thus nonignorable, and (15.21) is not valid.

When (15.19) does not hold, one might question the usefulness of $f(y_i|x_i)$ for analytic inference. If we wish to consider it we might try to model the distribution $f(y_i|x_i, z_i)$ and obtain $Pr(y_i|x_i, I_i = 1)$ by marginalization. This is usually difficult, and a second approach is a pseudo-likelihood method that utilizes the known sample inclusion probabilities (see Chapter 2). If (15.20) and thus (15.21) are valid, the score function for a parameter vector θ specifying $f(y_i|x_i)$ is

$$U_L(\theta) = \sum_{i=1}^{N} I_i \frac{\partial \log f(y_i|x_i)}{\partial \theta}. \quad (15.22)$$

If (15.20) does not hold we consider instead the pseudo-score function (e.g. Thompson, 1997, section 6.2)

$$U_W(\theta) = \sum_{i=1}^{N} \frac{I_i}{\pi_i} \frac{\partial \log f(y_i|x_i)}{\partial \theta}. \quad (15.23)$$

If $E_{Y_i|x_i}\{\partial \log f(Y_i|x_i)/\partial\theta\} = 0$ then $E\{U_W(\theta)\} = 0$, where the expectation is now over the I_i, Y_i pairs in the population, given the x_i, and estimation of θ can be based on the equation $U_W(\theta) = 0$. Nothing is assumed here about independence of the terms in (15.23); this must be addressed when developing a distribution theory for estimation or testing purposes.

Estimation based on (15.23) is sometimes suggested as a general preference with the argument that it is 'robust' to superpopulation model misspecification. But as noted, when (15.19) fails the utility of $f(y_i|x_i)$ is questionable; to study individual-level processes, every attempt should be made to obtain covariate information which makes (15.20) plausible. Skinner, Holt and Smith (1989) and Thompson (1997, chapter 6) provide general discussions of analytic inference from surveys.

The pseudo-score (15.23) can be useful when π_i in (15.18) depends on y_i. Hoem (1985, 1989) and Kalbfleisch and Lawless (1988) discuss examples of response-selective sampling in event history analysis. This is an important topic, for example when data are collected retrospectively. Lawless, Kalbfleisch and Wild (1999) consider settings where auxiliary design information is available on individuals not included in the sample; in that case more efficient alternatives to (15.23) can sometimes be developed.

15.5.   DURATION OR SURVIVAL ANALYSIS

Primary objectives of survival analysis are to study the distribution of a survival time T given covariates x, perhaps in some subgroup of the population. The hazard function for individual i is given by (15.7) for continuous time models. Discrete time models are often advantageous; in that case we denote the possible values of T as $1, 2, \ldots$ and specify discrete hazard functions

$$\lambda_i(t|x_i) = Pr(T_i = t | T_i \geq t, x_i). \tag{15.24}$$

Then (15.9) is then replaced with (e.g. Kalbfleisch and Prentice, 2002, section 1.2)

$$S(t|x_i) = \prod_{u=1}^{t-1} [1 - \lambda(u|x_i)]. \tag{15.25}$$

Following the discussion in Section 15.4, key questions are whether the sampling scheme is response selective, and whether survival times can be assumed independent. We consider these issues for non-parametric and parametric methodology.

15.5.1.   Non-parametric marginal survivor function estimation

Suppose that a survival distribution $S(t)$ is to be estimated, with no conditioning on covariates. Let the vector z include sample design information and denote $S_z(t) = Pr(T_i \geq t | Z_i = z)$. Assume for simplicity that Z is discrete and let $P_z = P(Z_i = z)$ and $\pi(z) = Pr(I_i = 1 | Z_i = z)$; note that P_z is part of the superpopulation model. Then

$$S(t) = Pr(T_i \geq t) = \sum_z P_z S_z(t) \tag{15.26}$$

and

$$Pr(T_i \geq t | I_i = 1) = \frac{\sum_z \pi(z) P_z S_z(t)}{\sum_z \pi(z) P_z}. \tag{15.27}$$

It is clear that (15.26) and (15.27) are the same if $\pi(z)$ is constant, i.e. the design is self-weighting. In this case the sampling design is ignorable in the sense that (15.20) holds. The design is also ignorable if $S_z(t) = S(t)$ for all z. In that case the scientific relevance of $S(t)$ seems clear. More generally, with $S(t)$ given by (15.26), its relevance is less obvious; although $S(t)$ is well defined as a population average, it may be of limited interest for analytic purposes.

Estimates of $S(t)$ are valuable for homogeneous subgroups in a population, and we consider non-parametric estimation by using discrete time. Following the set-up in Example 1, for individual $i \in s$ let τ_{i0} denote the start of observation, let $y_i = \min(t_i, \tau_{i1})$ denote either a failure time or censoring time, and let $\delta_i = I(y_i = t_i)$ denote a failure. If the sample design is ignorable then the log-likelihood contribution corresponding to (15.14) can be written as

$$\ell_i(\lambda) = \sum_{t=1}^{\infty} n_i(t)\{d_i(t)\log\lambda(t) + (1 - d_i(t))\log(1 - \lambda(t))\},$$

where $\lambda = (\lambda(1), \lambda(2), \ldots)$ denotes the vector of unknown $\lambda(t), n_i(t) = I(\tau_{i0} \leq t \leq y_i)$ indicates an individual is at risk of failure, and $d_i(t) = I(T_i = t, n_i(t) = 1)$ indicates that individual i was observed to fail at time t. If observations from different sampled individuals are independent then the score function $\sum \partial \ell_i / \partial \lambda$ has components

$$U(\lambda)_t = \sum_{i \in s} n_i(t) \left\{ \frac{d_i(t) - \lambda(t)}{\lambda(t)(1 - \lambda(t))} \right\}, \quad t = 1, 2, \ldots. \tag{15.28}$$

Solving $U(\lambda) = 0$, we get estimates

$$\hat{\lambda}(t) = \frac{\sum_{i \in s} n_i(t) d_i(t)}{\sum_{i \in s} n_i(t)} = \frac{d(t)}{n(t)}, \tag{15.29}$$

where $d(t)$ and $n(t)$ are the number of failures and number of individuals at risk at time t, respectively. By (15.25) the estimate of $S(t)$ is then the Kaplan–Meier estimate

$$\hat{S}(t) = \prod_{u=1}^{t-1} (1 - \hat{\lambda}(u)). \tag{15.30}$$

The estimating equation $U(\lambda) = 0$ also provides consistent estimates when the sample design is ignorable but the observations for individuals $i \in s$ are not mutually independent. When the design is nonignorable, using design weights in (15.28) has been proposed (e.g. Folsom, LaVange and Williams 1989). However, an important point about censoring or losses to follow-up should be noted. It is assumed that censoring is independent of failure times in the sense that $E\{d_i(t)|n_i(t) = 1\} = \lambda(t)$. If $S_z(t) \neq S(t)$ and losses to follow-up are related to z (which is plausible in many settings), then this condition is violated and even weighted estimation is inconsistent.

Variance estimates for the $\hat{\lambda}(t)$ or $\hat{S}(t)$ must take any association among sample individuals into account. If individuals are independent then standard maximum likelihood methods apply to (15.28), and yield asymptotic variance estimates (e.g. Cox and Oakes, 1984)

$$\hat{V}(\hat{\lambda}) = \text{diag}\left(\frac{\hat{\lambda}(t)(1 - \hat{\lambda}(t))}{n(t)} \right) \tag{15.31}$$

$$\hat{V}(\hat{S}(t)) = \hat{S}(t)^2 \sum_{u=1}^{t-1} \frac{\hat{\lambda}(u)}{n(u)(1-\hat{\lambda}(u))}. \tag{15.32}$$

If there is association among individuals various methods for computing design-based variance estimates could be utilized (e.g. Wolter, 1985). We consider ignorable designs and the following simple random groups approach. Assume that individuals $i \in s$ can be partitioned into C groups $c = 1, \ldots, C$ such that observations for individuals in different groups are independent, and the marginal distribution for T_i is the same across groups. Let s_c denote the sample individuals in group c and define

$$n_c(t) = \sum_{i \in s_c} n_i(t), \quad d_c(t) = \sum_{i \in s_c} d_i(t)$$

$$\hat{\lambda}_c(t) = \frac{d_c(t)}{n_c(t)}, t = 1, 2, \ldots; \quad c = 1, \ldots, C.$$

Then (see the Appendix) under some mild conditions a robust variance estimate for $\hat{\lambda}$ has entries (for $t = 1, \ldots, \tau; \ r = 1, \ldots, \tau$)

$$\hat{V}_R(\hat{\lambda})_{t,r} = \sum_{c=1}^{C} \frac{n_c(t)n_c(r)}{n(t)n(r)}(\hat{\lambda}_c(t) - \hat{\lambda}(t))(\hat{\lambda}_c(r) - \hat{\lambda}(r)), \tag{15.33}$$

where we take τ as an upper limit on T. The corresponding variance estimate for $\hat{S}(t)$ is

$$\hat{V}_R(\hat{S}(t)) = \hat{S}(t)^2 \sum_{c=1}^{C} \left\{ \sum_{u=1}^{t-1} \frac{n_c(u)(\hat{\lambda}_c(u) - \hat{\lambda}(u))}{n(u)(1-\hat{\lambda}(u))} \right\}^2. \tag{15.34}$$

The estimates $\hat{S}(t)$ and variance estimates apply to the case of continuous times as well, by identifying $t = 1, 2, \ldots$ with measured values of t and noting that $\hat{\lambda}(t) = 0$ if $d(t) = 0$. Williams (1995) derives a similar estimator by a linearization approach.

15.5.2. Parametric models

Marginal parametric models for T_i given x_i may be handled along similar lines. Consider a continuous time model with hazard and survivor functions of the form $\lambda(t|x_i; \ \theta)$ and $S(t|x_i; \ \theta)$, and assume that the sample design is ignorable. The contribution to the likelihood score function from data $(\tau_{i0}, y_i, \delta_i)$ for individual i is, from (15.14),

$$U_i(\theta) = \delta_i \frac{\partial \log \lambda(t_i|x_i; \ \theta)}{\partial \theta} - \frac{\partial}{\partial \theta} \int_{\tau_{i0}}^{y_i} \lambda(u|x_i; \ \theta) du. \tag{15.35}$$

We again consider association among observations via independent clusters $c = 1, \ldots, C$ within which association may be present. The estimating equation

$$U(\theta) = \sum_{c=1}^{C} \sum_{i \in s_c} U_i(\theta) = 0 \tag{15.36}$$

is unbiased, and under mild conditions the estimator $\hat{\theta}$ obtained from (15.36) is asymptotically normal with covariance matrix estimated consistently by

$$\hat{V}(\hat{\theta}) = I(\hat{\theta})^{-1} \widehat{\text{Var}}(U(\theta)) I(\hat{\theta})^{-1}, \tag{15.37}$$

where $I(\theta) = -\partial U(\theta)/\partial \theta'$ and

$$\widehat{\text{Var}}(U(\theta)) = \sum_{c=1}^{C} (\sum_{i \in s_c} U_i(\hat{\theta})) (\sum_{i \in s_c} U_i(\hat{\theta}))'. \tag{15.38}$$

The case where observations are mutually independent is covered by (15.38) with all clusters of size 1. An alternative estimator for $\hat{\theta}$ in this case is $\hat{V}(\hat{\theta}) = I(\hat{\theta})^{-1}$. Existing software can be used to fit parametric models, with some extra coding to evaluate (15.38). Huster, Brookmeyer and Self (1989) provide an example of this approach for clusters of size 2, and Lee, Wei and Amato (1992) use it with Cox's (1972) semi-parametric proportional hazards model.

An alternative approach for clustered data is to formulate multivariate models $S(t_1, \ldots, t_k)$ for the failure times associated with a cluster of size k. This is important when association among individuals is of substantive interest. The primary approaches are through the use of cluster-specific random effects (e.g. Clayton, 1978; Xue and Brookmeyer, 1996) or so-called copula models (e.g. Joe, 1997, Ch. 5). Hougaard (2000) discusses multivariate models, particularly of random effects type, in detail.

A model which has been studied by Clayton (1978) and others has joint survivor function for $T_1, \ldots, T_k$ of the form

$$S(t_1, \ldots, t_k) = \left\{ \sum_{j=1}^{k} S_j(t_j)^{-\phi^{-1}} - (k-1) \right\}^{-\phi}, \tag{15.39}$$

where the $S_j(t_j)$ are the marginal survivor functions and ϕ is an 'association' parameter. Specifications of $S_j(t_j)$ in terms of covariates can be based on standard parametric survival models (e.g. Lawless, 2002, Ch. 6). Maximum likelihood estimation for (15.39) is in principle straightforward for a sample of independent clusters, not necessarily equal in size.

15.5.3. Semi-parametric methods

Semi-parametric models are widely used in survival analysis, the most popular being the Cox (1972) proportional hazards model, where T_i has a hazard function of the form

$$\lambda(t|x_i) = \lambda_0(t) \exp(\beta' x_i), \tag{15.40}$$

where $\lambda_0(t) > 0$ is an arbitrary 'baseline' hazard function. In the case of independent observations, partial likelihood analysis, which yields estimates of β and non-parametric estimates of $\Lambda_0(t) = \int_0^t \lambda_0(u)du$, is standard and well known (c.g. Kalbfleisch and Prentice, 2002). The case where x_i in (15.40) varies with t is also easily handled. For clustered data, Lee, Wei and Amato (1992) have proposed marginal methods analogous to those in Section 15.5.2. That is, the marginal distributions of clustered failure times $T_1, \ldots, T_k$ are modelled using (15.40), estimates are obtained under the assumption of independence, but a robust variance estimate is obtained for $\hat{\beta}$. Lin (1994) provides further discussion and software. This methodology is extended further to include stratification as well as estimation of baseline cumulative hazard functions and survival probabilities by Spiekerman and Lin (1998) and Boudreau and Lawless (2001). Binder (1992) has discussed design-based variance estimation for $\hat{\beta}$ in marginal Cox models, for complex survey designs; his procedures utilize weighted pseudo-score functions. Lin (2000) extends these results and considers related model-based estimation. Software packages such as SUDAAN implement such analyses; Korn, Graubard and Midthune (1997) and Korn and Graubard (1999) illustrate this methodology. Boudreau and Lawless (2001) describe the use of general software like S-Plus and SAS for model-based analysis. Semi-parametric methods based on random effects or copula models have also been proposed, but investigated only in special settings (e.g. Klein and Moeschberger, 1997, Ch. 13).

15.6. ANALYSIS OF EVENT OCCURRENCES

Many processes involve several types of event which may occur repeatedly or in a certain order. For interesting examples involving cohabitation, marriage, and marriage dissolution see Trussell, Rodriguez and Vaughan (1992) and Hoem and Hoem (1992). It is not possible to give a detailed discussion, but we consider briefly the analysis of recurrent events and then methods for more complex processes.

15.6.1. Analysis of recurrent events

Objectives in analyzing recurrent or repeated events include the study of patterns of occurrence and of the relationship of fixed or time-varying covariates to event occurrence. If the exact times at which the events occur are available then individual-level models based on intensity function specifications such as (15.11) or (15.12) may be employed, with inference based on likelihood functions of the form (15.13). Berman and Turner (1992) discuss a convenient format for parametric maximum likelihood computations, utilizing a discretized version of (15.13). Adjustments to variance estimation to account for cluster samples can be accommodated as in Section 15.5.2.

Semi-parametric methods may be based on the same partial likelihood ideas as for survival analysis (Andersen *et al.*, 1993, Chs 4 and 6; Therneau and Hamilton, 1997). The book by Therneau and Grambsch (2000) provides many practical details and illustrations of this methodology. For example, under model (15.11) we may specify $g(x_i(t))$ parametrically, say as $g(x_i(t); \beta) = \exp(\beta' x_i(t))$, and leave the baseline intensity function $\lambda_0(t)$ unspecified. A partial likelihood for β based on n independent individuals is then

$$L_P(\beta) = \prod_{i=1}^{n} \prod_{j=1}^{m_i} \left\{ \frac{\exp(\beta' x_i(t_{ij}))}{\sum_{l=1}^{n} r_\ell(t_{ij})(\beta' x_\ell(t_{ij}))} \right\}, \tag{15.41}$$

where $r_\ell(t) = 1$ if and only if individual ℓ is under observation at time t. Therneau and Hamilton (1997) and Therneau and Grambsch (2000) discuss how Cox survival analysis software in S-Plus or SAS may be used to estimate β from (15.41), and model extensions to allow stratification on the number of prior events. The baseline cumulative intensity function $\Lambda_0(t) = \int_0^t \lambda_0(u) du$ can be estimated by

$$\hat{\Lambda}_0(t) = \sum_{(i,j):t_{ij}<t} \left\{ \frac{1}{\sum_{l=1}^{n} r_\ell(t_{ij}) \exp(\hat{\beta}' x_\ell(t_{ij}))} \right\}, \tag{15.42}$$

where $\hat{\beta}$ is obtained by maximizing (15.41). In the case where there are no covariates, (15.42) becomes the Nelson–Aalen estimator (Andersen *et al.*, 1993)

$$\hat{\Lambda}_{NA}(t) = \sum_{(i,j):t_{ij}<t} \left\{ \frac{1}{n(t_{ij})} \right\}, \tag{15.43}$$

where $n(t_{ij}) = \sum_{l=1}^{n} r_\ell(t_{ij})$ is the total number of individuals observed at time t_{ij}.

Lawless and Nadeau (1995) and Lawless (1995) show that this approach also gives robust procedures for estimating the mean functions for the number of events over $[0, t]$. They provide robust variance estimates, and estimates for $\hat{\beta}$ only may also be obtained from S-Plus or SAS programs for Cox regression (Therneau and Grambsch, 2000). Random effects can be incorporated in models for recurrent events (Lawless, 1987; Andersen *et al.*, 1993; Hougaard, 2000; Therneau and Grambsch, 2000).

In some settings exact event times are not provided and we instead have event counts for intervals such as weeks, months, or years. Discrete time versions of the methods above are easily developed. For example, suppose that $t = 1, 2, \ldots$ indexes time periods and let $y_i(t)$ and $x_i(t)$ represent the number of events and covariate vector for individual i in period t. Conditional models may be based on specifications such as

$$y_i(t) | H(t), x_i(t) \sim \text{Poisson}[\lambda_0(t) \exp(\beta' x_i^*(t))],$$

where $x_i^*(t)$ may include both information in $x_i(t)$ and information about prior events. In some cases $y_i(t)$ can equal only 0 or 1, and then analogous binary response models (e.g. Diggle, Liang and Zeger, 1994) may be used. Unobservable heterogeneity or overdispersion may be handled by the incorporation of random effects (e.g. Lawless, 1987). Robust methods in which the focus is not on conditional models but on marginal mean functions such as

$$E\{y_i(t)|x_i(t)\} = \lambda_0(t) \exp(\beta' x_i(t))$$ (15.44)

are simple modifications of methods described above (Lawless and Nadeau, 1995).

15.6.2. Multiple event types

If several types of events are of interest the intensity-based frameworks of Sections 15.2 and 15.3 may be used. In most situations the intensity functions for different types of events do not involve any common parameters. The likelihood functions based on terms of the form (15.13) then factor into separate pieces for each event type, meaning that models for each type can be fitted separately. Often the methodology for survival times or recurrent events serve as convenient building blocks. Therneau and Grambsch (2000) discuss how software for Cox models can be used for a variety of settings. Lawless (2002, Ch. 11) discusses the application of survival methods and software to more general event history processes. Rather than attempt any general discussion, we give two short examples which represent types of circumstances. In the first, methods based on ideas in Section 15.6.1 would be useful; in the second, survival analysis methods as in Section 15.5 can be exploited.

Example 3.   Use of medical facilities

Consider factors which affect a person's decision to use a hospital emergency department, clinic, or family physician for certain types of medical treatment or consultation. Simple forms of analysis might look at the numbers of times various facilities are used, against explanatory variables for an individual. Patterns of usage can also be examined: for example, do persons tend to follow emergency room visits with a visit to a family physician? Association between the uses of different facilities can be considered through the methods of Section 15.6.1, by using time-dependent covariates that reflect prior usage. Ng and Cook (1999) provide robust methods for marginal means analysis of several types of events.

Example 4.   Cohabitation and marriage

Trussell, Rodriguez and Vaughan (1992) discuss data on cohabitation and marriage from a Swedish survey of women aged 20–44. The events of main interest are unions (cohabitation or marriage) and the dissolution of unions, but the process is complicated by the fact that cohabitations or marriages may be first, second, third, and so on; and by the possibility that a given couple may cohabit, then marry. If we consider the durations of the sequence of 'states', survival analysis methodology may be applied. For example, we can consider the time to first marriage as a function of an individual's age and explanatory factors such as birth cohort, education history, cohabitation history, and employment, using survival analysis with time-varying covariates. The duration of first marriage can be similarly considered, with baseline covariates for the marriage partners, and post-marriage covariates such as the birth of children. After dissolution of a first marriage, subsequent marital unions and dissolutions may be considered likewise.

15.7. ANALYSIS OF MULTI-STATE DATA

Space does not permit a detailed discussion of analysis for multi-state models; we mention only a few important points.

If intensity-based models in continuous time (see (15.4)) are used then, provided complete information about changes of state and covariates is available, (15.13) is the basis for the construction of likelihood functions. At any point in time, an individual is at risk for only certain types of events, namely those which correspond to transitions out of the state currently occupied and, as in Section 15.6.2, likelihood functions typically factor into separate pieces for each distinct type of transition. Andersen *et al.* (1993) extensively treat Markov models and describe non-parametric and parametric methods. For the semi-Markov case where the transitions out of a state depend on the time since the state was entered, survival analysis methods are useful. If transitions to more than one state are possible from a given state, then the so-called 'competing risks' extension of survival analysis (e.g. Kalbfleisch and Prentice, 2002, Ch. 8; Lawless, 2002, Chs 10 and 11) is needed. Example 4 describes a setting of this type.

Discrete time models are widely used for analysis, especially when individuals are observed at discrete times $t = 0, 1, 2, \ldots$ that have the same meaning for different individuals. The basic model is a Markov chain with time-dependent transition probabilities

$$P_{ijk}(t|x_i^*(t)) = Pr(Y_i(t+1) = k | Y_i(t) = j, x_i^*(t)),$$

where x_i^* may include information on covariates and on prior event history. If an individual is observed at time points $t = a, a+1, \ldots, b$ then likelihood methods are essentially those for multinomial response regression models (e.g. Fahrmeir and Tutz, 1994; Lindsey, 1993). Analysis in continuous time is preferred if individuals are observed at unequally spaced time points or if there tend to be several transitions between observation points. If it is not possible to determine the exact times of transitions, then analysis is more difficult (e.g. Kalbfleisch and Lawless, 1989).

15.8. ILLUSTRATION

The US National Longitudinal Survey of Youth (NLSY) interviews persons annually. In this illustration we consider information concerning mothers who chose to breast-feed their first-born child and, in particular, the duration T of breast feeding (see e.g. Klein and Moeschberger, 1997, section 1.14).

In the NLSY, females aged 14 to 21 in 1979 were interviewed yearly until 1988. They were asked about births and breast feeding which occurred since the last interview, starting in 1983; in 1983 information about births as far back as 1978 was also collected. The data considered here are for 927 first-born children whose mothers chose to breast-feed them; duration times of breast feeding are measured in weeks. Covariates included

> Race of mother (Black, White, Other)
> Mother in poverty (Yes, No)
> Mother smoked at birth of child (Yes, No)
> Mother used alcohol at birth of child (Yes, No)
> Age of mother at birth of child (Years)
> Year of birth (1978–88)
> Education level of mother (Years of school)
> Prenatal care after third month (Yes, No).

A potential problem with these types of data is the presence of measurement error in observed duration times, due to recall errors in the date at which breast feeding concluded. We consider this below, but first report the results of duration analysis which assumes no errors of measurement. Given the presence of covariates related to design factors and the unavailability of cluster information, a standard (unweighted) analysis assuming independence of individual responses was carried out. Analysis using semi-parametric pro-portional hazard methods (Cox, 1972) and accelerated failure time models (Lawless, 2002, Ch. 6) suggests an exponential or Weibull regression model with covariates for education, smoking, race, poverty, and year of birth. Education in total years is treated as a continuous covariate below. Table 15.1 shows estimates and standard errors for the semi-parametric Cox model with hazard function of the form (15.40) and for a Weibull model with hazard function

$$\lambda(t|x_i) = \gamma \exp(\alpha + \beta' x_i)(t \exp(\alpha + \beta' x_i))^{\gamma-1}. \tag{15.45}$$

For the model (15.45) the estimate of γ was almost exactly one, indicating an exponential regression model and that the β in (15.40) and in (15.45) have the same meaning. In Table 15.1, binary covariates are coded so that $1 = $ Yes, $0 = $ No; race is coded by two variables (Black $= 1$, not $= 0$ and White $= 1$, not $= 0$); year of birth is coded as -4 (1978) to 4 (1986) and education level of mother is centred to have mean 0 in the sample.

The analysis indicates longer duration for earlier years of birth, for white mothers than for non-white, for mothers with more education, and for mothers in poverty. There was a mild indication of an education–poverty interaction (not shown), with more educated mothers living in poverty tending to have shorter durations.

Table 15.1   Analysis of breast feeding duration.

	Cox model		Weibull model	
Covariate	β estimate	se	β estimate	se
Intercept	—	—	-2.587	0.087
Black	-0.106	0.128	-0.122	0.127
White	-0.279	0.097	-0.325	0.097
Education	-0.058	0.020	-0.063	0.020
Smoking	0.250	0.079	0.280	0.078
Poverty	-0.184	0.093	-0.200	0.092
Year of birth	0.068	0.018	0.078	0.018

As noted, measurement error may be an issue in the recording of duration times, if the dates breast feeding terminated are in error; this will affect only observed duration times, and not censoring times. In some settings where recall error is potentially very severe, a decision may be made to use only current status data and not the measured durations (e.g. Holt, McDonald and Skinner, 1991). That is, since dates of birth are known from records, one could at each annual interview merely record whether or not breast feeding of an infant born in the past year had terminated or not. The 'observation' would then be that either $T_i \leq C_i$ or $T_i > C_i$, where C_i is the time between the date of birth and the interview. Diamond and McDonald (1992) discuss the analysis of such data.

If observed duration times are used, an attempt may be made to incorporate measurement error or to assess its possible effect. For the data considered here, the dates of births and interviews are not given so nothing very realistic can be done. As a result we consider only a very crude sensitivity analysis based on randomly perturbing observed duration times t_i to $t_i^{\text{new}} = t_i + u_i$. Two scenarios were considered by way of illustration: (i) u_i has probability mass 0.2 at each of the values -4, -2, 0, 2, and 4 weeks, and (ii) u_i has probability mass 0.3, 0.3, 0.2, 0.1, 0.1 at the values -4, -2, 0, 2, 4, respectively. These represent cases where measurements errors in t_i are symmetric about zero and biased downward (i.e. dates of cessation of breast feeding are biased downward), respectively. For each scenario, 10 sets of values for t_i^{new} were obtained (censored duration times were left unchanged) and new models fitted for each dataset. The variation in estimates and standard errors across the simulated datasets was negligible for practical purposes and indicates no concerns about the substantive conclusions from the analysis; in each case the standard error of $\hat{\beta}$ across the simulations was less than 20% of the standard error of $\hat{\beta}$ (which was stable across the simulated datasets).

15.9.   CONCLUDING REMARKS

For individual-level modelling and explanatory analysis of event history processes, sufficient information about events and covariates must be obtained.

This supports unweighted analysis based on conventional assumptions of non-selective sampling, but cluster effects in survey populations and samples may necessitate multivariate models or adjustments for association among sampled individuals. An approach based on the identification of clusters was outlined in Section 15.5 for the case of survival analysis, but in general, variance estimation methods used with surveys have received little study for event history data.

Many issues have not been discussed. There are difficult measurement problems associated with longitudinal surveys, including factors related to the behaviour of panel members. The edited volume of Kasprzyk *et al.* (1989) contains a number of excellent discussions. Tracking individuals closely enough and designing data collection and validation so that one can obtain relevant information on individual life histories prior to enrolment and between follow-up times are crucial. The following is a short list of areas where methodological development is needed.

• Methods for handling missing data (e.g. Little, 1992).
• Methods for handling measurement errors in responses (e.g. Hoem, 1985; Trivellato and Torelli, 1989; Holt, McDonald and Skinner, 1991; Skinner and Humphreys, 1999) and in covariates (e.g. Carroll, Ruppert and Stefanski 1995).
• Studies of response-related losses to follow-up and nonresponse, and the need for auxiliary data.
• Methods for handling response-selective sampling induced by retrospective collection of data (e.g. Hoem, 1985, 1989).
• Fitting multivariate and hierarchical models with incomplete data.
• Model checking and the assessment of robustness with incomplete data.

Finally, the design of any longitudinal survey requires careful consideration, with a prioritization of analytic vs. descriptive objectives and analysis of the interplay between survey data, individual-level event history modelling, and explanatory analysis. For investigation of explanatory or causal mechanisms, trade-offs between large surveys and smaller, more controlled longitudinal studies directed at specific areas deserve attention. Smith and Holt (1989) provide some remarks on these issues, and they have often been discussed in the social sciences (e.g. Fienberg and Tanur, 1986).

APPENDIX: ROBUST VARIANCE ESTIMATION FOR THE KAPLAN–MEIER ESTIMATE

We write (15.28) as

$$U(\lambda)_t = \sum_{c=1}^{C} \sum_{i \in s_c} \delta_i(t) \left\{ \frac{d_i(t) - \lambda(t)}{\lambda(t)(1 - \lambda(t))} \right\}, \quad t = 1, \ldots, \tau.$$

The theory of estimating equations (e.g. Godambe, 1991; White, 1994) gives

$$\hat{V}_R(\hat{\lambda}) = I(\hat{\lambda})^{-1}\hat{V}(U(\lambda))I(\hat{\lambda})^{-1} \qquad (A.1)$$

as an estimate of the asymptotic variance of $\hat{\lambda}$, where $I(\lambda) = -\partial U(\lambda)/\partial \lambda'$ and $\hat{V}(U(\lambda))$ is an estimate of the covariance matrix for $U(\lambda)$. Now

$$\text{cov}(U(\lambda)_t, U(\lambda)_r) = \sum_{c=1}^{C} E \sum_{i \in s_c} \sum_{j \in s_c} \frac{\delta_i(t)\delta_i(r)(d_i(t) - \lambda(t))(d_j(r) - \lambda(r))}{\lambda(t)(1 - \lambda(t))\lambda(r)(1 - \lambda(r))},$$

where we use the fact that $E\{d_i(t)|\delta_i(t) = 1, i \in s_c\} = \lambda(t)$. Under mild conditions this may be estimated consistently (as $C \to \infty$) by

$$\sum_{c=1}^{C} \sum_{i \in s_c} \sum_{j \in s_c} \frac{\delta_i(t)\delta_i(r)(d_i(t) - \hat{\lambda}(t))(d_j(r) - \hat{\lambda}(r))}{\hat{\lambda}(t)(1 - \hat{\lambda}(t))\hat{\lambda}(r)(1 - \hat{\lambda}(r))},$$

which after rearrangement using (A.1) gives (15.33).

ACKNOWLEDGEMENTS

The author would like to thank Christian Boudreau, Mary Thompson, Chris Skinner, and David Binder for helpful discussion. Research was supported in part by a grant from the Natural Sciences and Engineering Research Council of Canada.

CHAPTER 16

Applying Heterogeneous Transition Models in Labour Economics: the Role of Youth Training in Labour Market Transitions

Fabrizia Mealli and Stephen Pudney

16.1. INTRODUCTION

Measuring the impact of youth training programmes on the labour market continues to be a major focus of microeconometric research and debate. In countries such as the UK, where experimental evaluation of training programmes is infeasible, research is more reliant on tools developed in the literature on multi-state transitions, using models which predict simultaneously the timing and destination state of a transition. Applications include Ridder (1987), Gritz (1993), Dolton, Makepeace and Treble (1994) and Mealli and Pudney (1999).

In this chapter we describe an application of multi-state event history analysis, based not on a sample survey but rather on a 'census' of 1988 male school-leavers in Lancashire. Despite the fact that we are working with an extreme form of survey, there are several methodological issues, common in the analysis of survey data, that must be addressed. There are a number of important model specification difficulties facing the applied researcher in this area. One is the problem of scale and complexity that besets any realistic model. Active labour market programmes like the British Youth Training Scheme (YTS) and its successors (see Main and Shelly, 1998; Dolton, 1993) are embedded in the youth labour market, which involves individual transitions between several different states: employment, unemployment and various forms of education

Analysis of Survey Data Edited by R. L. Chambers and C. J. Skinner
© 2003 John Wiley & Sons, Ltd

or training. In principle, every possible type of transition introduces an additional set of parameters to be estimated, so the dimension of the parameter space rises with the square of the number of separate states, generating both computational and identification difficulties. This is the *curse of dimensionality* that afflicts many different applications of statistical modelling of survey data in economics, including demand analysis (Pudney, 1981) and discrete response modelling (Weeks, 1995). A second problem is generated by the *institutional features* of the training and education system. YTS places are normally limited in duration and college courses are also normally of standard lengths. Conventional duration modelling is not appropriate for such episodes, but flexible semi-parametric approaches (such as that of Meyer, 1990) may introduce far too many additional parameters in a multi-state context. A third important issue is the *persistence* that is generally found in observed sequences of individual transitions. There are clearly very strong forces tending to hold many individuals in a particular state, once that state has been entered. This is a consideration that motivates the widely used Goodman (1961) mover–stayer model, which captures an extreme form of persistence. A fourth problem is *sample attrition*. This is important in any longitudinal study, but particularly so for the youth labour market, where many of the individuals involved may have weak attachment to training and employment, and some resistance to monitoring by survey agencies.

Given the scale of the modelling problem, there is no approach which offers an ideal solution to all these problems simultaneously. In practice we are seeking a model specification and an estimation approach which gives a reasonable compromise between the competing demands of generality and flexibility on the one hand and tractability on the other.

An important economic focus of the analysis is the selection problem. It is well known that selection mechanisms may play an important role in this context: that people who are (self-)assigned to training may differ in terms of their unobservable characteristics, and these unobservables may affect also their subsequent labour market experience. If the fundamental role of training per se is to be isolated from the effects of the current pattern of (self-)selection, then it is important to account for the presence of persistent unobserved heterogeneity in the process of model estimation. Once suitable estimates are available, it then becomes possible to assess the impact of YTS using simulations which hold constant the unobservables that generate individual heterogeneity.

16.2. YTS AND THE LCS DATASET

With the rise of unemployment, and especially youth unemployment, in the UK during the 1980s, government provision of training took on an important and evolving role in the labour market. The one-year YTS programme, introduced in 1983, was extended in 1986 to two years; this was modified and renamed Youth Training in 1989. The system was subsequently decentralised with the introduction of local Training and Enterprise Councils. Later versions of the scheme were intended to be more flexible, but the system remains essentially one of two-year support for young trainees. Our data relate to the 1988 cohort of school-leavers and thus to the two-year version of YTS. In exceptional circumstances, YTS could last longer than two years, but for a large majority of participants in the programme, the limit was potentially binding. YTS participants may have had special status as trainees (receiving only a standard YTS allowance), or be regarded as employees in the normal sense, being paid the normal rate for the job. Thus YTS had aspects of both training and employment subsidy schemes. Additional funds were available under YTS for payment to the training providers (usually firms, local authorities and organisations in the voluntary sector) to modify the training programme for people with special training needs who needed additional support.

The data used for the analysis are drawn from a database held on the computer system of the Lancashire Careers Service (LCS), whose duties are those of delivering vocational guidance and a placement service of young people into jobs, training schemes or further education. It generates a wide range of information on all young people who leave school in Lancashire and on the jobs and training programmes undertaken. Andrews and Bradley (1997) give a more detailed description of the dataset.

The sample used in this study comprises 3791 males who entered the labour market with the 1988 cohort of school-leavers, and for whom all necessary information (including postcode, school identifier, etc.) was available. This group was observed continuously from the time they left school, aged 16, in the spring or summer of their fifth year of secondary school, up to the final threshold of 30 June 1992, and every change of participation state was recorded. We identify four principal states: continuation in formal education, which we refer to as college (C); employment (E); unemployment (U); and participation in one of the variants of the government youth training scheme (all referred to here as YTS). Note that 16- and 17-year-olds are not eligible for unemployment-related benefits, so unemployment in this context does not refer to registered unemployment, but is the result of a classification decision of the careers advisor. We have also classified a very few short, unspecified non-employment episodes as state U. There is a fifth state which we refer to as 'out of sample' (O). This is a catch-all classification referring to any situation in which either the youth concerned is out of the labour force for some reason, or the LCS has lost touch with him or her. Once state O is encountered in the record of any individual, the record is truncated at that point, so that it is an absorbing state in the sense that there can be no subsequent recorded transition out of state O. In the great majority of cases, a transition to O signifies the permanent loss of contact between the LCS and the individual, so that it is, in effect, the end of the observation period and represents the usual phenomenon of sample attrition. However, it is important that we deal with the potential endogeneity of attrition, so transitions into state O are modelled together with other transition types.

Many of the individual histories begin with a first spell corresponding to a summer 'waiting' period before starting a job, training or other education. We have excluded all such initial spells recorded by the LCS as waiting periods, and started the work history instead from the succeeding spell. After this adjustment is made, we observe for each individual a sequence of episodes, the final one uncompleted, and for each episode we have an observation on two endogenous variables: its duration and also (for all but the last) the destination state of the transition that terminates it. The data also include observations on explanatory variables such as age, educational attainment and a degree of detail on occupational category of the YTS place and its status (trainee, employee, special funding). Summary statistics for the sample are given in Table A1 in the appendix.

There are two obvious features revealed by inspection of the data, which give rise to non-standard elements of the model we estimate. The first of these is shown in Figure 16.1, which plots the smoothed empirical cdf of YTS spell durations. The cdf clearly shows the importance of the two-year limit on the length of YTS spells and the common occurrence of early termination. Nearly 30% of YTS spells finish within a year and nearly 50% before the two-year limit. Conventional transition model specifications cannot capture this feature, and we use below an extension of the limited competing risks (LCR) model introduced by Mealli, Pudney and Thomas (1996). Figure 16.2 plots the smoothed empirical hazard function of durations of college spells, and reveals another non-standard feature in the form of peaks at durations around 0.9 and 1.9 years (corresponding to educational courses lasting one and two

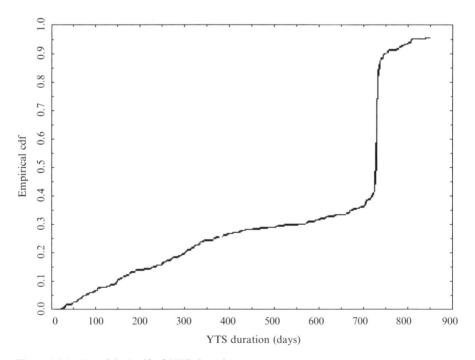

Figure 16.1 Empirical cdf of YTS durations.

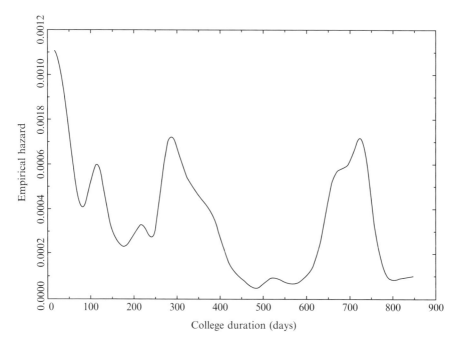

Figure 16.2   Smoothed empirical hazard function for transitions from college.

academic years). Again, we make an appropriate adaptation to our model to cope with this.

16.3.   A CORRELATED RANDOM EFFECTS TRANSITION MODEL

Longitudinal data of the type described above cover each individual from the beginning of his or her work history to an exogenously determined date at which the observation period ends. This generates for each sampled individual a set of m observed episodes (note that m is a random variable). Each episode has two important attributes: its duration; and the type of episode that succeeds it (the destination state). In our case there are four possible states that we might observe. We write the observed endogenous variables $r_0, \delta_1, r_1, \ldots, \delta_m$, where δ_j is the duration of the jth episode, and r_j is the destination state for the transition that brings it to an end. Thus the model can be regarded as a specification for the joint distribution of a set of m continuous variables (the δ_j) and m discrete variables (the r_j). For each episode there is a vector of observed explanatory variables, x_j, which may vary across episodes but which is assumed constant over time within episodes.

   The model we estimate in this study is a modified form of the conventional heterogeneous multi-spell multi-state transition model (see Pudney (1989), Lancaster (1990), and Chapter 15 by Lawless for surveys). Such models proceed by partitioning the observed work history into a sequence of episodes.

For the first spell of the sequence, there is a discrete distribution of the state variable r_0 with conditional probability mass function $Pr(r_0|x_0, v)$. Conditional on past history, each successive episode for $j = 1 \ldots m - 1$ is characterised by a joint density / mass function $f(\delta_j, r_j|x_j, v)$, where x_j may include functions of earlier state and duration variables, to allow for lagged state dependence. The term v is a vector of unobserved random effects, each element normalised to have unit variance; v is constant over time, and can thus generate strong serial dependence in the sequence of episodes (see Chapter 14 by Skinner and Holmes). Under our sampling scheme, the final observed spell is usually an episode of C, E, U or YTS, which is still in progress at the end of the observation period. For this last incomplete episode, the eventual destination state is unobserved, and its distribution is characterised by a survivor function $S(\delta_m|x_m, v)$ which gives the conditional probability of the mth spell lasting at least as long as δ_m. Conditional on the observed covariates $X = \{x_0 \ldots x_m\}$ and the unobserved effects v, the joint distribution of $r_0, \delta_1, r_1, \ldots, \delta_m$ is then

$$f(r_0, \delta_1, r_1 \ldots \delta_m|X, v) = Pr(r_0|x_0, v) \left[\prod_{j=1}^{m-1} f(\delta_j, r_j|x_j, v) \right] S(\delta_m|x_m, v). \quad (16.1)$$

For the smaller number of cases where the sample record ends with a transition to state O (in other words, attrition), there is no duration for state O and the last component of (16.1) is $S(\delta_m|x_m, v) \equiv 1$. There is a further complication for the still fewer cases where the record ends with a YTS $\rightarrow$ O transition, and this is discussed below.

The transition components of the model (the probability density / mass function f and the survivor function S) are based on the notion of a set of origin-and destination-specific transition intensity functions for each spell. These give the instantaneous probability of exit to a given destination at a particular time, conditional on no previous exit having occurred. Thus, for any given episode, spent in state k, the lth transition intensity function $\lambda_{k\ell}(t|z, v)$ is given by

$$Pr(r = k, \delta \in (t, t + dt)|\delta \geq t, x, v) = \lambda_{k\ell}(t|x, v)dt \quad (16.2)$$

where x and v are respectively vectors of observed and unobserved covariates which are specific to the individual but may vary across episodes for a given individual. Our data are constructed in such a way that an episode can never be followed by another episode of the same type, so the k, kth transition intensity λ_{kk} does not exist. The joint probability density / mass function of exit route, r, and realised duration, δ, is then constructed as

$$f(r, \delta|x, v) = \lambda_{kr}(\delta|x, v) \exp\left[-\sum_{\ell \neq k} I_{k\ell}(\delta|x, v) \right] \quad (16.3)$$

where $I_{k\ell}(\delta|x, v)$ is the k, lth integrated hazard:

$$I_{k\ell}(\delta|x, v) = \int_0^\delta \lambda_{k\ell}(t|x, v)dt. \quad (16.4)$$

Since the random effects v are unobserved, (16.1) cannot be used directly as the basis of an estimated model. However, if we assume a specific joint distribution function, $G(v)$, for the random effects, they can be removed by integration and estimation can then proceed by maximising the following log-likelihood based on (16.1) with respect to the model parameters:

$$\ln L = \sum_{i=1}^{n} \ln \left\{ \int Pr(r_0|x_0, v) \left[\prod_{j=1}^{m-1} f(\delta_j, r_j|x_j, v) \right] S(\delta_m|x_m, v) dG(v) \right\} \quad (16.5)$$

where the suffix $i = 1 \ldots n$ indexes the individuals in the sample.

It is important to realise that, for estimation purposes, the definition (16.2) of the transition intensity function is applicable to any form of continuous time multi-state transition process. It is also possible to think of such a process in terms of a competing risks structure, involving independently distributed latent durations for transition to each possible destination, with the observed duration and transition corresponding to the shortest of the latent durations. These two interpretations are observationally equivalent in the sense that it is always possible to construct a set of indpendent latent durations consistent with any given set of transition intensities. This aspect of the interpretation of the model therefore has no impact on estimation. However, when we come to simulating the model under assumptions of changed policy or abstracting from the biasing effects of sample attrition, then interpretation of the structure becomes important. For simulation purposes the competing risks interpretation has considerable analytical power, but at the cost of a very strong assumption about the structural invariance of the transition process. We return to these issues in Section 16.5 below.

The specifications we use for the various components of the model are described in the following sections.

16.3.1. Heterogeneity

We now turn to the specification of the persistent random effects. First, note that there has been some debate about the practical importance of heterogeneity in applied modelling. Ridder (1987) has shown that neglecting unobserved heterogeneity results in biases that are negligible, provided a sufficiently flexible baseline hazard is specified. However, his results apply only to the simple case of single-spell data with no censoring. In the multi-spell context where random effects capture persistence over time as well as inter-individual variation, and where there is a non-negligible censoring frequency, heterogeneity cannot be assumed to be innocuous. We opt instead for a model in which there is a reasonable degrees of flexibility in both the transition intensity functions and the heterogeneity distribution.

The same problem of dimensionality is found here as in the observable part of the model. Rather than the general case of 16 unobservables, each specific to a distinct origin–destination combination, we simplify the structure by using persistent heterogeneity to represent those unobservable factors which predispose individuals towards long or short stays in particular states. Our view is that this sort of state-dependent 'stickiness' is likely to be the main dimension of unobservable persistence – an assumption similar to, but less extreme than, the assumption underlying the familiar mover – stayer model (Goodman, 1961). Thus we use a four-factor specification, where each of the four random effects is constant over time and linked to a particular state of origin rather than destination. We assume that the observed covariates and these random effects enter the transition intensities in an exponential form, so that in general $\lambda_{k\ell}(t|x,v)$ can be expressed as $\lambda_{k\ell}^*(t; x, \omega_k v_k)$ where ω_k is a scale parameter. There is again a conflict between flexibility and tractability, in terms of the functional form of the distribution of the unobservables v_k. One might follow Heckman and Singer (1984) and Gritz (1993) by using a semi-parametric mass-point distribution, where the location of the mass points and the associated probabilities are treated as parameters to be estimated. Van den Berg (1997) has shown in the context of a two-state competing risks model that this specification has an advantage over other distributional forms (including the normal) in that it permits a wider range of possible correlations between the two underlying latent durations. However, in our four-dimensional setting, this would entail another great expansion of the parameter space. Since there is a fair amount of informal empirical experience suggesting that distributional form is relatively unimportant provided the transition intensities are specified sufficiently flexibly, we are content to assume a normal distribution for the v_k.

We introduce correlation across states in the persistent heterogeneity terms in a simple way which nevertheless encompasses the two most common forms used in practice. This is done by constructing the v_k from an underlying vector ζ as follows:

$$v_k = (1 - \rho)\zeta_k + \rho \sum_{p=1}^{4} \zeta_p \qquad (16.6)$$

where ρ is a single parameter controlling the general degree of cross-correlation. Under this specification, the correlation between any pair of heterogeneity terms, $\omega_k v_k$ and $\omega_\ell v_\ell$, is $2\text{sgn}(\omega_k \omega_\ell)\rho(1 + \rho)/(1 + 3\rho^2)$. Note that one of the scale parameters should be normalised with respect to its sign, since (with the ζ_k symmetrically distributed) the sample distribution induced by the model is invariant to multiplication of all the ω_k by -1. There are two important special cases of (16.6): $\rho = 0$ corresponds to the assumption of independence across (origin) states; $\rho = 1$ yields the one-factor specification discussed by Lindeboom and van den Berg (1994).

16.3.2. The initial state

Our model for the initial state indicator r_0 is a four-outcome multinomial logit (MNL) structure of the following form:

$$Pr(r_0 = k|x_0, v) = \frac{\exp(x_0\gamma_k + \psi_k v_k)}{\sum_{l=1}^{4} \exp(x_0\gamma_\ell + \psi_\ell v_\ell)} \qquad (16.7)$$

where γ_1 is normalised at 0. The parameters ψ_j are scale parameters that also control the correlation between the initial state and successive episodes.

16.3.3. The transition model

For a completely general model, 16 transition intensities should be specified – a practical impossibility, since this would lead to an enormously large parameter set. This dimensionality problem is very serious in applied work, and there are two obvious solutions, neither of which is ideal. The most common approach is to reduce the number of states, either by combining states (for example, college and training) into a single category, or by deleting individuals who make certain transitions (such as those who return to college or who leave the sample by attrition). The consequences of this type of simplification are potentially serious and obscure, since it is impossible to test the implicit underlying assumptions without maintaining the original degree of detail. In these two examples, the implicit assumptions are respectively: that transition rates to and from college and training are identical; and that the processes of transition to college or attrition are independent of all other transitions. Neither assumption is very appealing, so we prefer to retain the original level of detail in the data, and to simplify the model structure in (arguably) less restrictive and (definitely) more transparent ways.

We adopt as our basic model a simplified specification with separate intensity functions only for each exit route, letting the effect of the state of origin be captured by dummy variables included with the other regressors, together with a few variables which are specific to the state of origin (specifically SPECIAL and YTCHOICE describing the nature of the training placement in YTS spells, and YTMATCH, CLERK and TECH describing occupation and training match for E → U transitions).

The best choice for the functional form $\lambda_{k\ell}(t|x,v)$ is not obvious. Meyer (1990) has proposed a flexible semi-parametric approach which is now widely used in simpler contexts (see also Narendranathan and Stewart, 1993). It entails estimating the dependence of $\lambda_{k\ell}$ on t as a flexible step function, which introduces a separate parameter for each step. In our case, with 16 different transition types, this would entail a huge expansion in the dimension of the parameter space. Instead, we adopt a different approach. We specify a generic parametric functional form for $\lambda_{k\ell}(t|x,v)$, which is chosen to be reasonably flexible (in particular, not necessarily monotonic). However, we also exploit our a priori knowledge of the institutional features of the education and training systems to modify these functional forms to allow for the occurrence of 'standard' spell durations for many episodes of YTS and college. These modifications to the basic model are described below. The basic specification we use is the Burr form:

$$\lambda_{k\ell}(t|x,v) = \frac{\exp(z_k\beta_\ell + \omega_k v_k)\alpha_{k\ell}t^{\alpha_{k\ell}-1}}{1 + \sigma_{k\ell}^2\exp(z_k\beta_\ell + \omega_k v_k)t^{\alpha_{k\ell}}} \tag{16.8}$$

where z_k is a row vector of explanatory variables constructed from x in some way that may be specific to the state of origin, k. Note that the absence of a subscript k on β_ℓ is not restrictive: any form $z_k^* \beta_{k\ell}^*$ can be rewritten $z_k \beta_\ell$ by defining z_k appropriately, using origin-specific dummy variables in additive and multiplicative form. The form (16.8) is non-proportional and not necessarily monotonic, but it has the Weibull form as the special case $\sigma_{k\ell} = 0$. The parameters $\alpha_{k\ell}$ and $\sigma_{k\ell}$ are specific to the origin–destination combination k, l, and this gives the specification considerable flexibility. The Burr form has the following survivor function:

$$S_{k\ell}(\delta|x,v) = [1 + \sigma_{k\ell}^2 \exp(z_k \beta_\ell + \omega_k v_k) t^{\alpha_{k\ell}}]^{-1/\sigma_{k\ell}^2}. \qquad (16.9)$$

Note that the Burr model can be derived as a Weibull–gamma mixture, with the gamma heterogeneity spell-specific and independent across spells, but such an interpretation is not necessary and is not in any case testable without further a priori restrictions, such as proportionality.

16.3.4. YTS spells

There are two special features of YTS episodes that call for some modification of the standard transition model outlined above. One relates to attrition (YTS → O transitions). Given the monitoring function of the LCS for YTS trainees, it is essentially impossible for the LCS to lose contact with an individual while he or she remains in a YTS place. Thus a YTS → O transition must coincide with a transition from YTS to C, E or U, where the destination state is unobserved by the LCS. Thus a transition of this kind is a case where the observed duration in the $(m-1)$th spell (YTS) is indeed the true completed YTS duration, δ_{m-1}, but the destination state r_{m-1} is unobserved. For the small number of episodes of this kind, the distribution (16.3) is

$$f(\delta_{m-1}|x,v) = \sum_{\ell \neq k} \lambda_{k\ell}(\delta_{m-1}|x_{m-1},v) \exp\left[-\sum_{\ell \neq k} I_{k\ell}(\delta_{m-1}|x_{m-1},v)\right]. \qquad (16.10)$$

A second special feature of YTS episodes is the exogenous two-year limit imposed on them by the rules of the system. Mealli, Pudney and Thomas (1996) proposed a simple model for handling this complication which Figure 16.1 shows to be important in our data. The method involves making allowance for a discontinuity in the destination state probabilities conditional on YTS duration at the two-year limit. The transition structure operates normally until the limit is reached, at which point a separate MNL structure comes into play. Thus, for a YTS episode,

$$Pr(r = l|\delta = 2, x, v) = \frac{\exp(w\pi_\ell + \theta_\ell v_{YTS})}{\sum_p \exp(w\pi_p + \theta_p v_{YTS})} \qquad (16.11)$$

where w is a vector of relevant covariates. In the sample there are no cases at all of a college spell following a full-term YTS episode; consequently the summation in the denominator of (16.11) runs over only two alternatives, E and U. The π and θ parameters are normalised to zero for the latter.

16.3.5. Bunching of college durations

To capture the two peaks in the empirical hazard function for college spells, we superimpose two spikes uniformly across all transition intensity functions for college spells. Thus, for origin C and destinations $l =$ E, U, O, YTS, the modified transition intensities are

$$\lambda^*_{Cl}(t|x, v) = \lambda_{Cl}(t|x, v) \exp\left[\mu_1 A_1(t) + \mu_2 A_2(t)\right]$$

where $A_1(t)$ and $A_2(t)$ are indicator functions of $(0.8 \leq t \leq 1)$ and $(1.8 \leq t \leq 2)$ respectively.

16.3.6. Simulated maximum likelihood

The major computational problem involved in maximising the log-likelihood function (16.5) is the computation of the four-dimensional integral which defines each of the n likelihood components. The approach we take is to approximate the integral by an average over a set of Q pseudo-random deviates generated from the assumed joint standard lognormal distribution for ζ. We also make use of the antithetic variates technique to improve the efficiency of simulation, with all the underlying pseudo-random normal deviates reused with reversed sign to reduce simulation variance. Thus, in practice we maximise numerically the following approximate log-likelihood:

$$L(\theta) = \sum_{i=1}^{n} \ln \left(\frac{1}{Q} \sum_{q=1}^{Q} \frac{l_i(v^q) + l_i(-v^q)}{2} \right)$$

where $l_i(v^q)$ is the likelihood component for individual i, evaluated at a simulated value v^q for the unobservables.

This simulated ML estimator is consistent and asymptotically efficient as Q and $n \to \infty$ with $\sqrt{n}/Q \to 0$ (see Gourieroux and Monfort, 1996). Practical experience suggests that it generally works well even with small values of Q (see Mealli and Pudney (1996) for evidence on this). In this study, we use $Q = 40$. The asymptotic approximation to the covariance matrix of the estimated parameter vector $\hat{\theta}$ is computed via the conventional OPG (Outer Product of Gradient) formula, which gives a consistent estimate in the usual sense, under the same conditions on Q and n

16.4. ESTIMATION RESULTS

16.4.1. Model selection

Our preferred set of estimates is given in Tables A4–A7 in the appendix. These estimates are the outcome of a process of exploration which of necessity could not follow the 'general-to-specific' strategy that is usually favoured, since the most general specification within our framework would have approximately 450 parameters, with 16 separate random effects, and would certainly not be possible to estimate with available computing and data resources. Even after the considerable simplifications we have made, there remain 117 parameters in the model. Apart from the constraints imposed on us by this dimensionality problem, we have adopted throughout a conservative criterion, and retained in the model all variables with coefficient t-ratios in excess of 1.0. Some further explanatory variables were tried in earlier specifications, but found to be insignificant everywhere. These were all dummy variables, distinguishing those who: took a non-academic subject mix at school; were in a technical/craft occupation when in work; and had trainee rather than employee status when in YTS. Thus the sparse degree of occupational and training detail in the final model is consistent with the available sample information.

The relatively low frequencies of certain transition types have made it necessary to impose further restrictions to achieve adequate estimation precision. In particular, the $\alpha_{k\ell}$ and $\sigma_{k\ell}$ parameters for destination $l = O$ (sample attrition) could not be separately estimated, so we have imposed the restrictions $\alpha_{kO} = \alpha_{hO}$ and $\sigma_{kO} = \sigma_{hO}$ for all k, h.

16.4.2. The heterogeneity distribution

Table A7 gives details of the parameters underlying the joint distribution of the persistent heterogeneity terms appearing in the initial state and transition structures. Heterogeneity appears strongly significant in the initial state logit only for the exponents associated with employment and YTS. The transition intensities have significant origin-specific persistent heterogeneity linked to C, U and YTS. The evidence for heterogeneity associated with employment is weak, although a very conservative significance level would imply a significant role for it, and also in the logit that comes into force at the two-year YTS limit (θ_E in Table A6).

The estimated value of ρ implies a correlation of ± 0.30 between any pair of scaled heterogeneity terms, $\omega_k v_k$, in the transition part of the model. There is a positive correlation between the heterogeneity terms associated with the E, U and YTS states, but these are all negatively correlated with the heterogeneity term associated with college. This implies a distinction between those who are predisposed towards long college spells and those with a tendency towards long U and YTS spells. Note that the estimate of ρ is significantly different from both 0 and 1 at any reasonable significance level, so both the one-factor and independent stochastic structures are rejected by these results, although independence is clearly preferable to the single-factor assumption.

Wherever significant, there is a negative correlation between the random effect appearing in a branch of the initial state logit, $\psi_k v_k$, and the corresponding random effect in the transition structure, $\omega_k v_k$. This implies, as one might expect, that a high probability of starting in a particular state tends to be associated with long durations in that state.

The logit structure which determines exit route probabilities once the two-year YTS limit is reached involves a random effect which is correlated with the random effect in the transition intensities for YTS spells. However, this is of doubtful significance.

16.4.3. Duration dependence

The functional forms of the destination-specific transition intensities are plotted in Figures 16.3–16.6 conditional on different states of origin. In constructing these plots, we have fixed the elements of the vector of observed covariates x at the representative values listed in Table 16.1 below. The persistent origin-specific random effects, v, are fixed at their median values, 0. The relative diversity of

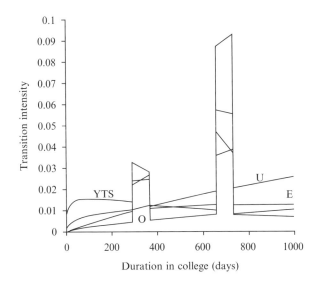

Figure 16.3   Estimated transition intensities for exit from education (C).

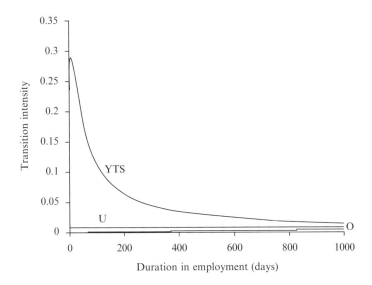

Figure 16.4   Estimated transition intensities for exit from employment (E).

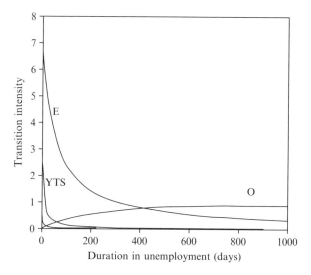

Figure 16.5   Estimated transition intensities for exit from unemployment (U).

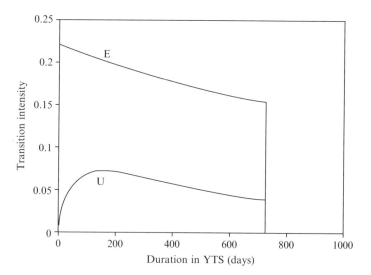

Figure 16.6   Estimated transition intensities for exit from YTS.

the functional forms across states of origin gives an indication of the degree of flexibility inherent in the structure we have estimated. There are several points to note here.

First, the estimated values of the 13 $\sigma_{k\ell}$ parameters are significant in all but four cases, implying that the restricted Weibull form would be rejected against the Burr model that we have estimated; thus, if we were prepared to assume proportional Weibull competing risks, this would imply a significant role for destination-specific Gamma heterogeneity uncorrelated across spells. Second, the transition intensities are not generally monotonic; an increasing then falling pattern is found for the transitions $C \rightarrow YTS$, $E \rightarrow YTS$ and $YTS \rightarrow U$. The aggregate hazard rate is non-monotonic for exits from college, employment and unemployment. Third, transition intensities for exit to college are very small for all states of origin except unemployment, where there is a sizeable intensity of transition into education at short unemployment durations. The generally low degree of transition into state C reflects the fact that, for most people, formal post-16 education is a state entered as first destination after leaving school, or not at all. However, the fact that there are unobservables common to both the initial state and transition parts of the model implies that the decision to enter college after school is endogenous and cannot be modelled separately from the transitions among the other three states. The discontinuity of exit probabilities of the two-year YTS limit is very marked (see Figure 16.7).

Figure 16.8 shows the aggregated hazard rates, $\lambda_k(t|x, v) = \sum_\ell \lambda_{k\ell}$, governing exits from each state of origin, k. The typical short unemployment duration- simply a high hazard rate for exits from unemployment, but declining strongly with duration, implying a heavy right hand tail for the distribution of unemployment durations. For the other three states of origin, the hazard rates are rather flatter, except for the one- and two-year peaks for college spells. Note that we cannot distinguish unambiguously between true duration dependence and the effects of non-persistent heterogeneity here, at least not without imposing restrictions such as proportionality of hazards.

16.4.4. Simulation strategy

The model structure is sufficiently complex that it is difficult to interpret the parameter estimates directly. Instead we use simple illustrative simulations to bring out the economic implications of the estimated parameter values. The

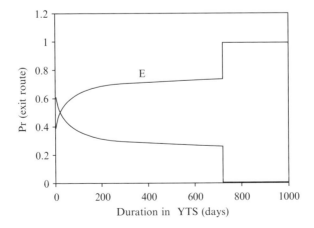

Figure 16.7 Probabilities of exit routes from YTS conditional on duration.

'base case' simulations are performed for a hypothetical individual who is average with respect to quantitative attributes and modal with respect to most qualitative ones. An exception to this is educational attainment, which we fix at the next-to-lowest category (GCSE2), to represent the group for whom YTS is potentially most important. Thus our representative individual has the characteristics listed in Table 16.1.

The treatment of state O (nonignorable attrition) is critical. We assume that, conditional on the persistent heterogeneity terms v_C, v_E, v_U, v_{YTS}, the labour market transition process is the outcome of a set of independent competing risks represented by the hazards $\lambda_{k\ell}$. Superimposed on this process is a fifth

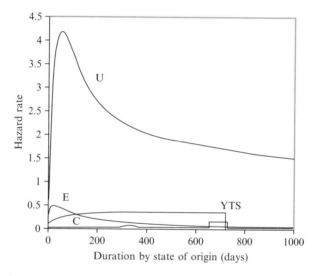

Figure 16.8   Aggregated hazard rates.

Table 16.1   Attributes of illustrative individual.

Attribute	Assumption used for simulations
Date of birth	28 February 1972
Ethnic origin	White
Educational attainment	One or more GCSE passes, none above grade D
Subject mix	Academic mix of school subjects
Health	No major health problem
School quality	Attended a school where 38.4% of pupils achieved five or more GCSE passes
Area quality	Lives in a ward where 77.9% of homes are owner-occupied
Local unemployment	Unemployment rate in ward of residence is 10.3%
Date of episode	Current episode began on 10 March 1989
Previous YTS	No previous experience of YTS
Occupation	When employed, is neither clerical nor craft/technical
Special needs	Has no special training needs when in YTS

independent risk of attrition which is relevant to the observation process but irrelevant to the labour market itself. In this sense, attrition is *conditionally* ignorable, by which we mean that it is independent of the labour market transition process after conditioning on the persistent unobservables. It is not *unconditionally* ignorable since the unobservables generate correlation between observed attrition and labour market outcomes, which would cause bias if ignored during estimation. However, conditional ignorability means that, once we have estimated the model and attrition process jointly (as we have done above), we can nevertheless ignore attrition in any simulation that holds fixed the values of the persistent unobservables. In these circumstances, ignoring attrition means marginalising with respect to the attrition process. Given the assumption of conditional independence of the competing risks, this is equivalent to deleting all hazards for exit to state O and simulating the remaining four-risk model. This procedure simulates the distribution of labour market event histories that would, if unhampered by attrition, be observed in the subpopulation of individuals with heterogeneity terms equal to the given fixed values v_k.

The simulation algorithm works as follows. For the representative individual defined in Table 16.1, 500 five-year work histories are generated via stochastic simulation of the estimated model.[1] These are summarised by calculating the average proportion of time spent in each of the four states and the average frequency of each spell type. To control for endogenous selection and attrition, we keep all the random effects fixed at their median values of zero, and reset all transition intensities into state O to zero. We then explore the effects of the covariates by considering a set of hypothetical individuals with slightly different characteristics from the representative individual. These explore the effects of ethnicity, educational attainment and the nature of the locality. For the last of these, we change the SCHOOL, AREA and URATE variables to values of 10%, 25% and 20% respectively.

Table 16.2 reveals a large impact for the variables representing ethnicity and educational attainment, in comparison with the variables used to capture the influence of social background. An individual identical to the base case, but from a non-white ethnic group (typically South Asian in practice), is predicted to have a much higher probability of remaining in full-time education (59% of the five-year period on average, compared with 20% for the reference white individual). However, for ethnic minority individuals who are not in education, the picture is gloomy. Non-whites have a much higher proportion of their non-college time (22% compared with 9%) spent unemployed, with a roughly comparable proportion spent in YTS.

The effect of increasing educational attainment at GCSE is to increase the proportion of time spent in post-16 education from 20% to 31% and 66% for the three GCSE performance classes used in the analysis. Improving GCSE

---

[1] The simulation process involves sampling from the type I extreme value distribution for the logit parts of the model, and from the distribution of each latent duration for the transition part. In both cases, the inverse of the relevant cdf was evaluated using uniform pseudo-random numbers.

Table 16.2   Simulated effects of the covariates for a hypothetical individual.

Simulated individual	Spell type	Proportion of time (%)	Proportion of non-college time	Frequency of spells (%)	Mean no. of spells
Base case	C	19.7	—	18.2	
(see Table 16.1)	E	56.9	70.9	39.0	2.59
	U	7.2	9.0	24.1	
	YTS	16.2	20.2	18.7	
Non-white	C	58.7	—	55.3	
	E	24.2	58.6	17.8	2.01
	U	9.3	22.5	18.8	
	YTS	7.8	18.9	8.0	
1–3 GCSEs at	C	30.6	—	29.9	
grade C or	E	47.7	68.7	33.1	2.16
better	U	9.3	13.4	23.2	
	YTS	12.4	17.9	13.8	
More than 3	C	65.6	—	64.8	
GCSEs at grade	E	24.3	70.6	17.2	1.55
C or better	U	4.5	13.1	12.0	
	YTS	5.6	16.3	6.0	
Major health	C	22.2	—	21.3	
problem	E	52.4	67.4	33.6	2.57
	U	4.8	6.2	22.3	
	YTS	20.6	26.5	22.8	
Poor school and	C	18.4	—	17.0	
area quality	E	52.8	64.7	35.9	2.75
	U	9.1	11.2	24.4	
	YTS	19.6	24.0	22.7	

Note: 500 replications over a five-year period; random effects fixed at 0.

performance has relatively little impact on the amount of time predicted to be spent in unemployment and its main effect is to generate a substitution of formal education for employment and YTS training.

There is a moderate estimated effect of physical and social disadvantage. Individuals identified as having some sort of (subjectively defined) major health problem are predicted to spend a greater proportion of their first five post-school years in college or YTS (43% rather than 36%) compared with the otherwise similar base case. This displaces employment (52% rather than 57%), but also reduces the time spent unemployed by about two and a half percentage points. In this sense, there is evidence that the youth employment system was managing to provide effective support for the physically disadvantaged, if only temporarily. After controlling for other personal characteristics, there is a significant role for local social influences as captured by the occupational, educational and housing characteristics of the local area, and the quality of the individual's school. Poor school and neighbourhood characteristics are associated with a slightly reduced prediction of time spent in college and employment, with a corresponding increase in unemployment and YTS tenure. Nevertheless, compared with race and education effects, these are minor influences.

16.4.5. The effects of unobserved heterogeneity

To analyse the effects of persistent heterogeneity specific to each state of origin, we conduct simulations similar to those presented in the previous paragraph. The results are shown in Figures 16.9–16.12. We consider the representative individual and then conduct the following sequence of stochastic simulations. For each state $k = \mathrm{C, E, U, YTS}$ set all the heterogeneity terms to zero except for one, v_k, whose value is varied over a grid of values in the range $[-2, 2]$ (covering approximately four standard deviations). At each point in the grid, 500 five-year work histories are simulated stochastically and the average proportion of time spent in each state is recorded. This is done for each of the four v_k, and the results plotted. The plots in Figures 16.9–16.12 show the effect of varying each of the heterogeneity terms on the proportion of time spent respectively in college, employment, unemployment and unemployment.

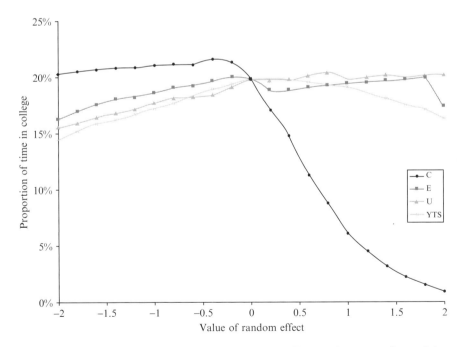

Figure 16.9 The effect of each state-specific random effect on the proportions of time spent in college.

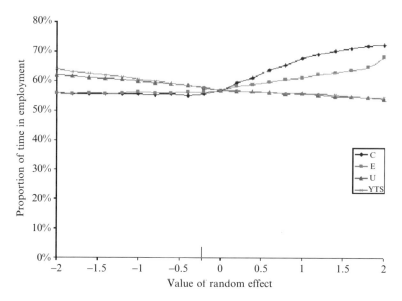

Figure 16.10 The effect of each state-specific random effect on the proportions of time spent in employment.

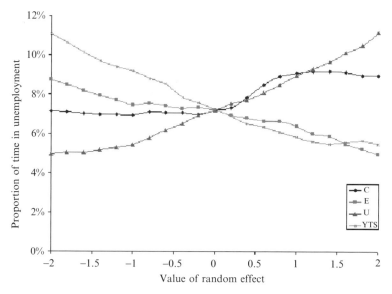

Figure 16.11 The effect of each state-specific random effect on the proportions of time spent in unemployment.

The striking feature of these plots is the large impact of these persistent unobservable factors on the average proportions of the five-year simulation period spent in each of the four states. This is particularly true for college,

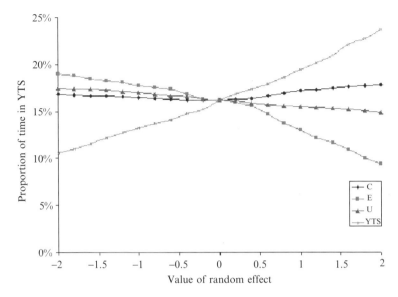

Figure 16.12   The effect of each state-specific random effect on the proportions of time spent in YTS.

where the proportion of time spent in education falls from over 20 % at $v_C = 0$ to almost zero at $v_C = 2$, with a corresponding rise in the time spent in employment and unemployment. The proportion of time spent unemployed (essentially the unemployment rate among individuals of the representative type) is strongly influenced by all four state-specific random effects, with a 6 percentage point variation in the unemployment rate.

16.5. SIMULATIONS OF THE EFFECTS OF YTS

We now bring out the policy implications of the model by estimating the average impact of YTS for different types of individual, again using stochastic simulation as a basis (Robins, Greenland and Hu (1999) use similar tools to estimate the magnitude of a causal effect of a time-varying exposure). A formal policy simulation can be conducted by comparing the model's predictions in two hypothetical worlds in which the YTS system does and does not exist. The latter (the 'counterfactual') requires the estimated model to be modified in such a way that YTS spells can no longer occur. The results and the interpretational problems associated with this exercise are presented in Section 16.5.2 below. However, first we consider the effects of YTS participation and and of early dropout from YTS, by comparing the simulated labour market experience of YTS participants and non-participants within a YTS world. For this we use the model as estimated, except that the 'risk' of attrition (transition to state O) is deleted.

16.5.1.   The effects of YTS participation

We work with the same set of reference individuals as in Sections 16.4.4–16.4.5 above. Again, the state-specific random effects are fixed at their median values of 0, so that the simulations avoid the problems of endogenous selection arising from persistent unobservable characteristics. This time the 500 replications are divided into two groups: the first one contains histories with no YTS spell and the second one histories with at least one YTS spell. We have then two groups of fictional individuals, identical except that the first happen by chance to have avoided entry into YTS, while the second have been through YTS: thus we compare two potential types of work history, only one of which could be observed for a single individual (Holland, 1986).

To make the comparison as equal as possible, we take the last three years of the simulated five-year history for the non-YTS group and the post-YTS period (which is of random length) for the YTS group. We exclude from each group those individuals for whom there is a college spell in the reference period, thus focusing attention solely on labour market participants.

Figure 16.13 shows, for the base case individual, the difference in simulated unemployment incidence for the two groups. At the median value of the random effects, the difference amounts to approximately 5 percentage points, so that YTS experience produces a substantially reduced unemployment risk. We have investigated the impact of unobservable persistent heterogeneity by repeating the simulations for a range of fixed values for each of the v_k. Figure 16.13 shows

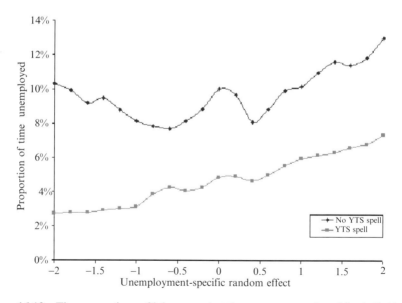

Figure 16.13   The proportions of labour market time spent unemployed for individuals with and without YTS episodes.

the plot for v_U; broadly similar patterns are found for the other v_k, suggesting that the beneficial effect of YTS participation is more or less constant across individuals with differing unobservable characteristics.

Table 16.3 shows the influence of observable characteristics, summarising the results of simulations for the base case and peturbations with respect to ethnicity, education and area/school quality. The beneficial effects of YTS participation are evident in all cases, but are particularly strong for members of ethnic minorities and for those with better levels of school examination achievement. Note that these are the groups with the highest probabilities of full-term YTS spells.

16.5.2. Simulating a world without YTS

The ultimate aim of this type of modelling exercise is to say something about the economic effects of implementing a training/employment subsidy scheme such as YTS. The obvious way to attempt this is to compare simulations of the model in two alternative settings: one (the 'actual') corresponding to the YTS scheme as it existed during the observation period; and the other (the 'counterfactual') corresponding to an otherwise identical hypothetical world in which YTS does not exist. There are well-known and obvious limits on what can be concluded from this type of comparison, since we have no direct way of knowing how the counterfactual should be designed. Note that this is not a

Table 16.3 Simulated effects of YTS participation on employment frequency and duration for hypothetical individuals.

Simulated individual	Replications with no YTS spell	
	% period in work	% spells in work
Base case	89.9	86.0
Non-white	65.3	60.5
1–3 GCSEs at grade C	85.1	83.0
> 3 GCSEs at grade C	88.3	85.4
Low school and area quality	86.5	84.0

Simulated individual	Replications containing a YTS spell			
	% post-YTS period in work	% post-YTS spells in work	Mean YTS duration	% YTS spells full term
Base case	95.1	89.2	1.47	51.0
Non-white	86.6	80.2	1.56	59.5
1–3 GCSEs at grade C	96.5	91.8	1.62	63.6
> 3 GCSEs at grade C	98.6	96.8	1.75	73.1
Low school and area quality	90.1	81.1	1.47	51.8

Note: 500 replications over a five-year period; random effects fixed at 0.

problem specific to the simulations presented in this chapter; any attempt to give a policy-oriented interpretation of survey-based results is implicitly subject to the same uncertainties.

The design of a counterfactual case requires assumptions about three major sources of interpretative error, usually referred to, rather loosely, as deadweight loss, displacement and scale effects. Deadweight loss refers to the possibility that YTS (whose objective is employment promotion) may direct some resources to those who would have found employment even without YTS. Since YTS has some of the characteristics of an employment subsidy, this is a strong possibility. It seems likely that if YTS had not existed during our observation period, then some of those who were in fact observed to participate in YTS would have been offered conventional employment instead, possibly on old-style private apprenticeships. Displacement refers to a second possibility that a net increase in employment for the YTS target group might be achieved at the expense of a reduction in the employment rate for some other group, presumably older, poorly qualified workers. Note, however, that displacement effects can also work in the other direction. For example, Johnson and Layard (1986) showed, in the context of a segmented labour market with persistent unsatisfied demand for skilled labour and unemployment amongst unskilled workers, that training programmes can simultaneously produce an earnings increase and reduced unemployment probability for the trainee (which might be detected by an evaluation study) and also make available a job for one of the current pool of unemployed. A third interpretative problem is that the aggregate net effect of a training programme may be non-linear in its scale, so that extrapolation of a micro-level analysis gives a misleading prediction of the effect of a general expansion of the scheme. This mechanism may work, for instance, through the effect of the system on the relative wages of skilled and unskilled labour (see Blau and Robins, 1987).

The evidence on these effects is patchy. Deakin and Pratten (1987) give results from a survey of British employers which suggests that roughly a half of YTS places may have either gone to those who would have been employed by the training provider anyway or substituted for other types of worker (with deadweight loss accounting for the greater part of this inefficiency). However, other authors have found much smaller effects (see Jones, 1988), and the issue remains largely unresolved. Blau and Robins (1987) found some empirical evidence of a non-linear scale effect, by estimating a significant interaction between programme size and its effects. The need for caution in interpreting the estimated effects of YTS participation is evident, but there exists no clear and simple method for adjusting for deadweight, displacement and scale effects.

The economic assumptions we make about the counterfactual have a direct parallel with the interpretation of the statistical transition model (see also Greenland, Robins and Pearl (1999) for a discussion on this). To say anything about the effects of removing the YTS programme from the youth labour market requires some assumption about how the statistical structure would change if we were to remove one of the possible states. The simulations we present in Table 16.4 correspond to the very simplest counterfactual case and,

Table 16.4   Simulated work histories for hypothetical individuals with and without the YTS scheme in existence.

Simulated individual	Spell type	Proportion of time for non-YTS world (%)	Increase compared with YTS world (% points)
Base case	C	24.0	+4.3
	E	58.1	+1.2
	U	17.9	+10.7
Non-white	C	65.3	+6.6
	E	21.7	−2.5
	U	13.0	+3.7
1–3 GCSEs at	C	38.8	+8.2
grade C	E	43.1	−4.6
	U	18.1	+8.8
> 3 GCSEs at	C	70.2	+4.6
grade C	E	22.3	−2.0
	U	7.4	+2.9
Major health	C	31.5	+9.3
problem	E	49.5	−2.9
	U	19.0	+14.2
Poor school and	C	25.7	+7.3
area quality	E	52.1	−0.7
	U	22.3	+13.2

Note: 500 replications over a five-year period; random effects fixed at 0.

equivalently, to the simplest competing risks interpretation. In the non-YTS world, we simply force the transition intensities for movements from any state into YTS, and the probability of YTS as a first destination, to be zero. The remainder of the estimated model is left unchanged, so that it generates transitions between the remaining three states. In other words, we interpret the model as a competing risks structure, in which the YTS 'risk' can be removed without altering the levels of 'hazard' associated with the other possible destination states. This is, of course, a strong assumption and avoids the issue of the macro-level effects which might occur if there really were an abolition of the whole state training programme.

As before, we work with a set of hypothetical individuals, and remove the effect of inter-individual random variation by fixing the persistent individual-specific random effects at zero. Table 16.4 then summarises the outcome of 500 replications of a stochastic simulation of the model. The sequence of pseudo-random numbers used for each replication is generated using a randomly selected seed specific to that replication; within replications, the same pseudo-random sequence is used for the actual and counterfactual cases. Note that the results are not directly comparable with those presented in Section 16.4.2 which compared YTS participants and non-participants, since we are considering here the whole five-year simulation period rather than the later part of it. We are also not focusing exclusively on the labour market, since we retain in the analysis individuals who are predicted by the simulations to remain in education. A third major difference is that the analysis of Section 16.4.2 did not consider the effects of differences in YTS participation frequency together with the effects of YTS per participant, whereas the simulations reported here will necessarily show bigger impacts of abolition for groups with high YTS participation rates.

On the basis of these results in Table 16.4, the effect of the YTS programme on employment frequencies is important but moderate: a fall of no more than 5 percentage points in the proportion of time spent in employment. Instead, the major impact of abolition is on time spent in education and in unemployment. With YTS abolished, the proportion of time spent in unemployment rises for most cases by between 6 and 14 percentage points, although the rise is necessarily much smaller for those with low probabilities of YTS participation (notably non-whites and those with good GCSE results). The simulated degree of substitution between continuing education and YTS is substantial, with the duration rising by 4–9 percentage points in every case. The rise is largest for individuals disadvantaged by ethnicity, health or social/educational background, but also for those with a modestly increased level of school examination achievement relative to the base case.

16.6. CONCLUDING REMARKS

We have estimated a large and highly complex transition model designed to address the formidable problems of understanding the role played by government training schemes in the labour market experience of school-leavers. The question 'what is the effect of YTS?' is a remarkably complex one, and we have looked at its various dimensions using stochastic simulation of the estimated model. Abstracting from endogenous (self-)selection into YTS, we have found evidence suggesting a significant improvement in subsequent employment prospects for those who do go through YTS, particularly in the case of YTS 'stayers'. This is a rather more encouraging conclusion than that of Dolton, Makepeace and Treble (1994), and is roughly in line with the earlier applied literature, based on less sophisticated statistical models. Our results suggest that, for the first five years after reaching school-leaving age, YTS appears mainly to have absorbed individuals who would otherwise have gone into unemployment or stayed on in the educational system. The employment promotion effect of YTS among 16–21-year-olds might in contrast be judged worthwhile but modest. Our estimated model is not intended to have any direct application to a period longer than the five-year simulation period we have used. However, arguably, these results do give us some grounds for claiming the existence of a positive longer term effect for YTS. The increased employment probabilities induced by YTS naturally occur in the late post-YTS part of the five-year history we have simulated. As a result, we can conclude that, conditional on observables and persistent unobservable characteristics, a greater proportion of individuals can be expected to reach age 21 in employment, if YTS has been available during the previous five years, than would otherwise be the case. On the reasonable assumption of a relatively high degree of employment stability after age 21, this suggests a strong, positive, long-term effect of YTS on employment probabilities.

APPENDIX

Table A1 Variables used in the models.

Variable	Definition	Mean
Time-invariant characteristics (mean over all individuals):		
DOB	Date of birth (years after 1 January 1960)	12.16
WHITE	Dummy = 1 if white; 0 if other ethnic origin	0.942
GCSE2	Dummy for at least 1 General Certificate of Secondary Education (GCSE) pass at grade D or E	0.263
GCSE3	Dummy for 1–3 GCSE passes at grade C or better	0.185
GCSE4	Dummy for at least 4 GCSE passes at grade C or better	0.413
ILL	Dummy for the existence of a major health problem	0.012
SCHOOL	Measure of school quality = proportion of pupils with at least 5 GCSE passes in first published school league table	0.384
AREA	Measure of social background = proportion of homes in ward of residence that are owner-occupied	0.779
Spell-specific variables (mean over all episodes):		
DATE	Date of the start of spell (years since 1 January 1988)	1.11
YTSYET	Dummy for existence of a spell of YTS prior to the current spell	0.229
YTSDUR	Total length of time spent on YTS prior to the current spell (years)	0.300
YTSLIM	Dummy = 1 if two-year limit on YTS was reached prior to the current spell	0.094
YTSMATCH	Dummy = 1 if current spell is in employment and there was a previous YTS spell in the same industrial sector	0.121
CLERICAL	Dummy = 1 if current spell is in clerical employment	0.036
TECH	Dummy = 1 if current spell is in craft/technical employment	0.135
STN	Dummy = 1 for YTS spell and trainee with special training needs	0.013
CHOICE	Dummy = 1 if last or current YTS spell in desired industrial sector	0.136
URATE	Local rate of unemployment (at ward level)	0.103

Table A2   Sample transition frequencies (%).

(a) Initial spell

State of origin	Destination state						Marginal
	C	E	U	YTS	Attrition	Incomplete	
C	—	7.7	12.3	4.8	4.9	70.3	47.4
E	3.0	—	17.8	15.4	0.9	62.9	10.6
U	9.1	27.2	—	57.6	5.5	0.7	28.2
YTS	1.6	73.9	11.9	—	10.1	2.5	13.8
Marginal	3.1	21.5	9.7	20.2	5.1	40.5	100

(b) All spells

State of origin	Destination state						Marginal
	C	E	U	YTS	Attrition	Incomplete	
C	—	8.1	13.4	4.9	4.8	68.8	25.1
E	0.7	—	13.2	5.8	0.4	79.9	30.5
U	5.3	37.8	—	41.2	13.3	2.4	24.5
YTS	1.3	70.6	16.5	—	8.2	3.4	19.9
Marginal	1.8	25.3	10.7	13.1	6.2	42.9	100

Table A3   Mean durations (years).

	C	E	U	YTS
Mean duration for completed spells	1.00	0.57	0.27	1.48
Mean elapsed duration for both complete and incomplete spells	2.95	2.17	0.31	1.52

Table A4   Estimates: initial state logit component (standard errors in parentheses).

Covariate	Destination state (relative to YTS)		
	C	E	U
Constant	1.575 (0.36)	0.321 (0.38)	3.835 (0.47)
WHITE	−2.580 (0.34)	—	−0.867 (0.37)
GCSE2	0.530 (0.16)	—	—
GCSE3	1.284 (0.20)	0.212 (0.22)	0.307 (0.16)
GCSE4	2.985 (0.20)	0.473 (0.24)	0.522 (0.17)
ILL	—	−2.311 (1.10)	—
SCHOOL	1.855 (0.31)	—	1.236 (0.35)
AREA	—	—	−1.818 (0.33)
URATE	—	−8.607 (2.57)	−12.071 (1.95)

Table A5(a)   Estimates: transition component (standard errors in parentheses).

Coefficient	Destination-specific transition intensities				
$(\hat{\beta}_\ell)$	C	E	U	O	YTS
Constant	−1.377 (1.85)	−5.924 (0.57)	−1.820 (0.71)	−6.926 (0.78)	−4.481 (1.11)
DATE	−8.090 (2.99)	—	—	0.795 (0.11)	−1.884 (0.14)
YTSYET	—	—	1.461 (0.43)	—	—
YTSDUR	—	0.762 (0.18)	−1.328 (0.46)	−0.198 (0.17)	—
YTSLIMIT	—	−2.568 (0.71)	−3.234 (0.75)	—	—
YTSMATCH	—	—	−0.610 (0.50)	—	—
CLERICAL	—	—	−0.865 (0.53)	—	—
STN	—	—	1.158 (0.41)	—	—
CHOICE	—	−0.335 (0.15)	—	—	—
WHITE	−1.919 (0.77)	1.433 (0.28)	−0.751 (0.32)	—	1.007 (0.29)
GCSE2	2.150 (0.61)	—	−0.666 (0.20)	—	0.437 (0.18)
GCSE3	2.369 (0.88)	−0.700 (0.17)	−1.233 (0.24)	−1.115 (0.33)	−1.036 (0.45)
GCSE4	3.406 (0.94)	−0.939 (0.18)	−2.046 (0.26)	−2.221 (0.32)	−1.642 (0.45)
ILL	—	—	−0.642 (0.39)	—	0.964 (0.75)
E	—	—	4.782 (0.59)	−0.962 (0.61)	3.469 (1.05)
U	5.530 (0.73)	6.654 (0.54)	—	5.079 (0.58)	6.066 (1.04)
YTS	—	3.558 (0.49)	2.853 (0.41)	—	—
U* (GCSE3/4)	−0.447 (0.72)	0.927 (0.22)	—	1.197 (0.36)	1.635 (0.45)
SCHOOL	—	0.233 (0.32)	−0.690 (0.47)	−1.389 (0.45)	1.451 (0.50)
AREA	1.512 (1.23)	—	−1.628 (0.51)	—	—
URATE	—	−3.231 (1.99)	−2.630 (2.62)	8.488 (3.32)	5.724 (3.20)
College μ_1			0.817 (0.16)		
College μ_2			1.516 (0.16)		

Table A5(b)   Estimates: Burr shape parameters (standard errors in parentheses).

	Destination				
Origin	C	E	U	O	YTS
			$\alpha_{k\ell}$		
C	—	1.341 (0.15)	1.852 (0.16)	1.636 (0.13)	1.167 (0.22)
E	0.356 (0.45)	—	1.528 (0.21)	1.636 (0.13)	1.190 (0.22)
U	2.667 (0.25)	1.601 (0.10)	—	1.636 (0.13)	1.722 (0.11)
YTS	0.592 (0.58)	1.427 (0.12)	1.100 (0.13)	1.636 (0.13)	—
			$\sigma_{k\ell}$		
C	—	2.494 (0.62)	1.171 (0.47)	0.555 (0.26)	5.547 (1.48)
E	0.414 (1.70)	—	4.083 (0.43)	0.555 (0.26)	5.465 (0.75)
U	2.429 (0.41)	1.652 (0.13)	—	0.555 (0.26)	1.508 (0.12)
YTS	5.569 (4.45)	1.018 (0.36)	1.315 (0.40)	0.555 (0.26)	—

Table A6 Estimates: YTS limit logit (standard errors in parentheses).

Parameter	Coefficients for state E
Constant	5.205 (1.25)
Heterogeneity (θ_E)	−1.493 (0.81)

Table A7 Estimates: coefficients of random effects and correlation parameter (standard errors in parentheses).

	State			
	C	E	U	YTS
Initial state logit (ψ_k)	0.075 (0.10)	1.106 (0.39)	0.160 (0.23)	0.521 (0.23)
Transition model (ω_k)	3.832 (0.33)	−0.248 (0.21)	−0.335 (0.13)	−0.586 (0.20)
Correlation parameter (ρ)			−0.224 (0.04)	

ACKNOWLEDGEMENTS

We are grateful to the Economic and Social Research Council for financial support under grants R000234072 and H519255003. We would also like to thank Martyn Andrews, Steve Bradley and Richard Upward for their advice and efforts in constructing the database used in this study. Pinuccia Calia provided valuable research assistance. We received helpful comments from seminar participants at the Universities of Florence and Southampton.

Incomplete Data

CHAPTER 17

Introduction to Part E

R. L. Chambers

17.1. INTRODUCTION

The standard theoretical scenario for survey data analysis is one where a sample is selected from a target population, data are obtained from the sampled population units and these data, together with auxiliary information about the population, are used to make an inference about either a characteristic of the population or a parameter of a statistical model for the population. An implicit assumption is that the survey data are "complete" as far as analysis is concerned and so can be represented by a rectangular matrix, with rows corresponding to sample units and columns corresponding to sample variables, and with all cells in the matrix containing a valid entry. In practice, however, complex survey design and sample nonresponse lead to survey data structures that are more complicated than this, and different amounts of information relevant to the parameters of interest are often available from different variables. In many cases this is because these variables have different amounts of "missingness". Skinner and Holmes provide an illustration of this in Chapter 14 where variables in different waves of a longitudinal survey are recorded for different numbers of units because of attrition.

The three chapters making up Part E of this book explore survey data analysis where relevant data are, in one way or another, and to different extents, missing. The first, by Little (Chapter 18), develops the Bayesian approach to inference in general settings where the missingness arises as a consequence of different forms of survey nonresponse. In contrast, Fuller (Chapter 19) focuses on estimation of a finite population total in the specific context of two-phase sampling. This is an example of "missingness by design" in the sense that an initial sample (the first-phase sample) is selected and a variable X observed. A subsample of these units is then selected (the second-phase sample) and another variable Y observed for the subsampled units. Values of Y are then missing for units in the first-phase sample who are not in the second-phase sample. As we show in the following section, this chapter is also relevant to the case of item nonresponse on a variable Y where imputation may make use of information on another survey variable X. Finally,

Analysis of Survey Data Edited by R. L. Chambers and C. J. Skinner

Steel, Tranmer and Holt (Chapter 20) consider the situation where either the unit level data of interest are missing completely, but detailed aggregates corresponding to these data are available, or a limited amount of survey information is available, but not sufficient to allow linkage of analysis and design variables in the modelling process.

In Chapter 18 Little builds on the Bayesian theory introduced in Chapter 4 to develop survey inference methods when data are missing because of nonresponse. Following standard practice he differentiates between unit nonresponse, where no data are available for survey non-respondents, and item nonresponse, where partial data are available for survey non-respondents. In the first case (unit nonresponse), the standard methodology is to compensate for this nonresponse by appropriately reweighting the responding sample units, while in the second case (item nonresponse) it is more usual to pursue an imputation strategy, and to fill in the "holes" in the rectangular complete data matrix by best guesses about the missing variable values. From a theoretical perspective, this distinction is unnecessary. Unit nonresponse is just an extreme form of item nonresponse, and both weighting (or more generally, estimation) and imputation approaches can be developed for this case. In Section 17.3 below we briefly sketch the arguments for either approach and relate them to the Bayesian approach advocated by Little.

Fuller's development in Chapter 19 adopts a model-assisted approach to develop estimation under two-phase sampling, using a linear regression model for Y in terms of X to improve estimation of the population mean of Y. Since two-phase sampling is a special case of item nonresponse, the same methods used to deal with item nonresponse can be applied here. In Section 17.4 below we describe the approach to estimation in this case. Since imputation-based methods are a standard way of dealing with item nonresponse, a natural extension is where the second-phase sample data provide information for imputing the "missing" Y-values in the first-phase sample.

The development by Steel, Tranmer and Holt (Chapter 20) is very different in its orientation. Here the focus is on estimation of regression parameters, and missingness arises through aggregation, in the sense that the survey data, to a greater or lesser extent, consist not of unit-level measurements but of aggregate measurements for identified groups in the target population. If these group-level measurements are treated as equivalent to individual measurements then analysis is subject to the well-known ecological fallacy (Cleave, Brown and Payne 1995; King, 1997; Chambers and Steel, 2001). Steel, Tranmer and Holt in fact discuss several survey data scenarios, corresponding to different amounts of aggregated data and different sources for the survey data, including situations where disaggregated data are available from conventional surveys and aggregated data are available from censuses or registers. Integration of survey data from different sources and with different measurement characteristics (in this case aggregated and individual) is fast becoming a realistic practical option for survey data analysts, and the theory outlined in Chapter 20 is a guide to some of the statistical issues that need to be faced when such an analysis is attempted. Section 17.5 provides an introduction to these issues.

17.2. AN EXAMPLE

To illustrate the impact of missingness on survey-data-based inference, we return to the maximum likelihood approach outlined in Section 2.2, and extend the normal theory application described in Example 1 there to take account of survey nonresponse.

Consider the situation where the joint population distribution of the survey variables Y and X and the auxiliary variable Z is multivariate normal, and we wish to estimate the unknown mean vector $\mu = (\mu_Y, \mu_X, \mu_Z)'$ of this distribution. As in Example 1 of Chapter 2 we assume that the covariance matrix

$$\Sigma = \begin{bmatrix} \sigma_{YY} & \sigma_{YX} & \sigma_{YZ} \\ \sigma_{XY} & \sigma_{XX} & \sigma_{XZ} \\ \sigma_{ZY} & \sigma_{ZY} & \sigma_{ZZ} \end{bmatrix}$$

of these variables is known and that sampling is noninformative given the population values of Z, so the population vector z_u of sample inclusion indicators $\{i_t; t = 1, \ldots, N\}$ can be ignored in inference. In what follows we use the same notation as in that example.

Now suppose that there is complete response on X, but only partial response on Y. Further, this nonresponse is noninformative given Z, so the respondent values of Y, the sample values of X and the population values of Z define the survey data. Let the original sample size be n, with n_{obs} respondents on Y and $n_{mis} = n - n_{obs}$ non-respondents. Sample means for these two groups will be denoted by subscripts of obs and mis below, with the vector of respondent values of Y denoted by y_{obs}. Let E_{obs} denote expectation conditional on the survey data y_{obs}, x_s, z_u and let E_s denote expectation conditional on the (unobserved) 'complete' survey data y_s, x_s, z_u. The score function for μ is then

$$sc_{obs}(\mu) = E_{obs}(E_s(sc(\mu))) = \Sigma^{-1} \left[n \begin{pmatrix} E_{obs}(\bar{y}_s) - \mu_Y \\ \bar{x}_s - \mu_X \\ \bar{z}_s - \mu_Z \end{pmatrix} + \frac{(N-n)}{\sigma_{ZZ}} \begin{pmatrix} \sigma_{YZ} \\ \sigma_{XZ} \\ \sigma_{ZZ} \end{pmatrix} (\bar{z}_{U-s} - \mu_Z) \right]$$

where $\bar{z}_{U-s}$ denotes the average value of Z over the $N - n$ non-sampled population units. We use the properties of the normal distribution to show that

$$E_{obs}(\bar{y}_s) = \frac{1}{n}[n_{obs}\bar{y}_{obs} + n_{mis}(\mu_Y + \lambda_X(\bar{x}_{mis} - \mu_X) + \lambda_Z(\bar{z}_{mis} - \mu_Z))]$$

where λ_X and λ_Z are the coefficients of X and Z in the regression of Y on *both* X and Z. That is,

$$\lambda_X = \frac{\sigma_{ZZ}\sigma_{YX} - \sigma_{YZ}\sigma_{XZ}}{\sigma_{ZZ}\sigma_{XX} - \sigma_{XZ}^2} \quad \text{and} \quad \lambda_Z = \frac{\sigma_{XX}\sigma_{YZ} - \sigma_{YX}\sigma_{XZ}}{\sigma_{ZZ}\sigma_{XX} - \sigma_{XZ}^2}.$$

The MLE for μ_Y is defined by setting the first component of $sc_{obs}(\mu)$ to zero and solving for μ_Y. This leads to an MLE for this parameter of the form

$$\hat{\mu}_Y = \bar{y}_{obs} + \frac{\sigma_{YZ}}{\sigma_{ZZ}}(\hat{\mu}_Z - \bar{z}_{obs}) - \frac{n_{mis}}{n_{obs}}\left\{(\hat{\mu}_Z - \bar{z}_{mis})\left(\lambda_Z - \frac{\sigma_{YZ}}{\sigma_{ZZ}}\right) + \lambda_X(\hat{\mu}_X - \bar{x}_{mis})\right\}.$$

The MLEs for μ_X and μ_Z are clearly unchanged from the regression estimators defined in (2.6). The sample information for μ is obtained by applying (2.5). This leads to

$$
\begin{aligned}
info_{obs}(\mu) &= E_{obs}(info(\mu)) - \mathrm{var}_{obs}(sc(\mu)) \\
&= E_{obs}(E_s(info(\mu))) - E_{obs}(\mathrm{var}_s(sc(\mu))) - \mathrm{var}_{obs}(E_s(sc(\mu))) \\
&= E_{obs}(info_s(\mu)) - \mathrm{var}_{obs}(sc_s(\mu)) \\
&= \Sigma^{-1}\left[N\left\{\Sigma - \left(1 - \frac{n}{N}\right)C\right\} - n_{mis}D\right]\Sigma^{-1}
\end{aligned}
$$

where

$$D = \begin{bmatrix} \sigma^2_{Y|XZ} & 0 & 0 \\ 0 & 0 & 0 \\ 0 & 0 & 0 \end{bmatrix}$$

and $\sigma^2_{Y|XZ}$ is the conditional variance of Y given X and Z.

Comparing the above expression for $info_{obs}(\mu)$ with that obtained for $info_s(\mu)$ in Example 1 in Chapter 2, we can see that the loss in information due to missingness in Y is equal to $n_{mis}\Sigma^{-1}D\Sigma^{-1}$. In particular, this implies that there is little loss of information if X and Z are good "predictors" of Y (so $\sigma^2_{Y|XZ}$ is close to zero).

17.3. SURVEY INFERENCE UNDER NON-RESPONSE

A key issue that must be confronted when dealing with missing data is the relationship between the missing data indicator, R, the sample selection indicator, I, the survey variable(s), Y, and the auxiliary population variable, Z. The concept of *missing at random* or MAR is now well established in the statistical literature (Rubin, 1976; Little and Rubin, 1987), and is strongly related to the concept of noninformative nonresponse that was defined in Chapter 2. Essentially, MAR requires that the conditional distribution, given the matrix of population covariates z_u and the (unobserved) 'complete' sample data matrix y_s, of the matrix of sample nonresponse indicators r_s generated by the survey nonresponse mechanism, be the same as that of the conditional distribution of this outcome given z_u and the observed sample data y_{obs}. Like the noninformative nonresponse assumption, it allows a useful factorisation of the joint distribution of the survey data that allows one to condition on just y_{obs}, i and z_u in inference.

In Chapter 18 Little extends this idea to the operation of the sample selection mechanism, defining *selection at random* or SAR where the conditional distribution of the values in i given z_u and the population vector y is the same as its distribution given **z** and the observed values y_{obs}. The relationship of this idea to noninformative sampling (see Chapters 1 and 2) is clear, and, as with that definition, we see that SAR allows one to ignore essentially the sample selection mechanism in inference about either the parameters defining the distribution of y_u, or population quantities defined by this matrix. In particular, probability sampling, where i_u is completely determined by z_u, corresponds to SAR and in this case the selection mechanism can be ignored in inference provided inference conditions on z_u. On the other hand, as Little points out, the MAR assumption is less easy to justify. However, it is very popular, and often serves as a convenient baseline assumption for inference under nonresponse. Under both MAR and SAR it is easy to see that inference depends only on the joint distribution of y_u and z_u, as the example in the previous section demonstrates. Following the development in this example, we therefore now assume that Y in fact is made up of two variables, Y and X, where Y has missingness, but X is complete. Obviously, under unit nonresponse X is "empty".

The prime focus of the discussion in Chapter 18 is prediction of the finite population mean $\bar{y}_U$. The minimum mean squared error predictor of this mean is its conditional expectation, given the sample data. Under SAR and MAR, these data correspond to y_{obs}, x_s and z_u. Let $\mu(x_t, z_t; \lambda)$ denote the conditional expectation of Y given $X = x_t$ and $Z = z_t$, with the corresponding expectation of Y given $Z = z_t$ denoted $\mu(z_t; \lambda, \phi)$. Here λ and ϕ are unknown parameters. For simplicity we assume distinct population units are mutually independent. This optimal predictor then takes the form

$$\hat{\bar{y}}_U^{opt} = N^{-1} \left(\sum_{obs} y_t + \sum_{mis} \mu(x_t, z_t; \lambda) + \sum_{U-s} \mu(z_t; \lambda, \phi) \right) \qquad (17.1)$$

where *obs* and *mis* denote the respondents and non-respondents in the sample and $U - s$ denotes the non-sampled population units. A standard approach is where X is categorical, defining so-called *response homogeneity* classes. When Z is also categorical, reflecting sampling strata, this predictor is equivalent to a weighted combination of response cell means for Y, where these cells are defined by the joint distribution of X and Z. These means can be calculated from the respondent values of Y, and substituted into (17.1) above, leading to a weighted estimator of $\bar{y}_U$, with weights reflecting variation in the distribution of the sample respondents across these cells. This estimator is often referred to as the poststratified estimator, and its use is an example of the *weighting* approach to dealing with survey nonresponse. Of course, more sophisticated models linking Y, X and Z can be developed, and (17.1) will change as a consequence. In Chapter 18 Little describes a random effects specification that allows "borrowing strength" across response cells when estimating the conditional expectations in (17.1). This decreases the variability of the cell mean estimates but at the potential cost of incurring model misspecification bias due to non-exchangeability of the cell means.

The alternative approach to dealing with nonresponse is *imputation*. The form of the optimal predictor (17.1) above shows us that substituting the unknown values in y_{mis} by our best estimate of their conditional expectations given the observed sample values in x_s and the population values in z_u will result in an estimated mean that is identical to the optimal complete data estimator of $\bar{y}_U$. One disadvantage with this method of imputation is that the resulting sample distribution of Y (including both real and imputed values) then bears no relation to the actual distribution of such values that would have been observed had we actually measured Y for the entire sample. Consequently other analyses that treat the imputed dataset as if it were actually measured in this way are generally invalid. One way to get around this problem is to adopt what is sometimes called a "regression hot deck" imputation procedure. Here each imputed Y-value is defined to be its estimated conditional expectation given its observed values of X and Z plus an imputed residual drawn via simple random sampling with replacement from the residuals $y_t - \mu(x_t, z_t; \hat{\lambda})$, $t \in obs$. The imputed Y-values obtained in this way provide a better estimate of the unobserved true distribution of this variable over the sample.

By using a standard iterated expectation argument, it is easy to see that this type of single-value imputation introduces an extra component of variability that is typically referred to as imputation variance. Furthermore, the resulting estimator (which treats the observed and imputed data values on an equal footing) is not optimal in the sense that it does not coincide with the minimum mean squared error predictor (17.1) above. This can be overcome by adopting a multiple imputation approach (Rubin, 1987), as Little demonstrates in Chapter 18. Properly implemented, with imputations drawn from the posterior distribution of y_{mis}, multiple imputation can also lead to consistent variance estimates, and hence valid confidence intervals. However, such a proper implementation is non-trivial, since most commonly used methods of imputation do not correspond to draws from a posterior distribution. An alternative approach that is the subject of current research is so-called controlled or calibrated imputation, where the hot deck sampling method is modified so that the imputation-based estimator is identical to the optimal estimator (17.1). In this case the variance of the imputation estimator is identical to that of this estimator.

17.4. A MODEL-BASED APPROACH TO ESTIMATION UNDER TWO-PHASE SAMPLING

In Chapter 19 Fuller considers estimation of $\bar{y}_U$ under another type of missingness. This is where values of Y are only available for a "second-phase" probability subsample s_2 of size n_2 drawn from the original "first-phase" probability sample s_1 of size n_1 that is itself drawn from the target population of size N. Clearly, this is equivalent to Y-values being missing for sample units in the probability subsample $s_1 - s_2$ of size $n_1 - n_2$ drawn from s_1. As above, we assume that values of another variable X are available for all units in s_1. However, following Fuller's development, we ignore the variable Z, assuming that all probabilistic quantities are conditioned upon it.

Consider the situation where X is bivariate, given by $X = (1\,D)$, so $x_t = (1\,d_t)$, with population mean $E_U(y_t|x_t) = \mu(x_t; \beta) = \beta_0 + d_t\gamma$, in which case $E_U(y_t) = \beta_0 + \eta_t(\phi)\gamma$, where $\eta_t(\phi)$ denotes the expected value of d_t. In this case (17.1) can be expressed

$$
\begin{aligned}
\hat{\bar{y}}_U^{opt} &= N^{-1}\left(\sum_{s_2} y_t + \sum_{s_1-s_2} \beta_0 + d_t\gamma + \sum_{U-s_1} \beta_0 + \eta_t(\phi)\gamma \right) \\
&= N^{-1}(n_2\bar{y}_2 + (N - n_2)\beta_0 + (n_1\bar{d}_1 - n_2\bar{d}_2 + N\bar{\eta}(\phi) - n_1\bar{\eta}_1(\phi))\gamma) \\
&= \beta_0 + \bar{\eta}(\phi)\gamma + N^{-1}(n_2(\bar{y}_2 - \beta_0 - \bar{d}_2\gamma) + n_1(\bar{d}_1 - \bar{\eta}_1(\phi))\gamma)
\end{aligned}
$$

Here $\bar{\eta}(\phi)$ and $\bar{\eta}_1(\phi)$ denote the population and first phase sample averages of the $\eta_t(\phi)$ respectively.

In practice the parameters β_0, γ and ϕ are unknown, and must be estimated from the sample data. Optimal estimation of these parameters can be carried out using, for example, the maximum likelihood methods outlined in Chapter 2. More simply, however, suppose sampling at both phases is noninformative and we assume a model for Y and D where these variables have first- and second-order population moments given by $E_U(Y|D) = \beta_0 + \gamma D, E_U(D) = \phi, \text{var}_U(Y|D = d) = \sigma^2$ and $\text{var}_U(D) = \omega^2$. Under this model the least squares estimators of β_0, γ, and ϕ are optimal. Note that the least squares estimators of β_0 and γ are based on the phase 2 sample data, while that of ϕ is the phase 1 sample mean of D. When these are substituted into the expression for $\hat{\bar{y}}_U^{opt}$ in the preceding paragraph it is easy to see that it reduces to the standard regression estimator $\bar{y}_{reg} = \bar{y}_2 + (\bar{d}_1 - \bar{d}_2)\hat{\gamma}$ of the population mean of Y. In fact, the regression estimator corresponds to the optimal predictor (17.1) when Y and D are assumed to have a joint bivariate normal distribution over the population and maximum likelihood estimates of the parameters of this distribution are substituted for their unknown values in this predictor. In contrast, in Chapter 19 Fuller uses sample inclusion probability-weighted estimators for β_0, γ and ϕ, which is equivalent to substituting Horvitz–Thompson pseudo-likelihood estimators of these parameters into the optimal predictor.

Variance estimation for this regression estimator can be carried out in a number of ways. In Chapter 18 Fuller describes a design-based approach to this problem. Here we sketch a corresponding model-based development. From this perspective, we aim to estimate the prediction variance $\text{var}_U(\bar{y}_{reg} - \bar{y})$ of this estimator. When the first-phase sample size n_1 is small relative to the population size N, this prediction variance is approximately the model-based variance of the regression estimator itself. Noting that the regression estimator can be written $\bar{y}_{reg} = \hat{\beta}_0 + \hat{\phi}\hat{\gamma}$, its model-based variance can be estimated using the Taylor series approximation

$$
\text{var}_U(\bar{y}_{reg}) = (1 \quad \hat{\phi} \quad \hat{\gamma})\, info_s^{-1}(\hat{\beta}_0, \hat{\gamma}, \hat{\phi})(1 \quad \hat{\phi} \quad \hat{\gamma})'
$$

where $info_s^{-1}(\hat{\beta}_0, \hat{\gamma}, \hat{\phi})$ is the inverse of the observed sample information matrix for the parameters β_0, γ and ϕ. This information matrix can be calculated using the theory outlined in Section 2.2. Equivalently, since $\bar{y}_{reg}$ is in fact the maximum likelihood estimator of the marginal expected value μ of Y, we can use the equivalence of two-phase sampling and item nonresponse to apply a simplified version of the analysis described in the example in Section 17.2 (dropping the variable Z) to obtain

$$info_s(\mu, \phi) = V^{-1}[n_1 V - (n_1 - n_2)W]V^{-1} = n_1 V^{-1} - (n_1 - n_2)V^{-1}WV^{-1}$$

where V denotes the variance–covariance matrix of Y and D,

$$V = \begin{bmatrix} \gamma^2\omega^2 + \sigma^2 & \gamma\omega^2 \\ \gamma\omega^2 & \omega^2 \end{bmatrix}$$

and

$$W = \begin{bmatrix} \sigma^2 & 0 \\ 0 & 0 \end{bmatrix}.$$

The parameters σ^2 and ω^2 can be estimated from the phase 2 residuals $y_t - \bar{y}_2 - (d_t - d_2)\hat{\gamma}$ and phase 1 residuals $d_t - \bar{d}_1$ respectively, and the model-based variance of the regression estimator then estimated by the [1,1] component of the inverse of the sample information matrix $info_s(\mu, \phi)$ above, with unknown parameters substituted by estimates.

Given the equivalence of two-phase estimation and estimation under item nonresponse, it is not surprising that a more general imputation-based approach to two-phase estimation can be developed. In particular, substituting the missing values of Y by estimates of their conditional expectations given their observed values of D will result in a phase 1 mean of both observed and imputed Y-values that is identical to the regression estimator.

As noted in Section 17.2, a disadvantage with this 'fixed value' imputation approach is that the phase 1 sample distribution of Y-values (both real and imputed) in our data will then bear no relation to the actual distribution of such values that would have been observed had we actually measured Y for the entire phase 1 sample. This suggests we adapt the regression hot deck imputation idea to two-phase estimation.

To be more specific, for $t \in s_1$, put $\tilde{y}_t$ equal to y_t if $t \in s_2$ and equal to $\hat{\beta}_0 + d_t\hat{\gamma} + \tilde{r}_t$ otherwise. Here $\tilde{r}_t$ is drawn at random, and with replacement, from the phase 2 residuals $r_u = y_u - \bar{y}_2 - (d_u - \bar{d}_2)\hat{\gamma}, u \in s_2$. The imputed mean of Y over the phase 1 sample is then

$$\tilde{\bar{y}}_1 = n_1^{-1} \sum_{s_1} \tilde{y}_t = \bar{y}_{reg} + n_1^{-1} \sum_{s_1-s_2} \tilde{r}_t = \bar{y}_{reg} + n_1^{-1} \sum_{s_2} \delta_t r_t \qquad (17.2)$$

where δ_t denotes the number of times unit $t \in s_2$ is a 'donor' in this process. Since the residuals are selected using simple random sampling it is easy to see that the values $\{\delta_t, t \in s_2\}$ correspond to a realisation of a multinomial random variable with $n_1 - n_2$ trials and success probability vector with all components equal to $1/n_2$. Consequently, when we use a subscript of s to denote conditioning on the sample data,

$$E_U(\tilde{\bar{y}}_1) = E_U\left(E_s(\tilde{\bar{y}}_1)\right)$$

$$= E_U\left(\bar{y}_{reg} + n_1^{-1}\sum_{s_2} E_s(\delta_t)r_t\right) = E_U\left(\bar{y}_{reg} + \left(\frac{1}{n_2} - \frac{1}{n_1}\right)\sum_{s_2} r_t\right)$$

and

$$\text{var}_U(\tilde{\bar{y}}_1) = \text{var}_U\left(E_s(\tilde{\bar{y}}_1)\right) + E_U\left(\text{var}_s(\tilde{\bar{y}}_1)\right)$$

$$= \text{var}_U\left(\bar{y}_{reg} + \left(\frac{1}{n_2} - \frac{1}{n_1}\right)\sum_{s_2} r_t\right)$$

$$+ E_U\left(\frac{1}{n_1}\left(\frac{1}{n_2} - \frac{1}{n_1}\right)\left[\sum_{s_2} r_t^2 - \frac{1}{n_2}\left(\sum_{s_2} r_t\right)^2\right]\right).$$

By construction, the phase 2 residuals sum to zero over s_2, and so the imputed phase 1 mean $\tilde{\bar{y}}_1$ and the regression estimator $\bar{y}_{reg}$ have the same expectation. However, the variance of $\tilde{\bar{y}}_1$ contains an extra component of variability proportional to the expected value of the sum of the squared phase 2 residuals. This is the imputation variance component of the 'total' variability of $\tilde{\bar{y}}_1$. Note that the model-based variance of this imputed mean is not estimated by calculating the variance of the $\tilde{y}_t$ over s_1 and dividing by n_1. From the above expression for the model-based variance of $\tilde{\bar{y}}_1$ we can see that the appropriate estimator of this variance is

$$\hat{V}(\tilde{\bar{y}}_1) = \hat{V}(\bar{y}_{reg}) + \frac{n_2 - 2}{n_1}\left(\frac{1}{n_2} - \frac{1}{n_1}\right)\hat{\sigma}^2$$

where $\hat{V}(\bar{y}_{reg})$ denotes the estimator of the model-based variance of the regression estimator that was developed earlier, and $\hat{\sigma}^2$ is the usual estimator of the error variance for the linear regression of Y on D in the phase 2 sample.

Controlled or calibrated imputation is where the imputations are still random, but now are also subject to the restriction that the estimator based on the imputed data recovers the efficient estimator based on the observed sample data. In Chapter 19 Fuller applies this idea in the context of two-phase sampling. Here the efficient estimator is the regression estimator, and the regression hot deck imputation scheme, rather than being based on simple random sampling, is such that $\tilde{\bar{y}}_1$ and the regression estimator $\bar{y}_{reg}$ coincide. From (17.2) one can easily see that a sufficient condition for this to be the case is where the sum of the imputed residuals $\tilde{r}_t$ is zero. Obviously, under this type of imputation the variance of the imputed mean and the variance of the regression estimator are the same, and no efficiency is lost by adopting an imputation-based approach to estimation.

17.5. COMBINING SURVEY DATA AND AGGREGATE DATA IN ANALYSIS

Many populations that are of interest in the social sciences have a natural "group" structure, reflecting social, political, demographic, geographic and economic differences between different population units in the groups. Probably the most common example is hierarchical grouping, with individuals grouped into households that are in turn grouped into neighbourhoods and so on. A key characteristic of the units making up such groups is that they tend to be more like other units from the same group than units from other groups. This intra-group correlation is typically reflected in the data collected in multistage social surveys that use the groups as sampling units. Analysis of these data is then multilevel in character, incorporating the group structure into the models fitted to these data. However, when the groups themselves are defined for administrative or economic reasons, as when dwellings are grouped into census enumeration districts in order to facilitate data collection, it seems inappropriate to define target parameters that effectively condition on these rather arbitrary groupings. In these cases there is a strong argument for "averaging" in some way over the groups when defining these parameters.

In Chapter 20, Steel, Tranmer and Holt adopt this approach. More formally, they define a multivariate survey variable W for each unit in the population, and an indicator variable C that indicates group membership for a unit. As usual, auxiliary information is defined by the vector-valued variable Z. Their population model then conditions on the group structure defined by C, in the sense that they use a linear mixed model specification to characterise group membership, with realised value w_t for unit t in group g in the population satisfying

$$w_t = \mu_{w|z} + \beta'_{w|z} z_t + v_g + \varepsilon_t. \qquad (17.3)$$

Here v_g is a random "group effect" shared by all units in group g and ε_t is a unit-specific random effect. Both these random effects are assumed to have zero expectations, constant covariance matrices $\Sigma^{(2)}_{w|z}$ and $\Sigma^{(1)}_{w|z}$ respectively, and to be uncorrelated between different groups and different units. The aim is to estimate the unconditional covariance matrix Σ of W in the population, given by

$$\Sigma = \Sigma^{(1)}_{w|z} + \Sigma^{(2)}_{w|z} + \beta'_{w|z} \Sigma_{zz} \beta_{w|z}. \qquad (17.4)$$

If W corresponds to a bivariate variable with components Y and X then the marginal regression of Y on X in the population is defined by the components of Σ, so this regression can be estimated by substitution once an estimate of Σ is available.

Steel, Tranmer and Holt describe a number of sample data scenarios, ranging from the rather straightforward one where unit-level data on W and Z, along with group identifiers, are available to the more challenging situation where the survey data just consist of group-level sample averages of W and Z. In the former case Σ can be estimated via substitution of estimates for the parameters of the right hand side of (17.4) obtained by using a multilevel modelling package to fit (17.3) to these data, while in the latter case, estimation of Σ from these group averages generally leads to aggregation bias (i.e. the ecological fallacy).

In an effort to minimise this bias, these authors consider an "augmented" data situation where a limited amount of unit-level values of Z are available in addition to the group averages. This situation, for example, occurs where a unit-level sample file, without group indicators, is released for selected census variables (including Z) and at the same time comprehensive group-level average information is available for a much wider set of census outputs (covering both W and Z). In Section 20.5 they note that an estimate $\hat{\Sigma}_{zz}$ of the population unit-level covariance of Z obtained from the unit-level data can be combined, using (17.4), with estimates of $\Sigma_{w|z} = \Sigma_{w|z}^{(1)} + \Sigma_{w|z}^{(2)}$ and $\beta_{w|z}$ calculated from the group-level average dataset to obtain an adjusted estimate of Σ. However, since both these group-level estimators are biased for their unit-level values, with, for example, the bias of the group level estimator of $\Sigma_{w|z}$ essentially increasing with the product of the average group size and $\Sigma_{w|z}^{(2)}$, it is clear that unless Z substantially "explains" the intra-group correlations in the population (i.e. $\Sigma_{w|z}^{(2)}$ is effectively zero) or the sample sizes underpinning the group averages are very small, this adjusted estimator of Σ will still be biased, though its bias will be somewhat smaller than the naive group-level estimator of Σ that ignores the information in Z.

Before closing, it is interesting to point out that the entire analysis so far has been under what Steel (1985) calls a "group dependence" model, in the sense that Σ, the parameter of interest, has been defined as the sum of corresponding group-level and unit-level parameters. Recollect that our target was the unit-level regression of Y on X in the population, and this in turn has been defined in terms of the components of Σ. Consequently this regression parameter depends on the grouping in the population. However, as mentioned earlier, such grouping structure may well be of no interest at all, in the sense that if population data on Y and X were available, we would define this parameter and estimate its value without paying any attention to the grouping structure of the population. This would be the case if the groups corresponded to arbitrary divisions of the population. Modifications to the theory described in Chapter 20 that start from this rather different viewpoint remain a topic of active research.

CHAPTER 18

Bayesian Methods for Unit and Item Nonresponse

Roderick J. Little

18.1. INTRODUCTION AND MODELING FRAMEWORK

Chapter 4 discussed Bayesian methods of inference for sample surveys. In this chapter I apply this perspective to the analysis of survey data subject to unit and item nonresponse. I first extend the Bayesian modeling framework of Chapter 4 to allow for missing data, by adding a model for the missing-data mechanism. In Section 18.2, I apply this framework to unit nonresponse, yielding Bayesian extensions of weighting methods. In Section 18.3, I discuss item nonresponse, which can be handled by the technique of multiple imputation. Other topics are described in the concluding section.

I begin by recapping the notation of Chapter 4:

$y_U = (y_{tj})$, $y_{tj} =$ value of survey variable j for unit t, $j = 1, \ldots, J$;
$t \in U = \{1, \ldots, N\}$
$Q = Q(y_U) =$ finite population quantity
$i_U = (i_1, \ldots, i_N) =$ sample inclusion indicators; $i_t = 1$ if unit t included, 0 otherwise
$y_U = (y_{inc}, y_{exc})$, $y_{inc} =$ included part of y_U, $y_{exc} =$ excluded part of y_U
$z_U =$ fully observed covariates, design variables.

Without loss of generality the sampled units are labeled $t = 1, \ldots, n$, so that y_{inc} is an $(n \times J)$ matrix. Now suppose that a subset of the values in y_{inc} is missing because of unit or item nonresponse. Write $y_{inc} = (y_{obs}, y_{mis})$, $y_{obs} =$ observed, $y_{mis} =$ missing, and, as in Sections 2.2 and 17.3, define the $(n \times J)$ missing-data indicator matrix:

$$r_s = (r_{tj}), \quad r_{tj} = \begin{cases} 1, & \text{if } y_{tj} \text{ observed;} \\ 0, & \text{if } y_{tj} \text{ missing.} \end{cases}$$

Analysis of Survey Data Edited by R. L. Chambers and C. J. Skinner
© 2003 John Wiley & Son, Ltd

Following the superpopulation model of Little (1982) and the Bayesian formulation in Rubin (1987) and Gelman et al. (1995), the joint distribution of y_U, i_U and r_s can then be modeled as

$$p(y_U, i_U, r_s|z_U, \theta, \phi, \psi) = p(y_U|z_U, \theta)p(i_U|z_U, y_U, \phi)p(r_s|z_U, i_U, y_{inc}, \psi).$$

The likelihood based on the observed data (z_U, y_{obs}, i_U, r_s) is

$$L(\theta, \phi, \psi|z_U, y_{obs}, i_U, r_s) \propto p(y_{obs}, i_U, r_s|z_U, \theta, \phi, \psi)$$

$$= \int p(y_U, i_U, r_s|z_U, \theta, \phi, \psi)dy_{exc}dy_{mis}.$$

Adding a prior distribution $p(\theta, \phi, \psi|z_U)$ for the parameters, the posterior distribution of the parameters given the data is

$$p(\theta, \phi, \psi|z_U, y_{obs}, i_U, r_s) \propto p(\theta, \phi, \psi|z_U)L(\theta, \phi, \psi|z_U, y_{obs}, i_U, r_s).$$

Conditions for ignoring the selection mechanism in the presence of nonresponse are the

Selection at random (SAR): $p(i_U|z_U, y_U, \phi) = p(i_U|z_U, y_{obs}, \phi)$ for all y_{exc}, y_{mis}.

Bayesian distinctness: $p(\theta, \phi, \psi|z_U) = p(\theta, \psi|z_U)p(\phi|z_U)$,

which are again satisfied for probability sampling designs. If the selection mechanism is ignorable, conditions for ignoring the missing-data mechanism are

Missing at random (MAR): $p(r_s|z_U, i_U, y_{inc}, \psi) = p(r_s|z_U, i_U, y_{obs}, \psi)$ for all y_{mis}.

Bayesian distinctness: $p(\theta, \psi|z_U) = p(\theta|z_U)p(\psi|z_U)$.

If the selection and missing-data mechanisms are ignorable, then

$$p(\theta|z_U, y_{obs}) = p(\theta|z_U, y_{obs}, r_s, i_U) \quad \text{and}$$
$$p(Q(y_U)|z_U, y_{obs}, r_s, i_U) = p(Q(y_U)|z_U, y_{obs}),$$

so the models for the selection and missing-data mechanisms do not affect inferences about θ and $Q(y_U)$.

The selection mechanism can be assumed ignorable for probability sample designs, provided the relevant design information is recorded and included in the analysis. In contrast, the assumption that the missing-data mechanism is ignorable is much stronger in many applications, since the mechanism is not in the control of the sampler, and nonresponse may well be related to unobserved values y_{mis} and hence not MAR. This fact notwithstanding, methods for handling nonresponse in surveys nearly all assume that the mechanism is ignorable, and the methods in the next two sections make this assumption. Methods for nonignorable mechanisms are discussed in Section 18.4.

18.2. ADJUSTMENT-CELL MODELS FOR UNIT NONRESPONSE

18.2.1. Weighting methods

The term 'unit nonresponse' is used to describe the case where all the items in the survey are missing, because of noncontact, refusal to respond, or some other reason. The standard survey approach is to discard the nonrespondents and analyze the respondents weighted by the inverse of the estimated probability of response. A common form of weighting is to form adjustment cells based on variables X recorded for respondents and nonrespondents, and then weight respondents in adjustment cell k by the inverse of the response rate in that cell; more generally the inverse of the response rate is multiplied by the sampling weight to obtain a weight that allows for differential probabilities of sampling and nonresponse. This weighting approach is simple and corrects for nonresponse bias if the nonresponse mechanism is MAR; that is, if respondents are a random sample of sampled cases within each adjustment cell.

Weighting for unit nonresponse is generally viewed as a design-based approach, but it can also be derived from the Bayesian perspective via fixed effects models for the survey outcomes within adjustment cells defined by X. The following example provides the main idea.

Example 1.   Unit nonresponse adjustments for the mean from a simple random sample

Suppose a simple random sample of size n is selected from a population of size N, and r units respond. Let X be a categorical variable defining K adjustment cells within which the nonresponse mechanism is assumed MAR. In cell k, let N_k be the population size and $P = (P_1, \ldots, P_K)$ where $P_k = N_k/N$ is the population proportion. Also let n_k and $p_k = n_k/n$ denote the number and proportion of sampled units and r_k the number of responding units in cell k. Let Y denote a survey variable with population mean $\bar{y}_{Uk}$ in cell k, and suppose interest concerns the overall mean

$$\bar{y}_U = \sum_{k=1}^{K} P_k \bar{y}_{Uk}.$$

Suppose initially that the adjustment-cell proportions P are known, as when X defines population post-strata. Let ξ_Y denote a full probability model for the values of Y. Then $\bar{y}_U$ can be estimated by the posterior mean given P, the survey data, and any known covariates, denoted collectively as D:

$$\hat{Y}_P = E(\bar{y}_U|P, D, \xi_Y) = \sum_{k=1}^{K} P_k E(\bar{y}_{Uk}|P_k, D, \xi_Y)$$

$$= \sum_{k=1}^{K} P_k(f_k \bar{y}_k + (1 - f_k)E(\bar{y}_k^*|P_k, D, \xi_Y)), \qquad (18.1)$$

where $\bar{y}_k$ is the respondent mean and $\bar{y}_k^*$ is the mean of the nonsampled and nonresponding units, and $f_k = r_k/N_k$. This equation emphasizes that the model provides a prediction of the mean of the nonincluded values. A measure of the precision of the estimate is the posterior variance

$$V_P = Var(\bar{y}_U|P, D, \xi_Y)$$

$$= \sum_{k=1}^{K} P_k^2(1 - f_k)^2(Var(\bar{y}_k^*|P, D, \xi_Y))$$

$$+ \sum_{k=1}^{K}\sum_{k' \neq k}^{K} P_k P_{k'}(1 - f_k)(1 - f_{k'})(Cov(\bar{y}_k^*, \bar{y}_{k'}^*|P, D, \xi_Y).$$

Now consider a fixed effects model for Y with noninformative constant prior for the means within the adjustment cells:

$$[y_t|X_t = k, \{\mu_k, \sigma_k^2\}] \sim_{ind} N(\mu_k, \sigma_k^2),$$

$$[\{\mu_k, \log \sigma_k^2\}] \sim const. \tag{18.3}$$

The posterior mean for this model given P is

$$E(\bar{y}_U|P, D) = \bar{y}_{PS} = \sum_{k=1}^{K} P_k \bar{y}_k, \tag{18.4}$$

the post-stratified respondent mean that effectively weights the respondents in cell k by rP_k/r_k. The posterior variance given $\{\sigma_k^2\}$ is

$$Var(\bar{y}_U|\{\sigma_k^2\}, P, D) = \sigma_{PS}^2 \equiv \sum_{k=1}^{K} P_k^2(1 - f_k)\sigma_k^2/r_k, \quad f_k = r_k/N_k,$$

which is equivalent to the version of the design-based variance that conditions on the respondent counts $\{r_k\}$ (Holt and Smith, 1979). When the variances $\{\sigma_k^2\}$ are not known, Bayes' inference under a noninformative prior yields t-type corrections that provide better calibrated inferences in small samples (Little, 1993a).

If P is unknown, let ξ_M denote a model for the cell counts and write $\xi = (\xi_Y, \xi_M)$. Then $\bar{y}_U$ is estimated by the posterior mean

$$\hat{Y} = E(\hat{Y}_P|D, \xi) \tag{18.5}$$

and the precision by the posterior variance

$$Var(\bar{y}_U|D, \xi) = E(Var(\bar{y}_U|P, D, \xi)) + Var(\hat{Y}_P|D, \xi), \tag{18.6}$$

where the second term represents the increase in variance from estimation of the cell proportions. The Bayesian formalism decomposes the problem into two elements that are mixed in the design-based approach, namely predicting the nonincluded values of Y in each cell, and estimating the cell proportions when they are not known. Random sampling implies the following model ξ_P for the cell counts conditional on n:

$$\{n_k|n\} \sim Mnml(n; P), \tag{18.7}$$

where $Mnml\,(n, P)$ denotes the multinomial distribution with index n. With a flat prior on $\log P$, the posterior mean of P_k is n_k/n and the posterior mean of $\bar{y}_U$ is

$$E(\bar{y}_U \mid data) = \bar{y}_{WC} = \sum_{k=1}^{K} p_k \bar{y}_k, \tag{18.8}$$

which is the standard weighting-class estimator, effectively weighting respondents in cell k by $(r/n)(n_k/r_k)$, where n_k/r_k is the inverse of the response rate in that cell (Oh and Scheuren, 1983). The posterior variance is

$$Var(\bar{y}_U \mid data) = \sum_{k=1}^{K} E(P_k^2 \mid data)(1 - f_k)\sigma_k^2/r_k + Var\left[\sum_{k=1}^{K} P_k \bar{y}_k \mid data\right]$$

$$= \sum_{k=1}^{K} (p_k^2 + p_k(1 - p_k)/(n + 1))(1 - f_k)\sigma_k^2/r_k$$

$$+ \sum_{k=1}^{K} p_k(\bar{y}_k - \bar{y}_{WC})^2/(n + 1), \tag{18.9}$$

which can be compared to the design-based estimate of mean squared error in Oh and Scheuren (1983).

Example 2.   Unit nonresponse adjustments for a stratified mean

For a more complex example that includes both sampling and nonresponse weights, let Z denote a stratifying variable, and suppose a stratified random sample of size n_j is selected from stratum $Z = j$, with selection probability $\pi_j = n_j/N_j$. As before, let X be a categorical variable defining K nonresponse adjustment cells, and in stratum j, cell k, let N_{jk} be the population size and $P = (P_{11}, \ldots, P_{JK})$ where $P_{jk} = N_{jk}/N$ is the population proportion. Also, let n_{jk} and $p_{jk} = n_{jk}/n$ denote the number and proportion of sampled units and r_{jk} the number of responding units in cell jk. Let Y denote a survey variable with population mean $\bar{y}_{Ujk}$ in cell jk, and suppose interest concerns the overall mean

$$\bar{y}_U = \sum_{j=1}^{J} \sum_{k=1}^{K} P_{jk} \bar{y}_{Ujk}.$$

As in the previous example, we consider a fixed effects model for Y with noninformative constant prior for the means within the adjustment cells:

$$[y_t \mid Z_t = j, X_t = k, \{\mu_{jk}, \sigma_{jk}^2\}] \sim_{ind} N(\mu_{jk}, \sigma_{jk}^2),$$

$$[\{\mu_{jk}, \log \sigma_{jk}^2\}] \sim const. \tag{18.10}$$

The posterior mean for this model given P is

$$E(\bar{y}_U \mid P, data) = \bar{y}_{PS} = \sum_{j=1}^{J} \sum_{k=1}^{K} P_{jk} \bar{y}_{jk}, \tag{18.11}$$

the post-stratified respondent mean that effectively weights the respondents in cell k by rP_{jk}/r_{jk}. If X is based on sample information, random sampling within the strata implies the following model ξ_P for the adjustment-cell counts conditional on n_k:

$$\{n_{jk}|n_j\} \sim Mnml(n_j; Q_j), \quad Q_j = (Q_{j1}, \cdots, Q_{jK}), \quad Q_{jk} = P_{jk}/P_j. \quad (18.12)$$

With a flat prior on $\log Q_j$, the posterior mean of Q_{jk} is n_{jk}/n_j and the posterior mean of $\bar{y}_U$ is

$$E(\bar{y}_U|data) = \bar{y}_{WC1} \equiv \sum_{j=1}^{J}\sum_{k=1}^{K} P_j(n_{jk}/n_j)\bar{y}_{jk} = \sum_{j=1}^{J}\sum_{k=1}^{K} w_{jk}r_{jk}\bar{y}_{jk}\Big/ \sum_{j=1}^{J}\sum_{k=1}^{K} w_{jk}r_{jk},$$

$$(18.13)$$

where respondents in cell jk are weighted by $w_{jk} = (N_j/n_j)(n_{jk}/r_{jk})(r/N)$, which is proportional to the inverse of the product of the sampling rate in stratum j and the response rate in cell jk.

Note that (18.13) results from cross-classifying the data by the strata as well as by the adjustment cells when forming nonresponse weights. An alternative weighting-class estimator assumes that nonresponse is independent of the stratifying variable Z within adjustment cells, and estimates the response rate in cell jk as $\hat{\phi}_k = r_{+k}/n_{+k}$. This yields

$$\bar{y}_{WC2} \equiv \sum_{j=1}^{J}\sum_{k=1}^{K} w^*_{jk}r_{jk}\bar{y}_{jk}\Big/ \sum_{j=1}^{J}\sum_{k=1}^{K} w^*_{jk}r_{jk}, \quad \text{where } w^*_{jk} = \left(\frac{r_{jk}/n_{jk}}{\hat{\phi}_k}\right)w_{jk}. \quad (18.14)$$

This estimator is approximately unbiased because the expected value of w^*_{jk} equals w_{jk}. Some authors advocate a variant of (18.14) that includes sample weights in the estimate of the response rate, that is

$$\hat{\phi}_k = \left(\sum_j r_{jk}/\pi_j\right)\left(\sum_j n_{jk}/\pi_j\right)^{-1}.$$

Both these estimates have the practical advantage of assigning weight $w^*_{jk} = 0$ to cells where there are no respondents ($r_{jk} = 0$), and hence the respondent mean is not defined. However, assigning zero weight seems unjustified when there are sample cases in these cells, that is $n_{jk} > 0$, and it seems hard to justify the implied predictions for the means $\bar{y}_{Ujk}$ in those cells; a modeling approach to this problem is discussed in the next section.

18.2.2. Random effects models for the weight strata

When the respondent sample size in a cell is zero the weighted estimates (18.4), (18.8), (18.11), or (18.13) are undefined, and when it is small the weight assigned to respondents can be very large, yielding estimates with poor precision. The classical randomization approach to this problem is to modify the weights, by pooling cells or by trimming (Potter, 1990). From the Bayesian perspective, the problem lies in the fixed effects model specification, which results in the mean in each cell being estimated by the cell respondent mean, with no borrowing of strength across cells. The solution lies in replacing the flat prior for the cell means by a proper prior that allows borrowing strength across cells.

In the simple random sampling setting, suppose the fixed effects model is replaced by

$$[y_t | X_t = k, \{\mu_k, \sigma_k^2\}] \sim_{ind} N(\mu_k, a_{1k}\sigma_k^2)$$
$$[\mu_k | P, C_k, \beta, \tau^2] \sim_{ind} N(\hat{y}_k, a_{2k}\tau^2), \hat{y}_k = g(P_k, C_k; \beta)$$
$$[\log \sigma_k^2 | \sigma^2, \phi^2] \sim_{ind} N(\log \sigma^2, \phi^2)$$
$$[\beta, \sigma^2, \phi^2, \tau^2] \sim \Pi, \tag{18.15}$$

where:

(a) $g(P_k, C_k; \beta)$ is a mean function of the proportion P_k and covariates C, indexed by unknown parameters β; and

(b) Π is a prior distribution for the hyperparameters $\beta, \sigma^2, \phi^2, \tau^2$, typically flat or weakly informative.

(c) Heterogeneous variance is modeled via known constants (a_{1k}, a_{2k}), $k = 1, \ldots, K$, and the hyperparameter ϕ^2, which is potentially unknown. The full variance structure combines models from various settings as special cases. Homogeneous variance is obtained by setting $a_{1k} = 1$ for all k and $\phi^2 = 0$. In previous work I and others have set a_{2k} equal to one, but alternative choices such as $a_{2k} = 1/P_k$ might also be considered.

Letting $\tau^2 \to \infty$ results in the fixed effects model for the cell means:

$$[y_t | X_t = k, \mu_k, \sigma_k^2] \sim_{ind} N(\mu_k, a_{1k}\sigma_k^2)$$
$$[\log \sigma_k^2 | \sigma^2, \phi^2] \sim_{ind} N(\log \sigma^2, \phi^2)$$
$$[\{\mu_k\}, \sigma^2, \phi^2] \sim [\{\mu_k\}][\sigma^2, \phi^2]$$
$$[\{\mu_k\}] \sim const, [\sigma^2, \phi^2] \sim \Pi, \tag{18.16}$$

which leads to the weighted estimators (18.4) or (18.8), depending on whether or not P is known. At the other extreme, setting $\tau^2 = \phi^2 = 0$ in (18.16) yields the direct regression form of the model:

$$[y_t | X_t = k, P_k, C_k, \beta, \sigma^2] \sim_{ind} N(\hat{y}_k, a_{1k}\sigma^2), \hat{y}_k = g(P_k, C_k; \beta),$$
$$[\beta, \log \sigma^2] \sim \Pi. \tag{18.17}$$

The posterior mean for this model for known P is the regression estimator:

$$E(\bar{y}_U | P, C, \beta, \sigma^2) = \sum_{k=1}^{K} P_k(f_k \bar{y}_k + (1 - f_k)\hat{y}_k), \tag{18.18}$$

which is potentially more efficient than (18.4) but is not necessarily design consistent, and is potentially vulnerable to bias from misspecification of the model (18.17).

For model (18.15) with $0 < \tau^2 < \infty$, the posterior mean conditional on $\{P, D, \beta, \{\sigma_k^2\}, \tau^2\}$ is

$$\hat{Y}_P^* = \sum_{k=1}^{K} P_k(f_k \bar{y}_k + (1 - f_k)(\lambda_k \bar{y}_k + (1 - \lambda_k)\hat{y}_k)), \quad \text{where}$$

$$\lambda_k = \frac{r_k a_{2k} \tau^2}{r_k a_{2k} \tau^2 + a_{1k} \sigma_k^2}. \tag{18.19}$$

The estimator (18.19) is not generally design unbiased, but it is design consistent, since as $r_k \to \infty$, $\lambda_k \to 1$ and (18.19) tends to the design-unbiased estimator (18.4). Design consistency is retained when the hyperparameters are replaced by consistent empirical Bayes' estimates. Alternatively, fully Bayes' inference can be applied based on priors on the prior distribution Π of the hyperparameters, yielding "Bayes' empirical Bayes" inferences that are also design consistent. These inferences take into account uncertainty in the hyperparameters, at the expense of an increase in computational complexity. For a comparison of these approaches in the setting of small-area estimation from surveys, see Singh, Stukel and Pfeffermann (1998).

The estimator (18.19) can be viewed as 'model assisted' in survey research terminology (for example, Cassel, Särndal and Wretman, 1977), since it represents a compromise between the 'design-based' estimator $\sum_{k=1}^{K} P_k \bar{y}_k$ and the 'model-based' estimator $\sum_{k=1}^{K} P_k \hat{y}_k$. However, rather than using a model to modify a design-based estimator, I advocate inference based directly on models that have robust design-based properties (Little, 1983b, 1991). That is, design information is used to improve the model, rather than using model information to improve design-based estimation. Conceptually, I prefer this approach since it lies within the model-based paradigm for statistical inference from surveys, while admitting a role for frequentist concepts in model selection (see, for example, Box, 1980; Rubin, 1984). A more pragmatic reason is that the assumptions of the estimation method are explicit in the form of the model.

Some possible forms of g for borrowing strength are

$$g(P_k; \beta) = \beta \tag{18.20}$$

which shrinks the post-stratum means to a constant value (Little, 1983b; Ghosh and Meeden, 1986);

$$g(P_k; \beta) = \beta_0 + \beta_1 P_k \tag{18.21}$$

which assumes a linear relationship between the mean and the post-stratum size;

$$g(C_k; \beta) = \beta_0 + \beta_1 C_k \tag{18.22}$$

where C_k is an ordinal variable indexing the post-strata (Lazzeroni and Little, 1998); or models that express the mean as a linear function of both P_k and C_k.

Elliott and Little (2000) study inferences based on (18.15) when $g(C_k, \beta)$ is given by (18.20), (18.22), or a smoothing spline. They show that these inferences compare favorably with weight trimming, where trimming points are either specified constants or estimated from the data.

18.3. ITEM NONRESPONSE

18.3.1. Introduction

Item nonresponse occurs when particular items in the survey are missing, because they were missed by the interview, or the respondent declined to answer particular questions. For item nonresponse the pattern of missing values is general complex and multivariate, and substantial covariate information is available to predict the missing values in the form of observed items. These characteristics make weighting adjustments unattractive, since weighting methods are difficult to generalize to general patterns of missing data (Little, 1988) and make limited use of the incomplete cases.

A common practical approach to item-missing data is imputation, where missing values are filled in by estimates and the resulting data are analyzed by complete-data methods. In this approach incomplete cases are retained in the analysis. Imputation methods until the late 1970s lacked an underlying theoretical rationale. Pragmatic estimates of the missing values were substituted, such as unconditional or conditional means, and inferences based on the filled-in data. A serious defect with the method is that it 'invents data.' More specifically, a single imputed value cannot represent all of the uncertainty about which value to impute, so analyses that treat imputed values just like observed values generally underestimate uncertainty, even if nonresponse is modeled correctly. Large-sample results (Rubin and Schenker, 1986) show that for simple situations with 30% of the data missing, single imputation under the correct model results in nominal 90% confidence intervals having actual coverages below 80%. The inaccuracy of nominal levels is even more extreme in multiparameter testing problems.

Rubin's (1987) theory of multiple imputation (MI) put imputation on a firm theoretical footing, and also provided simple ways of incorporating imputation uncertainty into the inference. Instead of imputing a single set of draws for the missing values, a set of Q (say $Q = 5$) datasets is created, each containing different sets of draws of the missing values from their predictive distribution given the observed data. The analysis of interest is then applied to each of the Q datasets and results are combined in simple ways, as discussed in the next section.

18.3.2. MI based on the predictive distribution of the missing variables

MI theory is founded on a simulation approximation of the Bayesian posterior distribution of the parameters given the observed data. Specifically, suppose the sampling mechanism is ignorable and let $y_s = (y_{tj})$ represent an $n \times J$ data matrix and r_s be an $n \times J$ missing-data indicator matrix with $r_{tj} = 1$ if y_{tj} is observed. Let y_{obs} be the observed data and y_{mis} the missing components of y_s. Let $p(y_s, r_s | z_U, \theta)$ be the distribution of y_s and r_s given design variables z_U, indexed by model parameters θ, and let $p(\theta | z_U)$ be a prior distribution for θ. The posterior distribution for θ given the observed data (z_U, y_{obs}, r_s) is related to the posterior distribution given hypothetical complete data (z_U, y_U, r_s) by the expression

$$p(\theta | z_U, y_{obs}, r_s) = \int p(\theta | z_U, y_s, r_s) p(y_{mis} | z_U, y_{obs}) \mathrm{d}y_{mis}.$$

MI approximates (or simulates) this expression as

$$p(\theta | z_U, y_{obs}, r_s) \cong \frac{1}{Q} \sum_{q=1}^{Q} p(\theta | z_U, y_s^{(q)}, r_s), \tag{18.23}$$

where $y_s^{(q)} = (y_{obs}, y_{mis}^{(q)})$ is an imputed dataset with missing values filled in by a draw $y_{mis}^{(q)}$ from the posterior predictive distribution of the missing data y_{mis} given the observed data z_U, y_{obs}, and r_s:

$$y_{mis}^{(q)} \sim p(y_{mis} | z_U, y_{obs}, r_s), \tag{18.24}$$

and $p(\theta | z_U, y_s^{(q)}, r_s)$ is the posterior for θ based on the filled-in dataset z_U and $y_s^{(q)}$. The posterior mean and covariance matrix of θ can be approximated similarly as

$$E(\theta | z_U, y_{obs}, r_s) \cong \bar{\theta} \equiv \frac{1}{Q} \sum_{q=1}^{Q} \hat{\theta}^{(q)}, \quad \text{where } \hat{\theta}^{(q)} = E(\theta | z_U, y_s^{(q)}, r_s); \tag{18.25}$$

$$Var(\theta | z_U, y_{obs}, r_s) \cong \frac{1}{Q} \sum_{q=1}^{Q} Var(\theta | z_U, y_s^{(q)}, r_s)$$

$$+ \frac{Q+1}{Q} \left[\frac{1}{Q-1} \sum_{q=1}^{Q} (\hat{\theta}^{(q)} - \bar{\theta})(\hat{\theta}^{(q)} - \bar{\theta})' \right]. \tag{18.26}$$

These expressions form the basis for MI inference of the filled-in datasets. Equation (18.25) indicates that the combined estimate of a parameter is obtained by averaging the estimates from each of the filled-in datasets. Equation (18.26) indicates that the covariance matrix of the estimate is obtained by averaging the covariance matrix from the filled-in datasets and adding the sample covariance matrix of the estimates $\hat{\theta}^{(q)}$ from each of the filled-in datasets; the $(Q+1)/Q$ factor is a small-sample correction that improves the approximation. The second component captures the added uncertainty from imputation that is missed by single-imputation methods.

Although the above MI theory is Bayesian, simulations show that inferences based on (18.23)–(18.26) have good frequentist properties, at least if the model and prior are reasonable. Another useful feature of MI is that the complete-data analysis does not necessarily need to be based on the model used to impute the missing values. In particular, the complete-data analysis might consist in computing estimates and standard errors using more conventional design-based methods, in which case the potential effects of model misspecification are confined to the imputation of the missing values. For more discussion of the properties of MI under model misspecification, see for example Rubin (1996), Fay (1996), Rao (1996), and the associated discussions.

An important feature of MI is that *draws* of the missing values are imputed rather than *means*. Means would be preferable if the objective was to obtain the best estimates of the missing values, but have drawbacks when the objective is to make inferences about parameters. Imputation of draws entails some loss of efficiency for point estimation, but the averaging over the K multiply-imputed datasets in (18.25) considerably reduces this loss. The gain from imputing draws is that it yields valid inferences for a wide range of estimands, including nonlinear functions such as percentiles and variances (Little, 1988).

The difficulty in implementing MI is in obtaining draws from the posterior distribution of y_{mis} given z_U, y_{obs} and r_s, which typically has an intractable form. Since draws from the posterior distribution of y_{mis} given z_U, y_{obs}, r_s, and θ are often easy to implement, a simpler scheme is to draw from the posterior distribution of y_{mis} given z_U, y_{obs}, r_s and $\bar{\theta}$, where $\bar{\theta}$ is an easily computed estimate of θ such as that obtained from the complete cases. This approach ignores uncertainty in estimating θ, and is termed *improper* in Rubin (1987). It yields acceptable approximations when the fraction of missing data is modest, but leads to overstatement of precision with large amounts of missing data. In the latter situations one option is to draw $\theta^{(q)}$ from its asymptotic distribution and then impute y_{mis} from its posterior distribution given z_U, y_{obs}, r_s, and $\theta^{(q)}$. A better but more computationally intensive approach is to cycle between draws $y_{mis}^{(i)} \sim P(y_{mis}|z_U, y_{obs}, r_s, \theta^{(i-1)})$ and $\theta^{(i)} \sim p(\theta|z_U, y_{obs}, y_{mis}^{(i)}, r_s)$, an application of the Gibbs sampler (Tanner and Wong, 1987; Tanner, 1996).

The formulation provided above requires specification of the joint distribution of y_s and r_s given z_U. As discussed in Section 18.1, if the missing data are missing at random (MAR) in that the distribution of r_s given y_s and z_U does not depend on y_{mis}, then inference can be based on a model for y_s alone rather than on a model for the joint distribution of y_s and r_s (Rubin 1976: Little and Rubin 2002). Specifically, (18.23)–(18.26) can replaced by the following:

$$p(\theta|z_U, y_{obs}) \cong \frac{1}{Q} \sum_{q=1}^{Q} p(\theta|z_U, y_s^{(q)}),$$

$$y_s^{(q)} = (y_{obs}, y_{mis}^{(q)}), y_{mis}^{(q)} \sim p(y_{mis}|z_U, y_{obs}),$$

$$E(\theta|z_U, y_{obs}) \cong \bar{\theta} \equiv \frac{1}{Q} \sum_{q=1}^{Q} \hat{\theta}^{(q)}, \quad \text{where } \hat{\theta}^{(q)} = E(\theta|z_U, y_s^{(q)}),$$

$$Var(\theta|z_U, y_{obs}) \cong \frac{1}{Q} \sum_{q=1}^{Q} Var(\theta|z_U, y_s^{(q)})$$

$$+ \frac{Q+1}{Q} \left[\frac{1}{Q-1} \sum_{q=1}^{Q} (\hat{\theta}^{(q)} - \theta)(\hat{\theta}^{(q)} - \theta)' \right],$$

where the conditioning on r_s has been dropped. This approach is called inference *ignoring the missing-data mechanism*. Since modeling the missing-data mechanism is difficult in many applications, and results are vulnerable to model misspecification, ignorable models are attractive. Whenever possible, surveys should be designed to make this assumption plausible by measuring information on at least a subsample of nonrespondents.

Example 3.   MI for the Third National Health and Nutrition Examination Survey

The Third National Health and Nutrition Examination Survey (NHANES-3) was the third in a series of periodic surveys conducted by the National Center for Health Statistics to assess the health and nutritional status of the US population. The NHANES-3 Survey began in 1988 and was conducted in two phases, the first in 1988–91 and the second in 1991–4. It involved data collection on a national probability sample of 39 695 individuals in the US population. The medical examination component of the survey dictated that it was carried out in a relatively small number (89) of localities of the country known as stands; stands thus form the primary sampling units. The survey was also stratified with oversampling of particular population subgroups.

This survey was subject to nonnegligible levels of unit and item nonresponse, in both its interview and its examination components. In previous surveys, nonresponse was handled primarily using weighting adjustments. Increasing levels of nonresponse in NHANES, and inconsistencies in analyses of NHANES data attributable to differing treatments of the missing values, led to the desire to develop imputation methods for NHANES-3 and subsequent NHANES surveys that yield valid inferences.

Variables in NHANES-3 can be usefully classified into three groups:

1.   Sample frame/household screening variables
2.   Interview variables (family and health history variables)
3.   Mobile Examination Center (MEC) variables.

The sample frame/household screening variables can be treated essentially as fully observed. Of all sampled individuals, 14.6% were unit nonrespondents who had only the sampling frame and household screening variables measured. The interview data consist of family questionnaire variables and health variables obtained for sampled individuals. These variables were subject to unit nonresponse and modest rates of item nonresponse. For example, self-rating of health status (for individuals aged 17 or over) was subject to an overall non-response rate (including unit nonresponse) of 18.8%, and family income had an overall nonresponse rate of 21.1%.

Missing data in the MEC variables are referred to here as examination nonresponse. Since about 8% of the sample individuals answered the interview questions but failed to attend the examination, rates of examination nonre-sponse were generally higher than rates of interview nonresponse. For example, body weight at examination had an overall nonresponse rate of 21.6%, systolic blood pressure an overall nonresponse rate of 28.1%, and serum cholesterol an overall nonresponse rate of 29.4%.

The three blocks of variables – screening, interview, examination – had an approximately monotone structure, with screening variables basically fully observed, questionnaire variables missing when the interview is not conducted, and examination variables missing when either (i) the interview is not con-ducted or (ii) the interview is conducted but the MEC examination does not take place. However, item nonresponse for interview data, and component and item-within component nonresponse for MEC data, spoil this monotone structure.

A combined weighting and multiple imputation strategy was adopted to create a public-use dataset consisting of over 70 of the main NHANES-3 variables (Ezzati and Khare, 1992; Ezzati-Rice et al., 1993, 1995; Khare et al., 1993). The dataset included the following information:

Basic demographics and geography: age, race/ethnicity, sex, household size, design stratum, stand, interview weight.
Other interview variables: alcohol consumption, education, poverty index, self-reported health, activity level, arthritis, cataracts, chest pain, heart attack, back pain, height, weight, optical health measures, dental health measures, first-hand and second-hand smoking variables.
Medical examination variables: blood pressure measures, serum cholesterol measures, serum triglycerides, hemoglobin, hematocrit, bone density mea-sures, size measures, skinfold measures, weight, iron, drusen score, macul-pathy, diabetic retinopathy, ferritin, mc measures, blood lead, red cell measures.

Many of the NHANES variables not included in the above set are recodes of included variables and hence easily derived.

As in previous NHANES surveys, unit nonrespondents were dropped from the sample and a nonresponse weight created for respondents to adjust for the fact they are no longer a random sample of the population. The nonresponse weights were created as inverses of estimated propensity-to-respond scores (Rubin, 1985), as described in Ezzati and Khare (1992). All other missing values were handled by multiple imputation, specifically creating five random draws from the predictive distribution of the missing values, based on a multivariate linear mixed model (Schafer, 1996). The database consists of the following six components:

1. *A core dataset* containing variables that are not subject to imputation (id, demographics, sampling weights, imputation flags) in fixed-width, space-delimited ASCII.
2. *Five versions* of a data file containing the observed data and the imputed values created as one draw from the joint predictive distribution of the missing values.
3. *SAS code* that will merge the core data with each of the imputed datasets, assign variables, names, etc., to produce five SAS datasets of identical size, with identical variable names.
4. *Sample analyses* using SUDAAN and Wesvar-PC to estimate means, proportions, quantiles, linear and logistic regression coefficients. Each analysis will have to be run five times.
5. *SAS code* for combining five sets of estimates and standard errors using Rubin's (1987) methods for multiple imputation inference, as outlined above.
6. *Documentation* written for a general audience that details (a) the history of the imputation project, (b) an overview of multiple imputation, (c) NHANES imputation models and procedures, (d) a summary of the 1994–5 evaluation study, (e) instructions on how to use the multiply-imputed database, and (f) caveats and limitations.

A separate model was fitted to sample individuals in nine age classes, with sample sizes ranging from 1410 to 8375 individuals. One reason for stratifying in this way is that the set of variables defined for individuals varies somewhat by age, with a restricted set applying to children under 17 years, and some variables restricted to adults aged over 60 years. Also, stratification on age is a simple modeling strategy for reflecting the fact that relationships between NHANES variables are known to vary with age.

I now describe the basic form of the model for a particular age stratum. For individual t in stand c, let y_{tc} be the $(1 \times J)$ vector of the set of items subject to missing data, and let x_{tc} be a fixed $(1 \times p)$ vector of design variables and items fully observed except for unit nonresponse. It is assumed that

$$
\begin{aligned}
y_{tc}|b_c &\sim N_J(x_{tc}\beta + b_c, \Sigma) \\
b_c &\sim N_J(0, \psi), \quad c = 1, \ldots, 89; t = 1, \ldots, n_c,
\end{aligned}
\tag{18.27}
$$

where β is a $(p \times J)$ vector of fixed effects, b_c is a $(1 \times J)$ vector of random stand effects with mean zero and covariance matrix $\psi = \text{diag}(\psi_1, \ldots, \psi_J)$, and Σ is an unstructured covariance matrix; conditioning on (β, Σ, ψ) in (18.27) is implicit. It is further assumed that the missing components of y_{tc} are missing at random and the parameters (β, Σ, ψ) are distinct from parameters defining the mechanism, so that the missing-data mechanism does not have to be modeled for likelihood inference (Rubin, 1976; Little and Rubin, 2002).

In view of the normality assumption in (18.27), most variables were transformed to approximate normality using standard power transformations. A few variables not amenable to this approach were forced into approximate normality by calculating the empirical cdf and corresponding quantiles of the standard normal distribution. The model (18.27) is a refinement over earlier imputation models in that stand is treated as a random effect rather than a fixed effect. This reduces the dimensionality of the model and allows for greater pooling of information across stands.

The Gibbs sampler was used to generate draws from the posterior distribution of the parameters and the missing values for the model in Section 18.2.1, with diffuse conjugate prior distributions on (β, Σ, ψ). S-Plus and Fortran code is available at Joseph Schafer's web site at http://www.psu.edu/~jls. In summary, given values from the tth iteration, the $(t+1)$th iteration of Gibbs' sampling involves the following five steps:

Draw $(b_c^{(i+1)}|y_{obs,\,tc}, y_{mis,\,tc}^{(i)}, \beta^{(i)}, \psi^{(i)}, \Sigma^{(i)}) \sim Normal, \quad c = 1, \dots, 89$

Draw $(\psi^{(i+1)}|y_{obs,\,tc}, y_{mis,\,tc}^{(i)}, \beta^{(i)}, \Sigma^{(i)}, \{b_c^{(i+1)}\}) \sim Inverse\ Wishart$

Draw $(\Sigma^{(i+1)}|y_{obs,\,tc}, y_{mis,\,tc}^{(i)}, \beta^{(i)}, \{b_c^{(i+1)}\}, \psi^{(i+1)}) \sim Inverse\ Wishart$

Draw $(\beta^{(i+1)}|y_{obs,\,tc}, y_{mis,\,tc}^{(i)}, \{b_c^{(i+1)}\}, \psi^{(i+1)}, \Sigma^{(i+1)}) \sim Normal$

Draw $(y_{mis,\,tc}^{(i+1)}|y_{obs,\,tc}, \{b_c^{(i+1)}\}, \psi^{(i+1)}, \Sigma^{(i+1)}, \beta^{(i+1)} \sim Normal, \quad c = 1, \dots, 89.$

Here $y_{obs,\,tc}$ consists of the set of observed items in the vector y_{tc} and $y_{mis,\,tc}$, the set of missing items. More details of the forms of these distributions are given in Schafer (1996).

The Gibbs sampler for each age stratum was run as a single chain and converged rapidly, reflecting the fact that the model parameters and random effects were well estimated. After an initial run-in period, draws of the missing values were taken at fixed intervals in the chain, and these were transformed back to their original scales and rounded to produce the five sets of imputations.

18.4. NONIGNORABLE MISSING DATA

The models discussed in the previous two sections assume the missing data are MAR. Nonignorable, non-MAR models are needed when missingness depends on the missing values. For example, suppose a participant in an income survey refused to report an income amount because the amount itself is high (or low). If missingness of the income amount is associated with the amount, after controlling for observed covariates (such as age, education, or occupation) then the mechanism is not MAR, and methods for imputing income based on MAR models are subject to bias. A correct analysis must be based on the full likelihood from a model for the joint distribution of y_s and r_s. The standard likelihood asymptotics apply to nonignorable models providing the parameters are identified, and computational tools such as the Gibbs sampler also apply to this more general class of models.

Suppose the missing-data mechanism is nonignorable, but the selection mechanism is ignorable, so that a model is not required for the inclusion indicators i_U. There are two broad classes of models for the joint distribution of y_s and r_s (Little and Rubin, 2002, Ch. 11; Little, 1993b). *Selection* models model the joint distribution as

$$p(y_s, r_s | z_U, \theta, \psi) = p(y_s | z_U, \theta) p(r_s | z_U, y_{inc}, \psi), \qquad (18.28)$$

as in Section 18.1. *Pattern-mixture* models specify

$$p(y_s, r_s | z_U, \gamma, \pi) = p(y_s | z_U, r_s, \gamma) p(r_s | z_U, \pi), \qquad (18.29)$$

where γ and π are unknown parameters, and the distribution of y_s is conditioned on the missing data pattern r_s. Equations (18.28) and (18.29) are simply two different ways of factoring the joint distribution of y_s and r_s. When r_s is independent of y_s the two specifications are equivalent with $\theta = \gamma$ and $\psi = \pi$. Otherwise (18.28) and (18.29) generally yield different models.

Pattern-mixture models (18.29) seem more natural when missingness defines a distinct stratum of the population of intrinsic interest, such as individuals reporting 'don't know' in an opinion survey. However, pattern-mixture models can also provide inferences for parameters θ of the complete-data distribution, by expressing the parameters of interest as functions of the pattern-mixture model parameters γ and π (Little, 1993b). An advantage of the pattern-mixture modeling approach over selection models is that assumptions about the form of the missing-data mechanism are sometimes less specific in their parametric form, since they are incorporated in the model via parameter restrictions. This point is explained for specific normal pattern-mixture models in Little (1994) and Little and Wang (1996).

Most of the literature on nonignorable missing data has concerned selection models of the form (18.28), for univariate nonresponse. An early example is the probit selection model.

Example 4.   Probit selection model

Suppose Y is scalar and incompletely observed, $X_1, \ldots, X_p$ represent design variables and fully observed survey variables, and interest concerns the parameters β of the regression of Y on $X_1, \ldots, X_p$. A normal linear model is assumed for this regression, that is

$$(y_t | x_{t1}, \ldots, x_{tp}) \sim N\left(\beta_0 + \sum_{j=1}^{p} \beta_j x_{tj}, \sigma^2\right). \qquad (18.30)$$

The probability that Y is observed given Y and $X_1, \ldots, X_p$ is modeled as a probit regression function:

$$\Pr(R_t = 1 | y_t, x_{t1}, \ldots, x_{tp}) = \Phi\left[\psi_0 + \sum_{j=1}^{p} \psi_j x_{tj} + \psi_{p+1} y_t\right], \qquad (18.31)$$

where Φ denotes the probit function. When $\psi_{p+1} \neq 0$, this probability is a monotonic function of the values of Y, and the missing-data mechanism is nonignorable. If, on the other hand, $\psi_{p+1} = 0$ and $(\psi_0, \ldots, \psi_p)$ and (β, σ^2) are distinct, then the missing-data mechanism is ignorable, and maximum likelihood estimates of (β, σ^2) are obtained by least squares linear regression based on the complete cases.

Amemiya (1984) calls (18.31) a Type II Tobit model, and it was first introduced to describe selection of women into the labor force (Heckman, 1976). It is closely related to the logit selection model of Greenlees, Reece and Zieschang (1982), which is extended to repeated-measures data in Diggle and Kenward (1994). This model is substantively appealing, but problematic in practice, since information to simultaneously estimate the parameters of the missing-data mechanism and the parameters of the complete-data model is usually very limited, and estimates are very sensitive to misspecification of the model (Little, 1985; Stolzenberg and Relles, 1990). The following example illustrates the problem.

Example 5.   Income nonresponse in the current population survey

Lillard, Smith and Welch (1982, 1986) applied the probit selection model of Example 4 to income nonresponse in four rounds of the Current Population Survey Income Supplement, conducted in 1970, 1975, 1976, and 1980. In 1980 their sample consisted of 32 879 employed white civilian males aged 16–65 who reported receipt (but not necessarily amount) of W = wages and salary earnings and who were not self-employed. Of these individuals, 27 909 reported the value of W and 4970 did not. In the notation of Example 4, Y is defined to equal $(W^{\delta-1})/\delta$, where δ is a power transformation of the kind proposed in Box and Cox (1964). The predictors X were chosen as education (five dummy variables), years of market experience (four linear splines), probability of being in first year of market experience, region (south or other), child of household head (yes, no), other relative of household head or member of secondary family (yes, no), personal interview (yes, no), and year in survey (1 or 2). The last four variables were omitted from the earnings equation (18.30); that is, their coefficients in the vector β were set equal to zero. The variables education, years of market experience, and region were omitted from the response equation (18.31); that is, their coefficients in the vector ψ were set to zero.

Lillard, Smith and Welch (1982) fit the probit selection model (18.30) and (18.31) for a variety of other choices of δ. Their best-fitting model, $\hat{\delta} = 0.45$, predicted large income amounts for nonrespondents, in fact 73% larger on average than imputations supplied by the Census Bureau, which used a hot deck method that assumes ignorable nonresponse. However, this large adjustment is founded on the normal assumption for the population residuals from the $\gamma = 0.45$ model, and on the specific choice of covariates in (18.30) and (18.31). It is quite plausible that nonresponse is ignorable and the unrestricted residuals follow the same (skewed) distribution as that in the respondent sample. Indeed, comparisons of Census Bureau imputations with IRS income amounts from matched CPS/IRS files do not indicate substantial underestimation (David *et al.*, 1986).

Rather than attempting to simultaneously estimate the parameters of the model for Y and the model for the missing-data mechanism, it seems preferable to conduct a sensitivity analysis to see how much the answers change for various assumptions about the missing-data mechanism. Examples of this approach for pattern-mixture models are given in Rubin (1977), Little (1994), Little and Wang (1996), and Scharfstein, Robins and Rotnitsky (1999). An alternative to simply accepting high rates of potentially nonignorable missing data for financial variables such as income is to use special questionnaire formats that are designed to collect a bracketed observation whenever a respondent is unable or unwilling to provide an exact response to a financial amount question. Heeringa, Little and Raghunathan (2002) describe a Bayesian MI method for multivariate bracketed data on household assets in the Health and Retirement Survey. The theoretical underpinning of these methods involves the extension of the formulation of missing-data problems via the joint distribution of y_U, i_U and r_s in Section 18.1 to more general incomplete data problems involving coarsened data (Heitjan and Rubin, 1991; Heitjan, 1994). Full and ignorable likelihoods can be defined for this more general setting.

18.5. CONCLUSION

This chapter is intended to provide some indication of the generality and flexibility of the Bayesian approach to surveys subject to unit and item nonresponse. The unified conceptual basis of the Bayesian paradigm is very appealing, and computational tools for implementing the approach are becoming increasingly available in the literature. What is needed to convince practitioners are more applications such as that described in Example 3, more understanding of useful baseline "reference" models for complex multistage survey designs, and more accessible, polished, and well-documented software: for example, SAS (2001) now has procedures (PROC MI and PROC MIANALYZE) for creating and analysing multiply-imputed data. I look forward to further developments in these directions in the future.

ACKNOWLEDGEMENTS

This research is supported by grant DMS-9803720 from the National Science Foundation.

CHAPTER 19

Estimation for Multiple Phase Samples

Wayne A. Fuller

19.1. INTRODUCTION

Two-phase sampling, also called double sampling, is used in surveys of many
types, including forest surveys, environmental studies, and official statistics.
The procedure is applicable when it is relatively inexpensive to collect information on a vector denoted by x, relatively expensive to collect information on
the vector y of primary interest, and x and y are correlated. In the two-phase
sample, the vector x is observed on a large sample and (x, y) is observed on a
smaller sample. The procedure dates from Neyman (1938). Rao (1973)
extended the theory and Cochran (1977, Ch. 12) contains a discussion of the
technique. Hidiroglou (2001) presents results for different configurations for
the two phases.

We are interested in constructing an estimator for a large vector of characteristics using data from several sources and/or several phases of sampling. We
will concentrate on the use of the information at the estimation stage, omitting
discussion of use at the design stage.

Two types of two-phase samples can be identified on the basis of sample
selection. In one type, a first-phase sample is selected, some characteristics of
the sample elements are identified, and a second-phase sample is selected using
the characteristics of the first-phase units as controls in the selection process.
A second type, and the type of considerable interest to us, is one in which a
first-phase sample is selected and a rule for selection of second-phase units is
specified as part of the field procedure. Very often the selection of second-phase
units is not a function of first-phase characteristics. One example of the second
type is a survey of soil properties conducted by selecting a large sample of
points. At the same time the large sample is selected, a subset of the points, the
second-phase sample, is specified. In the field operation, a small set of data is
collected from the first-phase sample and a larger set is collected from the

Analysis of Survey Data Edited by R. L. Chambers and C. J. Skinner
© 2003 John Wiley & Sons, Ltd

second-phase sample. A second example of the second type is that of a population census in which most individuals receive a short form, but a subsample receives a long from with more data elements.

The sample probability-of-selection structure of the two-phase sample is sometimes used for a survey containing item nonresponse. Item nonresponse is the situation in which respondents provide information for some, but not all, items on the questionnaire. The use of the two-phase model for this situation has been discussed by Särndal and Swensson (1987) and Rao and Sitter (1995). A very similar situation is a longitudinal survey in which respondents do not respond at every point of the survey. Procedures closely related to multiple phase estimation for these situations have been discussed by Fuller (1990, 1999). See also Little and Rubin (1987).

Our objective is to produce an easy-to-use dataset that meets several criteria. Generally speaking, an easy-to-use dataset is a file of complete records with associated weights such that linear estimators are simple weighted sums. The estimators should incorporate all available information, and should be design consistent for a wide range of population parameters at aggregate levels, such as states. The dataset will be suitable for analytic uses, such as comparison of domains, the computation of regression equations, or the computation of the solutions to estimating equations. We also desire a dataset that produces reasonable estimates for small areas, such as a county. A model for some of the small-area parameters may be required to meet reliability objectives for the small areas.

19.2. REGRESSION ESTIMATION

19.2.1. Introduction

Our discussion proceeds under the model in which the finite population is a sample from an infinite population of (x_t, y_t) vectors. It is assumed that the vectors have finite superpopulation fourth moments. The members of the finite population are indexed with integers $U = \{1, 2, \ldots, N\}$. We let (μ_x, μ_y) denote the mean of the superpopulation vector and let $(\bar{x}_U, \bar{y}_U)$ denote the mean of the finite population.

The set of integers that identify the sample is the set s. In a two-phase sample, we let s_1 be the set of elements in the first phase and s_2 be the set of elements in the second phase. Let there be n_1 units in the first sample and n_2 units in the second sample. When we discuss consistency, we assume that n_1 and n_2 increase at the same rate.

Assume a first-phase sample is selected with selection probabilities π_{1t}. A second-phase sample is selected by a procedure such that the total probability of selection is π_{2t}. Thus, the total probability of being selected for the second-phase sample can be written

$$\pi_{2t} = \pi_{1t}\pi_{2t|1}, \tag{19.1}$$

where $\pi_{2t|1}$ is the conditional probability of selecting unit t at the second phase given that it is selected in the first phase.

Recall the two types of two-phase samples discussed in the introduction. In the first type, where the second-phase sample is selected on the basis of the characteristics of the first-phase sample, it is possible that only the first-phase probabilities and the conditional probabilities of the second-phase units given the first-phase *sample* are known. In such a situation, the total probability of selection for the second phase is

$$\pi_{2t} = \sum_{s_1 \ni t} \pi_{1t} \pi_{2t|s_1} \tag{19.2}$$

where $\pi_{2t|s_1}$ is the conditional probability of selecting unit t in the second phase given that the first-phase sample is the sample s_1, and the sum is over those first-phase samples containing element t. If $\pi_{2t|s_1}$ is known only for the observed s_1, then π_{2t} cannot be calculated. On the other hand, in the second type of two-phase samples, the probabilities π_{1t} and π_{2t} are known.

19.2.2. Regression estimation for two-phase samples

The two-phase regression estimator of the mean of y is

$$\bar{y}_{reg} = \bar{x}_1 \hat{\beta}, \tag{19.3}$$

where

$$\hat{\beta} = \left(\sum_{t \in s_2} x_t' \lambda_t x_t\right)^{-1} \sum_{t \in s_2} x_t' \lambda_t y_t, \tag{19.4}$$

and $\bar{x}_1$ is the first-phase mean of x,

$$\bar{x}_1 = \left(\sum_{i \in s_1} \pi_{1t}^{-1}\right)^{-1} \sum_{i \in s_1} \pi_{1t}^{-1} x_t, \tag{19.5}$$

and λ_t are weights. We assume that a vector that is identically equal to one is in the column space of the x_t. Depending on the design, there are several choices for the weights in (19.4). If the total probability of selection is known, it is possible to set $\lambda_t = \pi_{2t}^{-1}$. If π_{2t} is not known, $\lambda_t = (\pi_{1t}\pi_{2t|s_1})^{-1}$ is a possibility. In some cases, such as simple random sampling at both phases, the weights are equivalent.

If the regression variable that is identically equal to one is isolated and the x-vector written as

$$x_t = (1, d_t), \tag{19.6}$$

the regression estimator can be written in the familiar form

$$\bar{y}_{reg} = \bar{y}_2 + (\bar{d}_1 - \bar{d}_2)\hat{\gamma}, \tag{19.7}$$

where

$$\gamma = \left[\sum_{t \in s_2}(d_t - \bar{d}_2)' \lambda_t (d_t - \bar{d}_2)\right]^{-1} \sum_{t \in s_2}(d_t - \bar{d}_2)' \lambda_t y_t \qquad (19.8)$$

and

$$(\bar{y}_2, \bar{d}_2) = \left(\sum_{t \in s_2} \lambda_t\right)^{-1} \sum_{t \in s_2} \lambda_t (y_t, d_t).$$

In the form (19.7) we see that each of the two choices for λ_t produces a consistent estimator of the population mean.

The estimator (19.3) is the regression estimator in which the overall mean (total) of x is used in estimation. Given that the individual x-values for the first phase are known, extensive information, such as the knowledge of which units are the smallest 10%, can be used to construct the x-variables used in the regression estimator.

Using conditional expectations, the variance of the two-phase estimator (19.3) is

$$V\{\bar{y}_{reg}\} = V\{E[\bar{y}_{reg}|s_1]\} + E\{V[\bar{y}_{reg}|s_1]\}. \qquad (19.9)$$

If we approximate the variance by the variance of the $O_p(n^{-1/2})$ terms in the Taylor expansion, we have

$$V\{E[\bar{y}_{reg}|s_1]\} = V\{\bar{y}_1\}, \qquad (19.10)$$

and

$$V\{\bar{y}_{reg}|s_1\} = V\left\{\left(\sum_{t \in s_2} \pi_{2t|s_1}^{-1}\right)^{-1} \sum_{t \in s_2} \pi_{2t|s_1}^{-1} e_{1,t}\ \Big|s_1\right\}, \qquad (19.11)$$

where $e_{1,t} = y_t - x_t \hat{\beta}_1$,

$$\hat{\beta}_1 = \left(\sum_{t \in s_1} x_t' \pi_{1t}^{-1} x_t\right)^{-1} \sum_{t \in s_1} x_t' \pi_{1t}^{-1} y_t,$$

$$\bar{y}_1 = \left(\sum_{t \in s_1} \pi_{1t}^{-1}\right)^{-1} \sum_{t \in s_1} \pi_{1t}^{-1} y_t,$$

and $\bar{y}_1$ is the (unobservable) mean for all units in the first-phase sample.

19.2.3. Three-phase samples

To extend the results to three-phase estimation, assume a third-phase sample of size n_3 is selected from a second-phase sample of size n_2, which is itself a sample of a first-phase sample of size n_1 selected from the finite population. Let s_1, s_2, s_3 be the set of indices in the phase 1, phase 2, and phase 3 samples, respectively.

A relatively common situation is that in which the first-phase sample is the entire population.

We assume the vector $(1, x)$ is observed in the phase 1 sample, the vector $(1, x, y)$ is observed in the phase 2 sample, and the vector $(1, x, y, z)$ is observed in the phase 3 sample. Let $\bar{x}_1$ be the weighted mean for the phase 1 sample and let a two-phase estimator of the mean of y be

$$\hat{\mu}_y = \bar{y}_2 + (\bar{x}_1 - \bar{x}_2)\hat{\beta}_2, \tag{19.12}$$

where

$$\hat{\beta}_2 = \left(\sum_{t \in s_2} (x_t - \bar{x}_1)' \lambda_{2t} (x_t - \bar{x}_1)\right)^{-1} \sum_{t \in s_2} (x_t - \bar{x}_1)' \lambda_{2t} (y_t - \bar{y}_1),$$

and λ_{2t} are weights. The estimator $\hat{\mu}_y$ is an estimator of the finite population mean, as well as of the superpopulation mean. Then a three-phase estimator of the mean of z is

$$\hat{\mu}_z = (\bar{x}_1 - \bar{x}_3, \hat{\mu}_y - \bar{y}_3)(\hat{\theta}_1', \hat{\theta}_2')', \tag{19.13}$$

where

$$\begin{pmatrix} \hat{\theta}_1 \\ \hat{\theta}_2 \end{pmatrix} = \left[\sum_{t \in s_3} (x_t - \bar{x}_3, y_t - \bar{y}_3)' \lambda_{3t} (x_t - \bar{x}_3, y_t - \bar{y}_3)\right]^{-1} B_{Rz}$$

and

$$B_{Rz} = \sum_{t \in s_3} (x_t - \bar{x}_3, y_t - \bar{y}_3)' \lambda_{3t} (z_t - \bar{z}_3)'.$$

We assume that the weights λ_{2t} and λ_{3t} are such that

$$\left(\sum_{t \in s_2} \lambda_{2t}\right)^{-1} \sum_{t \in s_2} \lambda_{2t}(x_t, y_t)$$

is consistent for $(\bar{x}_U, \bar{y}_U)$ and

$$\left(\sum_{t \in s_3} \lambda_{3t}\right)^{-1} \sum_{t \in s_3} \lambda_{3t}(x_t, y_t, z_t)$$

is consistent for $(\bar{x}_U, \bar{y}_U, \bar{z}_U)$. If all three samples are simple random samples from a normal superpopulation, the estimator (19.13) is the maximum likelihood estimator. See Anderson (1957).

The variance of the three-phase estimator can be obtained by repeated application of conditional expectation arguments:

$$\begin{aligned} V\{\hat{\mu}_z\} &= V\{E[\hat{\mu}_z|s_1]\} + E\{V[\hat{\mu}_z|s_1]\} \\ &= V\{\bar{z}_1\} + E\{V[E(\hat{\mu}_z|s_2)|s_1]\} + E\{V[(\hat{\mu}_z|s_2)|s_1]\} \\ &= V\{\bar{z}_1\} + E\{V[\bar{z}_2|s_1]\} + E\{V(\hat{\mu}_z|s_2)\}. \end{aligned} \tag{19.14}$$

19.2.4. Variance estimation

Several approaches can be considered for the estimation of the variance of a
two-phase sample. One approach is to attempt to estimate the two terms of
(19.9) using an estimator of the form

$$\hat{V}\{\bar{y}_{reg}\} = \hat{V}\{\bar{y}_1\} + \hat{V}\{\bar{y}_{reg}|s_1\}. \tag{19.15}$$

The estimated conditional variance, the second term on the right of (19.15), can
be constructed as the standard variance estimator for the regression estimator
of $\bar{y}_1$ given s_1. If the first-phase sample is a simple random nonreplacement
sample and the x-vector defines strata, then a consistent estimator of $V\{\bar{y}_1\}$
can be constructed as

$$\hat{V}\{\bar{y}_1\} = N^{-1}(N - n_1)n_1^{-1}\hat{S}^2,$$

where

$$\hat{S}^2 = \left(\sum_{t \in s_2} \pi_{1t}^{-1}\pi_{2t|s_1}^{-1}\right)^{-1} \sum_{t \in s_2} \pi_{1t}^{-1}\pi_{2t|s_1}^{-1}(y_t - \bar{y}_\pi)^2$$

and

$$\bar{y}_\pi = \left(\sum_{t \in s_2} \pi_{1t}^{-1}\pi_{2t|s_1}^{-1}\right)^{-1} \sum_{t \in s_2} \pi_{1t}^{-1}\pi_{2t|s_1}^{-1} y_t.$$

A general estimator of $V\{\bar{y}_1\}$ can be constructed by recalling that the Horvitz–
Thompson estimator of $V\{\bar{y}_1\}$ of (19.10) can be written in the form

$$\tilde{V}\{\bar{y}_1\} = \sum_{(t, u) \in s_1} \sum \pi_{1tu}^{-1}(\pi_{1tu} - \pi_{1t}\pi_{1u})\pi_{1t}^{-1}y_t\pi_{1u}^{-1}y_u, \tag{19.16}$$

where π_{1tu} is the probability that unit t and unit u appear together in the phase
1 sample. Using the observed phase 2 sample, the estimated variance (19.15)
can be estimated with

$$\hat{V}\{\bar{y}_1\} = \sum_{(t, u) \in s_2} \sum \pi_{1tu}^{-1}\pi_{2tu|s_1}^{-1}(\pi_{1tu} - \pi_{1t}\pi_{1u})\pi_{1t}^{-1}y_t\pi_{1u}^{-1}y_u, \tag{19.17}$$

provided $\pi_{2tu|s_1}$ is positive for all (t, u). This variance estimator is discussed by
Särndal, Swensson and Wretman (1992, ch. 9). Expression (19.17) is not always
easy to implement. See Kott (1990, 1995), Rao and Shao (1992), Breidt and
Fuller (1993), Rao and Sitter (1995), Rao (1996), and Binder (1996) for discussions of variance estimation.

   We now consider another approach to variance estimation. We can expand
the estimator (19.3) in a first-order Taylor expansion to obtain

$$\bar{x}_1\hat{\beta} = \bar{x}_U\beta_U + (\bar{x}_1 - \bar{x}_U)\beta_U + \bar{x}_U(\hat{\beta} - \beta_U) + O_p(n^{-1}),$$

where $\bar{x}_U$ is the finite population mean, and β_U is the finite population regression coefficient

$$\beta_U = \left(\sum_{t \in U} x_t' x_t \right)^{-1} \sum_{t \in U} x_t' y_t.$$

The n without a subscript is used as the index for a sequence of samples in which n_1 and n_2 increase at the same rate. Then the variance of the approximating random variable has variance

$$V\{\bar{y}_{reg}\} \doteq \beta_U' V\{\bar{x}_1\} \beta_U + \beta_U' C\{\bar{x}_1, \hat{\beta}\} \bar{x}_U' + \bar{x}_U C\{\hat{\beta}, \bar{x}_1\} \beta_U + \bar{x}_U V\{\hat{\beta}\} \bar{x}_U', \tag{19.18}$$

where $C\{\bar{x}_1, \hat{\beta}\}$ is the covariance matrix of the two vectors. In many designs, it is reasonable to assume that $C\{\bar{x}_1, \hat{\beta}\} = 0$, and variance estimation is much simplified. See Fuller (1998).

The variances in (19.18) are unconditional variances. The variance of $\bar{x}$ can be estimated with a variance formula appropriate for the complete first-phase sample. The component $V\{\hat{\beta}\}$ is the more difficult component to estimate in this representation. If the second-phase sampling can be approximated by Poisson sampling, then $V\{\hat{\beta}\}$ can be estimated with a two-stage variance formula. In the example of simple random sampling at the first phase and stratified sampling for the second phase, the diagonal elements of $V\{\hat{\beta}\}$ are the variances of the stratum means. In this case, use of the common estimators produces estimators (19.15) and (19.18) that are nearly identical.

19.2.5. Alternative representations

The estimator (19.3) can be written in a number of different ways. Because it is a linear estimator, we can write

$$\bar{y}_{reg} = \sum_{t \in s_2} w_{2t} y_t \tag{19.19}$$

where

$$w_{2t} = \bar{x}_1 \left(\sum_{u \in s_2} x_u' \lambda_u x_u \right)^{-1} x_t' \lambda_t.$$

We can also write the estimator as a weighted sum of the predicted values. Let the first-phase weights be

$$w_{1t} = \left(\sum_{u \in s_1} \pi_{1u}^{-1} \right)^{-1} \pi_{1t}^{-1}. \tag{19.20}$$

Then we can write

$$\bar{y}_{reg} = \sum_{t \in s_1} w_{1t} \hat{y}_t, \tag{19.21}$$

where

$$\hat{y}_t = x_t\hat{\beta}.$$

Under certain conditions we can write the estimator as

$$\bar{y}_{reg} = \sum_{t \in s_1} w_{1t}\ddot{y}_t, \tag{19.22}$$

where

$$\ddot{y}_t = y_t \qquad \text{if } t \in s_2$$
$$= \hat{y}_t \qquad \text{if } t \in s_1 \cap s_2^c$$

and $s_1 \cap s_2^c$ is the set of units in s_1 but not in s_2. The representation (19.22) holds if $\lambda_t = \pi_{2t}^{-1}$ and $\pi_{1t}^{-1}\pi_{2t}^{-1}$ is in the column space of the matrix $X = (x_1', x_2', \ldots, x_{n_2}')'$. It will also hold if $\lambda_t = (\pi_{1t}\pi_{2t|s_1})^{-1}$ and $\pi_{2t|s_1}$ is in the column space of the matrix X. The equivalence of (19.3) and (19.22) follows from the fact that

$$\sum_{t \in s_2} (y_t - \hat{y}_t)\lambda_t x_t = 0,$$

under the conditions on the column space of X.

We have presented estimators for the mean. If total estimation is of interest all estimators are multiplied by the population size, if it is known, or by $\sum_{t \in s_1} \pi_{1t}^{-1}$ if the population size is unknown.

The estimator (19.19), based on a dataset of n_2 elements, and the estimator (19.22), based on n_1 elements, give the same estimated value for a mean or total of y. However, the two formulations can be used to create different estimators of other parameters, such as estimators of subpopulations.

To construct estimator (19.3) for a domain D, we would define a new second-phase analysis variable

$$y_{Dt} \quad = y_t \qquad \text{if } t \in D$$
$$= 0 \qquad \text{otherwise.}$$

Then, one would regress y_{Dt} on x_t and compute the regression estimator, where the regression estimator is of the form (19.19) with the weights w_{2t} applied to the variable y_{Dt}. The design-consistent estimators for some small subpopulations may have very large variances because some of the samples will contain few or no final phase sample elements for the subpopulation.

The estimator for domain D associated with (19.22) is

$$\hat{Y}_{gd} = \sum_{t \in D \cap s_1} w_{1t}\ddot{y}_t, \tag{19.23}$$

where the summation is over first-phase units that are in domain D. The estimator (19.23) is sometimes called the *synthetic estimator* for the domain. An extreme, but realistic, case is a subpopulation containing no phase 2 elements, but containing phase 1 elements. Then, the estimator defined by (19.19) is zero, but the estimator (19.22) is not. The fact that estimator (19.22) produces nonzero estimates for any subpopulation with a first-phase x is very appealing.

However, the domain estimator (19.23) is a model-dependent estimator. Under the model in which

$$y_t = x_t\beta + e_t \tag{19.24}$$

and the e_t are zero-mean random variables, independent of x_u for all t and u, the estimator (19.23) is unbiased for the subpopulation sum. Under randomization theory, the estimator (19.23) is biased and it is possible to construct populations for which the estimator is very biased.

The method of producing a dataset with $(\hat{y}_t, x_t)$ as data elements will be satisfactory for some subpopulation totals. However, the single set of $\hat{y}$-values cannot be used to construct estimates for other characteristics of the distribution of y, such as the quantiles or percentiles of y. This is because the estimator of the cumulative distribution function evaluated at ς is the mean of an analysis variable that is one if y_t is less than ς and is zero otherwise. The creation of a dataset that will produce valid quantile estimates has been discussed in the imputation literature and as a special topic by others. See Rubin (1987), Little and Rubin (1987), Kalton and Kasprzyk (1986), Chambers and Dunstan (1986), Nascimento Silva and Skinner (1995), Brick and Kalton (1996), and Rao (1996). The majority of the imputation procedures require a model specification.

19.3. REGRESSION ESTIMATION WITH IMPUTATION

Our objective is to develop an estimation scheme that uses model estimation to improve estimates for small areas, but that retains many of the desirable properties of design-based two-phase estimation.

Our model for the population is

$$y_t = x_t\beta + e_t, \tag{19.25}$$

where the e_t are independent zero-mean, finite variance, random variables, independent of x_u for all t and u. We subdivide the population into subunits called *cells*. Assume that cells exist such that

$$e_t \sim II(0, \sigma_g^2) \qquad \text{for } t \in C_g, \tag{19.26}$$

where C_g is the set of indices in the gth cell, and the notation is read 'distributed independently and identically.' A common situation is that in which the x-vector defines a set of imputation cells. Then, the model (19.25)–(19.26) reduces to the assumption that

$$y_{gu} \sim II(\mu_g, \sigma_g^2). \tag{19.27}$$

That is, units in the cell, whether observed or not, are independently and identically distributed. The single assumption (19.27) can be obtained through assumptions on the nature of the parent population and the selection process.

Consider a two-phase sample in which the x-vectors are known for the entire first phase. Let the weighted sample regression be that defined in (19.3) and assume the λ_t and the selection probabilities are such that

$$\sum_{t \in s_2} (y_t - x_t \hat{\beta}) \pi_{2t}^{-1} = 0. \tag{19.28}$$

Assume that a dataset is constructed as

$$\tilde{y}_t = y_t \qquad \text{if } t \text{ is observed in phase 2,}$$
$$= x_t \hat{\beta} + \hat{e}_{rt} \quad \text{if } t \text{ is observed in phase 1 but not in phase 2,} \tag{19.29}$$

where $\hat{e}_{rt}$ is a deviation chosen at random from the set of regression deviations

$$\hat{e}_u = y_u - x_u \hat{\beta}, \qquad u = 1, 2, \ldots, n_2, \tag{19.30}$$

in the same cell as the recipient. The element contributing the $\hat{e}_{rt}$ is called the *donor* and the element t receiving the $\hat{e}_{rt}$ is called the *recipient*. Let the x-vector define the set of imputation cells and assume the π_{2t} are constant within every cell. Then the outlined procedure is equivalent to selection of an observation within the cell to use as the donor. The regression estimator of the cell mean is the weighted sample mean for the cell, when the x-vector is a set of indicator variables defining the cells.

To construct a design-consistent estimator when units within a cell have been selected with unequal probability, we select the donors within a cell with probability proportional to

$$\varsigma_t = \left[\sum_{u \in s_2} \pi_{1u}^{-1} (\pi_{2u|s_1}^{-1} - 1) \right]^{-1} \pi_{1t}^{-1} (\pi_{2t|s_1}^{-1} - 1). \tag{19.31}$$

Then,

$$E\left\{ \sum_{t \in s_1} \pi_{1t}^{-1} e_{rt} | s_2 \right\} = E\left\{ \sum_{t \in s_2} \pi_{1t}^{-1} e_{rt} | s_2 \right\} + E\left\{ \sum_{t \in s_1 \cap s_2^c} \pi_{1t}^{-1} e_{rt} | s_2 \right\}$$
$$= \sum_{u \in s_2} \pi_{1u}^{-1} e_u + \sum_{t \in s_1 \cap s_2^c} \pi_{1t}^{-1} \sum_{u \in s_2} \varsigma_u e_u. \tag{19.32}$$

Now

$$E\left\{ \sum_{u \in s_2} \pi_{1u}^{-1} \pi_{2u|s_1}^{-1} (1 - \pi_{2u|s_1}^{-1}) | s_1 \right\} = \sum_{u \in s_1} \pi_{1u}^{-1} (1 - \pi_{2u|s_1}^{-1})$$
$$= E\left\{ \sum_{t \in s_1 \cap s_2^c} \pi_{1t}^{-1} | s_1 \right\}. \tag{19.33}$$

Therefore, the expectation in (19.32) is approximately equal to the weighted sum

$$\sum_{t \in s_2} \left\{ \pi_{1t} \pi_{2t|s_1} \right\}^{-1} e_t.$$

See Rao and Shao (1992) and Platek and Gray (1983) for discussions of imputation with unequal probabilities.

The estimated total using the dataset defined by (19.29) is

$$\hat{Y}_{Ir} = \sum_{t \in s_1} \pi_{1t}^{-1} \tilde{y}_t = \sum_{t \in s_1} \pi_{1t}^{-1} x_t \hat{\beta} + \sum_{t \in s_1} \pi_{1t}^{-1} \hat{e}_{rt}, \qquad (19.34)$$

where $\ddot{e}_{rt} = \dot{e}_t$ if $t \in s_2$. If the first phase is the entire population,

$$\hat{Y}_{Ir} = \sum_{t=1}^{N} \tilde{y}_t = \sum_{t=1}^{N} x_t \hat{\beta} + N \bar{e}_r, \qquad (19.35)$$

where $\bar{e}_r$ is the mean of the imputed regression deviations.

Under random-with-replacement imputation, assuming the original sample is a single-phase nonreplacement sample with $\pi_t = nN^{-1}$, and ignoring the variance of $\hat{\beta}$, the variance of $\hat{Y}_{Ir}$ in (19.35) is

$$V\{\hat{Y}_{Ir}\} \doteq (N - n)n^{-1}[N + n]\sigma_e^2$$
$$= N(N - n)n^{-1}\sigma_e^2 + (N - n)\sigma_e^2 , \qquad (19.36)$$

where the term $(N - n)\sigma_e^2$ is introduced by the random imputation. With random imputation that restricts the sample sum of the imputed deviations to be (nearly) equal to zero, the imputation variance term is (nearly) removed from (19.36).

Under the described imputation procedure, the estimator of the total for the small-area D_u is

$$\hat{Y}_{Iru} = \sum_{t \in D_u \cap s_1} \pi_{1t}^{-1} \tilde{y}_t, \qquad (19.37)$$

where $D_u \cap s_1$ is the set of first-phase indexes that are in small area D_u. The estimator $\hat{Y}_{Iru}$ is the synthetic estimator plus an imputation error,

$$\hat{Y}_{Iru} = \sum_{t \in D_u \cap s_1} \pi_{1t}^{-1} x_t \hat{\beta} + \sum_{t \in D_u \cap s_1} \pi_{1t}^{-1} \hat{e}_{rt}. \qquad (19.38)$$

Under the regression model with simple random sampling, the variance expression (19.36) holds for the small area with (N_u, n_u) replacing (N, n), where N_u and n_u are the population and sample number of elements, respectively, in small area D_u. Note that the variance expression remains valid for $n_u = 0$.

Expression (19.35) for the grand total defines an estimator that is randomization valid and does not require the model assumptions associated with (19.25) and (19.26). That is, for simple random sampling we can write

$$V\{\hat{Y}_{Ir}|U\} = N(N - n)n^{-1}\sigma_{e.}^2 + (N - n)\sigma_{e.}^2 + o(n^{-1}N^2), \qquad (19.39)$$

where

$$\sigma_{e.}^2 = E\{(y_t - x_t B)^2\}$$

and

$$B = (E\{x_t' x_t\})^{-1} E\{x_t' y_t\}.$$

On the other hand, to write the variance of the small-area estimator as

$$V\{\hat{Y}_{Iru}\} = N_u(N_u - n_u)n_u^{-1}\sigma_{eu}^2 + (N_u - n_u)\sigma_{eu}^2, \qquad (19.40)$$

where σ_{eu}^2 is the variance of the e in the small area, requires the model assumptions. Under certain conditions it is possible to use the imputed values to estimate the variance of the first-phase mean. For example, if the first-phase stratum contains n_{h1} phase 1 elements and n_{h2} phase 2 elements, if the subsampling is proportional, and if imputation is balanced so that the variance of a phase 2 imputed mean is equal to the variance of the simple phase 2 mean, then

$$E\{\hat{V}[\bar{y}_{hI1}]\} = V\{\bar{y}_{hI1}\}\left[1 - (n_{h1} - 1)^{-1}n_{h2}^{-1}(n_{h1} - n_{h2})\right], \qquad (19.41)$$

where $\bar{y}_{hI1}$ is the first-phase mean for stratum h based on imputed data, $\bar{y}_{h1}$ is the (unobservable) first-phase mean for stratum h, n_{h1} is the number of first-phase units in stratum h, n_{h2} is the number of second-phase units in the first-phase stratum h, and $\hat{V}[\bar{y}_{hI1}]$ is the estimated variance of the phase 1 stratum mean computed treating imputed data as real observations. Thus, if there are a large number of second-phase units in the first-phase stratum, the bias of $V[\bar{y}_{h1}]$ will be small.

In summary, we have described a two-phase estimation scheme in which estimates of population totals are the standard randomization two-phase estimators subject to imputation variance. The estimates for small areas are synthetic estimators constructed under a model, subject to the consistency-of-total restriction.

19.4. IMPUTATION PROCEDURES

The outlined regression imputation procedure of Section 19.3 yields a design-consistent estimator for the total of the y-variable. Estimates of other parameters, such as quantiles, based on the imputed values (19.29) are not automatically design consistent for general x-variables.

Two conditions are sufficient for the imputation procedure to give design-consistent estimates of the distribution function. First, the vector of x-variables is a vector of indicator variables that defines a set of mutually exclusive and exhaustive subdivisions of the first-phase sample. Second, the original selection probabilities are used in the selection of donors. The subdivisions defined by the x-vectors in this case are sometimes called *first-phase strata*. If the sampling rates are constant within first-phase strata, simple selection of donors can be used.

EXAMPLE   *319*

In many situations, it is operationally feasible to use ordinary sampling techniques to reduce or eliminate the variance associated with the selection of donors. See Kalton and Kish (1984) and Brick and Kalton (1996). The simplest situation is equal probability sampling, where the first-phase sample is an integer multiple, m, of the second-phase sample. If every second-phase unit is used as a donor exactly m times, then there is zero imputation variance for a total. It is relatively easy, in practical situations, to reduce the imputation variance to that due to the rounding of the number of times a second-phase unit is used as a donor. In general, the number of times an element is used as a donor should be proportional to its weight.

An important procedure that can be used to reduce the imputation variance is to use fractional imputation, described by Kalton and Kish (1984). In this procedure k values obtained from k donors are used to create k imputed values for each first-phase element with no second-phase observation. Each of the k imputed values is given a weight, such that the procedure is unbiased. If the original sample and the second-phase sample are equal probability samples, the weight is the original first-phase weight divided by k. A fully efficient fractional imputation scheme uses every second-phase unit in a cell as a donor for every first-phase unit that is not a second-phase unit. The number of imputations required by this procedure is determined by the number of elements selected and the number not selected in each cell.

We call the imputation procedure controlled if it produces a dataset such that all estimates for y-characteristics at the aggregate level are equal to the standard (weighted phase 2) two-phase regression estimator. Controlled imputation can also be used to produce estimates based on imputed data for designated small areas that are nearly equal to the synthetic small-area estimates.

In practice, the first-phase can contain many potential control variables and the formation of first-phase strata will require selecting a subset of variables and/or defining categories for continuous variables. Thus, implementation of the imputation step can require a great deal of effort in a large-scale survey. Certainly the imputed dataset must be passed through an edit operation. Given a large dimensional dataset with many control variables, some of the original observations will be internally inconsistent and some created imputed observations will be internally inconsistent. The objective of the imputation and edit operations is to reduce the number of inconsistencies, and to produce a final dataset that closely approximates reality.

19.5. EXAMPLE

To illustrate some of the properties of alternative estimation procedures for two-phase samples, we use data from the National Resources Inventory (NRI), a large national survey of land cover and land use conducted by the Natural Resources Conservation Service of the US Department of Agriculture in cooperation with the Statistical Laboratory of Iowa State University.

The study is a panel study and data have been collected for the years 1982, 1987, 1992, and 1997. The sample covers the entire United States. The primary sampling units are segments of land, where segments range in size from 40 acres to 640 acres. The design is specified by state and the typical design is stratified with area strata and two segments selected in each stratum. There are about 300 000 segments in the sample. In most segments, either two-or three-sample point locations are specified. In two states there is one sample point in each segment. There are about 800 000 points in the sample. See Nusser and Goebel (1997) for a more complete description of the survey.

In the study, a mutually exclusive and exhaustive land classification called *coveruse* is used. In each segment two types of information are collected: segment information and point information. Segment information includes a determination of the total number of urban acres in the segment, as well as the acres in roads, water, and farmsteads. Information collected at points is more detailed and includes a determination of coveruse, as well as information on erosion, land use, soil characteristics, and wildlife habitat. The location of the points within the segment was fixed at the time of sample selection. Therefore, for example, a segment may contain urban land but none of the three points fall on urban land. The segment data are the first-phase data and the point data are the second-phase data.

We use some data on urban land from the 1997 survey. The analysis unit for the survey is a state, but we carry out estimation for a group of counties. Because the data have not been released, our data are a collection of counties and do not represent a state or any real area. Our study area is composed of 89 counties and 6757 segments.

Imputation is used to represent the first-phase (segment) information. In the imputation operation an imputed point is created for those segments with acres of urban land but no point falling on the urban land. The characteristics for attributes of the point, such as coveruse in previous years, are imputed by choosing a donor that is geographically close to the recipient. The weight assigned to the imputed point is equal to the inverse of the segment selection probability multiplied by the acres of the coveruse. See Breidt, McVey and Fuller (1996–7) for a description of the imputation procedure. There are 394 real points and 219 imputed points that were urban in both 1992 and 1997 in the study area.

The sample in our group of counties is a stratified sample of segments, where a segment is 160 acres in size. The original stratification is such that each stratum contains two segments. In our calculations, we used collapsed strata that are nearly the size of a county. Calculations were performed using PC CARP.

We grouped the counties into seven groups that we call *cells*. These cells function as second-phase strata and are used as the cells from which donors are selected. This represents an approximation to the imputation procedure actually used.

In Section 19.3, we discussed using the imputed values in the estimation of the first-phase part of the variance expression. Let $\hat{V}_1\{\hat{\tau}_y\}$ be the estimated

EXAMPLE   *321*

variance of the total of a y-characteristic computed treating the imputed points as real points. This estimator is a biased estimator of the variance of the mean of a complete first-phase sample of y-values because a value from one segment can be used as a donor for a recipient in another segment in the same first-phase stratum. Thus, the segment totals are slightly correlated and the variance of a stratum mean is larger than the segment variance divided by the number of segments in the stratum. Because of the large number of segments per variance estimation stratum, about 70, the bias is judged to be small. See expression (19.41).

The variance of the weighted two-phase estimator of a total for a characteristic y was estimated by a version of (19.15),

$$\hat{V}\{\hat{\tau}_y\} = \hat{V}_1\{\hat{\tau}_y\} + \sum_{c=1}^{7} \hat{\tau}_{x1c}^2 \hat{V}\{\overline{x}_{2c}^{-1}\overline{y}_{2c}|s_1\}, \tag{19.42}$$

where $\hat{\tau}_y$ is the estimated total of y, $\hat{\tau}_{x1c}$ is the phase 1 estimator of the total of x for cell c, x is the indicator for urban in 1992, and $\overline{x}_{2c}$ is the estimated mean of cell c for the second phase.

The second term on the right side of (19.42) is the estimator of the conditional variance of the regression estimator given the first-phase sample. In this example, the regression estimator of the mean is the stratified estimator using the first-phase estimated stratum (cell) sizes $\hat{\tau}_{x1c}$. Thus, the regression estimator of the variance is equivalent to the stratified estimator of variance.

Table 19.1 contains estimated coefficients of variation for the NRI sample. This sample is an unusual two-phase sample in the sense that the number of two-phase units is about two-thirds of the number of first-phase units. In the more common two-phase situation, the number of second-phase units is a much smaller fraction of the number of first-phase units.

The variance of total urban acres in a county using imputed data is simply the variance of the first-phase segment data. However, if only the observed points are used to create a dataset, variability is added to estimates for counties. There were 15 counties with no second-phase points, but all counties had some urban segment data. Thus, the imputation procedure produces positive estimates for all counties. Crudely speaking, the imputation procedure makes a 100% improvement for those counties.

Table 19.1   Estimated properties of alternative estimators for a set of NRI counties.

Procedure	Coefficient of variation		Fraction of zero county estimates
	Total urban acres	County urban acres[a]	
Two-phase (weighted)	0.066	0.492	0.169
Two-phase (imputed)	0.066	0.488	0.000

[a] Average of 12 counties with smallest values.

Because of the number of zero-observation and one-observation counties, it is difficult to make a global efficiency comparison at the county level. The column with heading 'County urban acres' contains the average coefficient of variation for the 12 counties with the smallest coefficients of variation. All of these counties have second-phase points. There is a modest difference between the variance of the two procedures because the ratio of the total for the county to the total for the cell is well estimated.

In the production NRI dataset, a single imputation is used. Kim and Fuller (1999) constructed a dataset for one state in which two donors were selected for each recipient. The weight for each imputed observation is then one-half of the original observation. Kim and Fuller (1999) demonstrate how to construct replicates using the fractionally imputed data to compute an estimated variance.

CHAPTER 20

Analysis Combining Survey and Geographically Aggregated Data

D. G. Steel, M. Tranmer and D. Holt

20.1. INTRODUCTION AND OVERVIEW

Surveys are expensive to conduct and are subject to sampling errors and a variety of non-sampling errors (for a review see Groves, 1989). An alternative sometimes used by researchers is to make use of aggregate data available from administrative sources or the population census. The aggregate data will often be in the form of means or totals for one or more sets of groups into which the population has been partitioned. The groups will often be geographic areas or may be institutions, for example schools or hospitals.

Use of aggregate data alone to make inferences about individual-level relationships can introduce serious biases, leading to the ecological fallacy (for a review see Steel and Holt, 1996a). Despite the potential of ecological bias, analysis of aggregate data is an approach employed in social and geographical research. Aggregate data are used because individual-level data cannot be made available due to confidentiality requirements, as in the analysis of electoral data (King, 1997) or because aggregate data are readily available at little or no cost.

Achen and Shively (1995) noted that the use of aggregate data was a popular approach in social research prior to the Second World War but the rapid development of sample surveys led to the use of individual-level survey data as the main approach since the 1950s. More recently researchers have been concerned that the analysis of unit-level survey data traditionally treats the units divorced from their context. Here the concern is with the atomistic fallacy. The context can be the geographic area in which individuals live but can also be other groupings.

Survey data and aggregate data may be used together. One approach is contextual analysis in which the characteristics of the group in which an

Analysis of Survey Data Edited by R. L. Chambers and C. J. Skinner

individual is located are included in the analysis of individual-level survey data (Riley, 1964; Boyd and Iversen, 1979). A more recent response to the concern of analysing individuals out of context has been the development of multi-level modelling (Goldstein, 1987, 1995). Here the influence of the context is included through group-level random effects and parameters that may be due to the effect of unmeasured contextual variables, although they may also be acting as proxies for selection effects. Measured contextual or group-level variables may also be included in multi-level modelling. Analytical group-level variables, which are calculated from the values of individuals in the group, can be included and are often the means of unit-level variables, but may also include other statistics calculated from the unit-level variables such as a measure of spread. There may also be global group-level variables included, which cannot be derived from the values of units in the group, for example the party of the local member of parliament (Lazarsfeld and Menzel, 1961).

Standard multi-level analysis requires unit-level data, including the indicators of which group each unit belongs, to be available to the analyst. If all units in selected groups are included in the sample then analytical group-level variables can be calculated from the individual-level data. However, if a sample is selected within groups then sample contextual variables can be used, but this will introduce measurement error in some explanatory variables. Alternatively the contextual variables can be obtained from external aggregate data, but the group indicators available must allow matching with the aggregate data. Also, if global group-level variables are to be used the indicators must allow appropriate matching. In many applications the data available will not be the complete set required for such analyses.

Consider a finite population U of N units. Associated with the tth unit in the population are the values of variables of interest y_t. We will focus on situations in which the primary interest is in the conditional relationship of the variables of interest on a set of explanatory variables, the values of which are x_t. Assume that the population is partitioned into M groups and c_t indicates the group to which the tth unit belongs. The gth group contains N_g units. For convenience we will assume only one set of groups, although in general the population may be subject to several groupings, which may be nested but need not be. The complete finite population values are given by

$$(y_t, x_t, c_t), t = 1, \ldots, N.$$

In this situation there are several different approaches to the type of analysis and targets of inference.

Consider a basic model for the population values for $t \in g$

$$y_t = \mu_y + v_{y_g} + \epsilon_{y_t} \tag{20.1}$$

$$x_t = \mu_x + v_{x_g} + \epsilon_{x_t} \tag{20.2}$$

where $v_g = (v'_{y_g}, v'_{x_g})' \sim N(0, \Sigma^{(2)})$ and $\epsilon_t = (\epsilon'_{y_t}, \epsilon'_{x_t})' \sim N(0, \Sigma^{(1)}), g = 1, \ldots, M$ and $t = 1, \ldots, N_g$. Define $\Sigma = \Sigma^{(1)} + \Sigma^{(2)}$. Independence between all random vectors is assumed. The elements of the matrices are as follows:

$$\Sigma^{(l)} = \begin{pmatrix} \Sigma_{yy}^{(l)} & \Sigma_{yx}^{(l)} \\ \Sigma_{xy}^{(l)} & \Sigma_{xx}^{(l)} \end{pmatrix}, \; l = 1, 2, \quad \text{and} \quad \Sigma = \begin{pmatrix} \Sigma_{yy} & \Sigma_{yx} \\ \Sigma_{xy} & \Sigma_{xx} \end{pmatrix}.$$

Associated population regression cofficients are defined by

$$\beta_{yx}^{(l)} = \Sigma_{xx}^{(l)-1} \Sigma_{xy}^{(l)} \quad \text{and} \quad \beta_{yx} = \Sigma_{xx}^{-1} \Sigma_{xy}.$$

Consideration has to be given as to whether the targets of inference should be parameters that characterize the marginal relationships between variables across the whole population or parameters that characterize relationships within groups, conditional on the group effects. Whether the groups are of interest or not depends on the substantive situation. Holt (1989) distinguished between a disaggregated analysis, aimed at relationships taking into account the groups, and an analysis that averages over the groups, which is referred to as an aggregated analysis. A disaggregated analysis will often be tackled using multi-level modelling. For groups of substantive interest such as schools or other institutions there will be interest in determining the effect of the group and a disaggregated analysis will be appropriate. When the groups are geographically defined both approaches can be considered. The groups may be seen as a nuisance and so interest is in parameters characterizing relationships marginal to the groups. This does not assume that there are not influences operating at the group level but that we have no direct interest in them. These group-level effects will impact the relationships observed at the individual level, but we have no interest in separating purely individual- and group-level effects.

We may be interested in disaggregated parameters such as $\Sigma^{(1)}, \Sigma^{(2)}$ and functions of them such as $\beta_{yx}^{(1)}, \beta_{yx}^{(2)}$. For these situations we must consider how to estimate the variance and covariance components at individual and group levels. In situations where interest is in marginal relationships averaged over the groups then the target is Σ and functions of it such as β_{yx}. Estimation of the components separately may or may not be necessary depending on whether individual or aggregate data are available and the average number of units per group in the dataset, as well as the relative sizes of the effects at each level. We will consider how estimation can proceed in the different cases of data availability and the bias and variance of various estimators that can be used.

Geographical and other groups can often be used in the sample selection, for example through cluster and multi-stage sampling. This is done to reduce costs or because there is no suitable sampling frame of population units available. An issue is then how to account for the impact of the population structure, that is group effects in the statistical analysis. There is a large literature on both the aggregated approach and the disaggregated approach (see Skinner, Holt and Smith (1989), henceforth abbreviated as SHS).

If there is particular interest in group effects, namely elements of $\Sigma^{(2)}$, then that can also be taken into account in the sample design by making sure that multiple units are selected in each group. Cohen (1995) suggested that when costs depend on the number of groups and the number of individuals in the sample then in many cases an optimal number of units to select per group is approximately proportional to the square root of the ratio of unit-level to group-level variance components. This is also the optimal choice of sample units to select per primary sampling unit (PSU) in two-stage sampling with a simple linear cost function (Hansen, Hurwitz and Madow, 1953). Longford (1993) noted that for a fixed sample size the optimal choice is one plus the ratio of the unit-level to group-level variance components.

The role of group-level variables has to be considered carefully. Many analyses do not include any group-level variables. However, there is often the need to reflect the influence of group effects through measured variables. This can be done to explain random group-level effects and assess the impact of group-level variables that are of direct substantive interest. Such variables can be measures of the context, but they may also reflect within-group interactions and may also act as a proxy for selection effects. Inclusion of group-level variables requires either collection of this information from the respondent or collection of group indicators, for example address or postcode, to enable linking with the aggregate group-level data.

We will consider situations where the interest is in explaining variation in an individual-level variable. There are cases when the main interest is in a group-level response variable. However, even then the fact that groups are composed of individuals can be important, leading to a structural analysis (Riley, 1964).

In this chapter we consider the potential value of using aggregate and survey data together. The survey data available may not contain appropriate group indicators or may not include multiple units in the same group, which will affect the ability to estimate effects at different levels. We consider how a small amount of survey data may be used to eliminate the ecological bias in analysing aggregate data. We also show how aggregate data may be used to improve analysis of multi-level populations using survey data, which may or may not contain group indicators. Section 20.2 considers the situations that can arise involving individual-level survey data and aggregate data. The value of different types of information is derived theoretically in Section 20.3. In Section 20.4 a simulation study is reported that demonstrates the information in each source. Section 20.5 considers the use of individual-level data on auxiliary variables in the analysis of aggregate data.

20.2. AGGREGATE AND SURVEY DATA AVAILABILITY

Consider two data sources, the first based on a sample of units, s_0, from which unit-level data are available and the second on the sample, s_1, from which aggregate data are available. To simplify the discussion and focus on the role of survey and aggregate data we will suppose that the sample designs used to select s_0 and s_1 do not involve any auxiliary variables, and depend at most on the group indicators c_t. Hence the selection mechanisms are ignorable for model-based inferences about the parameters of interest. We also assume that there is no substantive interest in the effect of measured group-level variables, although the theory can be extended to allow for such variables.

The sample s_k contains m_k groups and n_k units with an average of $\bar{n}_k = n_k/m_k$ sample units per selected group and the sample size in the gth group is n_{kg}. An important case is when the aggregate data are based on the whole population, that is $s_1 = U$, for example aggregate data from the population census or some administrative system, so $n_{1g} = N_g$. The group means based upon sample s_k are denoted $\bar{y}_{kg}$ and $\bar{x}_{kg}$. The overall sample means are

$$\bar{y}_k = \frac{1}{n_k} \sum_g n_{kg} \bar{y}_{kg} \quad \text{and} \quad \bar{x}_k = \frac{1}{n_k} \sum_g n_{kg} \bar{x}_{kg}.$$

We assume that we are concerned with analytical inference so that we can investigate relationships that may apply to populations other than the one actually sampled (Smith, 1984). Thus we will take only a purely model-based approach to analysis and suppose that the values in the finite population are generated by some stochastic process or drawn from some superpopulation. We wish to make inferences about parameters of the probability distributions associated with this stochastic process as discussed in SHS, section 1.6.

We will consider three types of data:

(1) $d_0 = \{(y_t, x_t, c_t), t \in s_0\}$,
(2) $d_0^* = \{(y_t, x_t), t \in s_0\}$,
(3) $d_1 = \{(\bar{y}_g, \bar{x}_g, n_{1g}), g \in s_1\}$.

Case (1) corresponds to unit-level survey data with group indicators. If $m_0 < n_0$ then there are multiple observations in at least some groups and so it will be possible to separate group- and unit-level effects. Alternatively if an analysis of marginal parameters is required it is possible to estimate the variances of descriptive and analytical statistics taking the population structure into account (see Skinner, 1989). If $m_0 = n_0$ then it will not usually be possible to separate group- and unit-level effects. For an analysis of marginal parameters, this is not a problem and use of standard methods, such as ordinary least squares (OLS) regression, will give valid estimates and inferences. In this case the group identifiers are of no use. If $\bar{n}_0 = n_0/m_0$ is close to one then the estimates of group and individual-level effects will have high variances. In such a situation an analysis of marginal parameters will not be affected much by the clustering and so there will be little error introduced by ignoring it. The issues associated with designing surveys for multi-level modelling are summarized by Kreft and De Leeuw (1998, pp. 124–6).

Case (2) corresponds to unit-level survey data without group indicators. This situation can arise in several ways. For example, a sample may be selected by telephone and no information is recorded about the area in which the selected household is located. Another example is the release of unit record files by statistical agencies for which no group identifiers are released to help protect confidentiality. The Australian Bureau of Statistics (ABS) releases such data for many household surveys in this way. An important example is the release of data for samples of households and individuals from the population census, which is done in several countries, including Australia, the UK and the USA.

Case (3) corresponds to purely aggregate data. As noted before, analysis based soley on aggregate data can be subject to considerable ecological bias.

We will consider what analyses are possible in each of these cases, focusing on the bias and variance of estimators for a model with normally distributed variables. We then consider how aggregate and unit-level data can be used together in the following situations of available data;

(4) d_0 and d_1,
(5) d_0^* and d_1.

Case (4) corresponds to having unit-level survey data containing group indicators and aggregate data. Case (5) corresponds to having unit-level survey data without group indicators and aggregate data. For these cases the impact of different relative values of $m_0, n_0, \bar{n}_0$ and $m_1, n_1, \bar{n}_1$ will be investigated.

20.3. BIAS AND VARIANCE OF VARIANCE COMPONENT ESTIMATORS BASED ON AGGREGATE AND SURVEY DATA

We will consider various estimates that can be calculated in the different cases of data availability, focusing on covariances and linear regression coefficients. It is easier to consider the case when the number of sample units per group is the same across groups, so $n_{0g} = \bar{n}_0$ for all $g \in s_0$ and $n_{1g} = \bar{n}_1$ for all $g \in s_1$. This case is sufficient to illustrate the main conclusions concerning the roles of unit and aggregate data in most cases. Hence we will mainly focus on this case, although results for more general cases are available (Steel, 1999).

Case (1) Data available: $d_0 = \{(y_t, x_t, c_t), t \in s_0\}$, individual-level data with group indicators. Conventional multi-level modelling is carried out using a unit-level dataset that has group indicators. Maximum likelihood estimation (MLE) can be done using iterated generalized least squares (IGLS) (Goldstein, 1987), a Fisher fast scoring algorithm (Longford, 1987) or the EM algorithm (Bryk and Raudenbush, 1992). These procedures have each been implemented in specialized computer packages. For example, the IGLS procedures are implemented in MLn.

Estimates could also be obtained using moment or ANOVA-type estimates. Let $w_t = (y_t', x_t')'$. Moment estimators of $\Sigma^{(1)}$ and $\Sigma^{(2)}$ would be based on the between- and within-group covariance matrices

$$S_0^{(2)} = \frac{1}{m_0 - 1} \sum_{g \in s_0} n_{0g}(\bar{w}_{0g} - \bar{w}_0)(\bar{w}_{0g} - \bar{w}_0)'$$

$$S_0^{(W)} = \frac{1}{n_0 - m_0} \sum_{g \in s_0} (n_{0g} - 1)S_{0g}$$

where

$$S_{0g} = \frac{1}{n_{0g} - 1} \sum_{t \in s_{0,g}} (w_t - \bar{w}_{0g})(w_t - \bar{w}_{0g})'.$$

[handwritten margin notes: "m = groups", "n = units", "grand mean", "between group variance", "group mean", "with group variance ?", "within group", "matrix ?"]

For the model given by (20.1) and (20.2) Steel and Holt (1996a) showed

$$E[S_0^{(2)}] = \Sigma^{(1)} + \bar{n}_0^* \Sigma^{(2)}, \quad E[S_{0g}] = \Sigma^{(1)}, \quad \text{so } E[S_0^{(W)}] = \Sigma^{(1)}$$

where

$$\bar{n}_0^* = \bar{n}_0 \left(1 - \frac{C_0^2}{m_0 - 1}\right), \quad \text{and} \quad C_0^2 = \frac{1}{m_0} \sum_{g \in s_0} (n_{0g} - \bar{n}_0)^2 / \bar{n}_0^2.$$

Unbiased estimates of $\Sigma^{(1)}$ and $\Sigma^{(2)}$ can be obtained:

$$\tilde{\Sigma}_1^{(1)} = S_0^{(W)}, \tilde{\Sigma}_1^{(2)} = \frac{1}{\bar{n}_0^*}(S_0^{(2)} - S_0^{(W)}).$$

The resulting unbiased estimator of Σ is

$$\tilde{\Sigma}_1 = \tilde{\Sigma}_1^{(1)} + \tilde{\Sigma}_1^{(2)} = \frac{1}{\bar{n}_0^*} S_0^{(2)} + S_0^{(W)} \left(1 - \frac{1}{\bar{n}_0^*}\right).$$

We will focus on the estimation of the elements of $\Sigma^{(l)}$, that is $\Sigma_{ab}^{(l)}$. Define the function of $\bar{n}_0$, $\phi_{ab}(\bar{n}_0) = (\Sigma_{aa}^{(1)} + \bar{n}_0 \Sigma_{aa}^{(2)})(\Sigma_{bb}^{(1)} + \bar{n}_0 \Sigma_{bb}^{(2)}) + (\Sigma_{ab}^{(1)} + \bar{n}_0 \Sigma_{ab}^{(2)})^2$ and the parameter $\theta_{ab} = (\Sigma_{aa}^{(1)} \Sigma_{bb}^{(1)} + \Sigma_{ab}^{(1)^2})$. The finite sample variances of the estimators are

$$var(\tilde{\Sigma}_{1ab}^{(1)}) = \frac{\theta_{ab}}{n_0 - m_0} \tag{20.3}$$

$$var(\tilde{\Sigma}_{1ab}^{(2)}) = \frac{1}{\bar{n}_0^2} \left[\frac{\phi_{ab}(\bar{n}_0)}{m_0 - 1} + \frac{\theta_{ab}}{n_0 - m_0}\right] \tag{20.4}$$

$$var(\tilde{\Sigma}_{1ab}) = \frac{1}{\bar{n}_0^2} \left[\frac{\phi_{ab}(\bar{n}_0)}{m_0 - 1} + \frac{\theta_{ab}(\bar{n}_0 - 1)}{m_0}\right]. \tag{20.5}$$

These results can be used to give an idea of the effect of using different values of n_0 and m_0.

Maximum likelihood estimates $\hat{\Sigma}^{(1)}, \hat{\Sigma}^{(2)}$ can be obtained using IGLS. The asymptotic variances of these estimates are the same as the moment estimators in the balanced case.

Case (2) Data available: $d_0^* = \{(y_t, x_t), t \in s_0\}$, individual-level data without group indicators. In this situation we can calculate

$$S_0^{(1)} = \frac{1}{n_0 - 1} \sum_{t \in s_0} (w_t - \bar{w}_0)(w_t - \bar{w}_0)',$$

the covariance matrix calculated from the unit-level data ignoring the groups. Steel and Holt (1996a) show

$$E[S_0^{(1)}] = \Sigma^{(1)} + \left(1 - \frac{\bar{n}_0^+ - 1}{n_0 - 1}\right) \Sigma^{(2)} \tag{20.6}$$

where $\bar{n}_0^+ = \bar{n}_0(1 + C_0^2)$.

The elements of $S_0^{(1)}$ have

$$var(S_{0ab}^{(1)}) = \frac{1}{n_0 - 1}\left[\phi_{ab}(\bar{n}_0)\frac{m_0 - 1}{n_0 - 1} + \theta_{ab}\frac{(n_0 - m_0)}{n_0 - 1}\right]. \qquad (20.7)$$

The term $(\bar{n}_0^+ - 1)/(n_0 - 1)$ is $O(m_0^{-1})$, so provided m_0 is large we can estimate Σ with small bias. However, the variance of the estimates will be underestimated if the clustering is ignored, and we assumed that the variance of $S_{0ab}^{(1)}$ was $\phi_{ab}(1)/(n_0 - 1)$ instead of (20.7).

Case (3)   Data available: $d_1 = \{(\bar{y}_g, \bar{x}_g, n_{1g}), g \in s_1\}$, aggregate data only. This is the situation that is usually tackled by ecological analysis. There are several approaches that can be attempted. From these data we can calculate $S_1^{(2)}$, which has the properties

$$E[S_1^{(2)}] = \Sigma^{(1)} + \bar{n}_1^* \Sigma^{(2)} = \Sigma + (\bar{n}_1^* - 1)\Sigma^{(2)} \qquad (20.8)$$

$$var(S_{1ab}^{(2)}) = \frac{\phi_{ab}(\bar{n}_1)}{m_1 - 1}. \qquad (20.9)$$

As an estimate of Σ, $S_1^{(2)}$ has expectation that is a linear combination of $\Sigma^{(1)}$ and $\Sigma^{(2)}$ with the weights given to the two components leading to a biased estimate of Σ and functions of it, unless $\bar{n}_1^* = 1$. If $\bar{n}_1^*$ is large then the estimator will be heavily biased, as an estimator of Σ, towards $\Sigma^{(2)}$. To remove the bias requires estimation of $\Sigma^{(1)}$ and $\Sigma^{(2)}$, even if the marginal parameters are of interest.

As noted by Holt, Steel and Tranmer (1996) as an estimate of $\Sigma^{(2)}$, $(1/\bar{n}_1^*)S_1^{(2)}$ has bias $(1/\bar{n}_1^*)\Sigma^{(1)}$, which may be small if the group sample sizes are large. However, even if the parameter of interest is $\Sigma^{(2)}$ or functions of it, the fact that there is some contribution from $\Sigma^{(1)}$ needs to be recognized and removed.

Consider the special case when y_t and x_t are both scalar and

$$S_1^{(2)} = \begin{pmatrix} S_{1yy}^{(2)} & S_{1yx}^{(2)} \\ S_{1xy}^{(2)} & S_{1xx}^{(2)} \end{pmatrix}.$$

The ecological regression coefficients relating y to x are $B_{1yx}^{(2)} = S_{1xx}^{(2)-1} S_{1xy}^{(2)}$, and

$$E[B_{1yx}^{(2)}] = (1 - a_1^*)\beta_{yx}^{(1)} + a_1^*\beta_{yx}^{(2)}$$

where $a_1^* = \bar{n}_1^*\delta_{xx}/[1 + (\bar{n}_1^* - 1)\delta_{xx}]$. The effect of aggregation is to shift the weight given to the population regression parameters towards the group level. Even a small value of δ_{xx} can lead to a considerable shift if $\bar{n}_1^*$ is large.

If there are no group-level effects, so $\Sigma^{(2)} = 0$, which is equivalent to random group formation, then $S_1^{(2)}$ is unbiased for Σ and $\phi_{ab}(\bar{n}_1) = \theta_{ab}$. Steel and Holt (1996b) showed that in this case there is no ecological fallacy for linear statistics, and parameters such as means, variances, regression and correlation coefficients can be unbiasedly estimated from group-level data. They also showed that the variances of statistics are mainly determined by the number of groups in the analysis. A regression or correlation analysis of m groups has the same statistical properties as an analysis based on m randomly selected individuals irrespective of the sample sizes on which the group means are based.

When only aggregate data are available estimates of the variance components can be obtained in theory. The key to separating $\Sigma^{(1)}$ and $\Sigma^{(2)}$ is the form of heteroscedasticity for the group means implied by models (20.1) and (20.2), that is

$$var(\bar{w}_{1g}) = \frac{\Sigma^{(1)}}{n_{1g}} + \Sigma^{(2)}. \tag{20.10}$$

A moments approach can be developed using group-level covariance matrices with different weightings. An MLE approach can also be developed (Goldstein, 1995). The effectiveness of these approaches depends on the variation in the group sample sizes, n_{1g}, and the validity of the assumed form of the heteroscedasticity in (20.10). If the covariance matrices of either ϵ_t or v_g are not constant but related in an unknown way to n_g then this approach will have problems.

Maximum likelihood estimates can be obtained using IGLS. The key steps in the IGLS procedure involve regression of the elements of $(\bar{w}_{1g} - \hat{\mu}_w)(\bar{w}_{1g} - \hat{\mu}_w)'$ against the vector of values $(1, 1/n_{1g})'$, where $\hat{\mu}_w$ is the current estimate of $\mu_w = (\mu'_y, \mu'_x)'$. The resulting coefficient of $1/n_g$ estimates $\Sigma^{(2)}$ and that of 1 estimates $\Sigma^{(1)}$, giving the estimates $\hat{\Sigma}_3^{(1)}$ and $\hat{\Sigma}_3^{(2)}$. If there is no variation in the group sample sizes, n_{1g}, then the coefficients cannot be estimated; hence we will consider cases when they do vary. The asymptotic variance of these estimates can be derived as (Steel, 1999)

$$var(\hat{\Sigma}_{3aa}^{(1)}) = \sum_{g \in s_1} n_{1g}^2 (\phi_{aa}(\bar{n}_{1g}))^{-1} \left[\sum_{g,h \in s_1} n_{1h}(n_{1h} - n_{1g})(\phi_{aa}(\bar{n}_{1g}))^{-1}(\phi_{aa}(\bar{n}_{1h}))^{-1} \right]^{-1}$$

$$var(\hat{\Sigma}_{3aa}^{(2)}) = \sum_{g \in s_1} (\phi_{aa}(\bar{n}_{1g}))^{-1} \left[\sum_{g,h \in s_1} n_{1h}(n_{1h} - n_{1g})(\phi_{aa}(\bar{n}_{1g}))^{-1}(\phi_{aa}(\bar{n}_{1h}))^{-1} \right]^{-1}.$$

An idea of the implications of these formulae can be obtained by considering the case when $n_{1g}\Sigma_{aa}^{(2)}$ is large compared with $\Sigma_{aa}^{(1)}$ so $\phi_{aa}(\bar{n}_{1g}) \approx n_{1g}\Sigma_{aa}^{(2)}$. Then

$$var(\hat{\Sigma}_{3aa}^{(1)}) = \frac{2\Sigma_{aa}^{(2)}}{m\bar{n}_{-1}^2} C_{-1}^2$$

$$var(\hat{\Sigma}_{3aa}^{(2)}) = \frac{2\Sigma_{aa}^{(2)}(1 + C_{-1}^2)}{mC_{-1}^2}$$

where

$$\bar{n}_{-1} = \frac{1}{m} \sum_{g \in s_1} \frac{1}{n_{1g}}$$

and C_{-1}^2 is the square of the coefficient of variation of the inverses of the group sample sizes. This case would often apply when the aggregate data are means based on a complete enumeration of the group, so that $n_{1g} = N_g$.

The other extreme is when $n_{1g}\Sigma_{aa}^{(2)}$ is small compared with $\Sigma_{aa}^{(1)}$, so that the variation in $\phi_{aa}(\bar{n}_{1g})$ is much less than n_g and can be approximated by $\phi_{aa}(\bar{n}_1)$. In this case

$$var(\widehat{\Sigma}_{3aa}^{(1)}) = \frac{\phi_{aa}(\bar{n}_1)(1 + C^2)}{mC^2}$$

$$var(\widehat{\Sigma}_{3aa}^{(2)}) = \frac{\phi_{aa}(\bar{n}_1)}{m\bar{n}_1^2 C^2},$$

where C^2 is the square of the coefficient of variation of the group sample sizes.

Unless the group sample sizes vary extremely we would expect both C^2 and C_{-1}^2 to be less than one and often they will be much less than one. In both situations we would expect the variances to be much larger than if we had been able to use the corresponding individual-level data. Also we should expect the estimate of $\Sigma_{aa}^{(1)}$ to be particularly poor.

Case (4)   Data available: d_0 and d_1, individual-level data with group indicators and aggregate data. As estimates can be produced using only d_0, the interest here is: what does the aggregate data add? Looking at (20.3) and (20.4) we see that the variances of the estimates of the level 1 variance components based solely on d_0 will be high when $n_0 - m_0$ is small. The variances of the estimates of the level 2 variance components based solely on d_0 will be high when $n_0 - m_0$ or $m_0 - 1$ is small. In these conditions the aggregate data based on s_1 will add useful information to help separate the effects at different levels if it gives data for additional groups or if it refers to the same groups but provides group means based on larger sample sizes than s_0.

It has been assumed that the group labels on the unit-level data enable the linking of the units with the relevant group mean in d_1. In some cases the group labels may have been anonymized so that units within the same group can be identified, but they cannot be matched with the group-level data. This situation can be handled as described in case (5).

Case (5)   Data available: d_0^* and d_1, individual-level data without group indicators and aggregate data. In practice there are two situations we have in mind. The first is when the main data available are the group means and we are interested in how a small amount of unit-level data can be used to eliminate the ecological bias. So this means that we will be interested in n_0 smaller than m_1. The other case is when the unit-level sample size is large but the lack of group indicators means that the unit- and group-level effects cannot be separated. A small amount of group-level data can then be used to separate the effects. So this means that m_1 is smaller than n_0. In general we can consider the impact of varying m_0, n_0, m_1, n_1.

Define $\tilde{n} = \bar{n}_1^* + [(\bar{n}_0^+ - 1)/(n_0^{-1})]$. Moment estimators of $\Sigma^{(1)}$ and $\Sigma^{(2)}$ can be based on (20.6) and (20.8) giving

$$\tilde{\Sigma}_5^{(1)} = \frac{1}{\tilde{n}-1} \left[\bar{n}_1^* S_0^{(1)} + \left(\frac{\bar{n}_0^+ - 1}{n_0 - 1} - 1 \right) S_1^{(2)} \right] \tag{20.11}$$

$$\tilde{\Sigma}_5^{(2)} = \frac{1}{\tilde{n}-1} (S_1^{(2)} - S_0^{(1)}). \tag{20.12}$$

The resulting estimator of Σ is

$$\tilde{\Sigma}_5 = \tilde{\Sigma}_5^{(1)} + \tilde{\Sigma}_5^{(1)} = \frac{1}{\tilde{n}-1} \left[\frac{\bar{n}_0^+ - 1}{n_0 - 1} S_1^{(2)} + (\bar{n}_1^* - 1) S_0^{(1)} \right]. \tag{20.13}$$

These estimators are linear combinations of $S_0^{(1)}$ and $S_1^{(2)}$ and so the variances of their components are determined by $var(S_{0ab}^{(1)})$, $var(S_{1ab}^{(2)})$ and $Cov(S_{0ab}^{(1)}, S_{1ab}^{(2)})$. We already have expressions for $var(S_{0ab}^{(1)})$ and $var(S_{1ab}^{(2)})$ given by (20.7) and (20.9) respectively. The covariance will depend on the overlap between the samples s_0 and s_1. If the two samples have no groups in common the covariance will be zero. At the other extreme when $s_0 = s_1$,

$$Cov(S_{0ab}^{(1)}, S_{1ab}^{(2)}) = \frac{\phi_{ab}(\bar{n}_1)}{n_0 - 1}$$

and the variances given in case (1) apply. A practical situation in between these two extremes occurs when we have aggregate data based on a census or large sample and we select s_0 to supplement the aggregate data by selecting a sub-sample of the groups and then a subsample is selected in each selected group, using the same sampling fraction, f_2, so $m_0 = f_1 m_1, \bar{n}_0 = f_2 \bar{n}_1$ and $n_0 = f n_1$ where $f = f_1 f_2$. In this situation

$$Cov(S_{0ab}^{(1)}, S_{1ab}^{(2)}) = f_2(f_1 m_1 - 1) \frac{\phi_{ab}(\bar{n}_1)}{(n_0 - 1)(m_1 - 1)} = f \, var(S_1^{(2)}) \frac{m_1 - 1}{n_0 - 1}.$$

For given sample sizes in s_0 and s_1 the variances of the estimators of the variance components are minimized when the covariance is maximized. For this situation this will occur when f_2 is minimized by selecting as many groups as possible. If the same groups are included in the two samples and s_0 is a subsample of s_1 selected using the same sampling fraction, f_2, in each group, so $f_1 = 1, f_2 = f, n_0 = f n_1$ and $m_0 = m_1$, then

$$Cov(S_{0ab}^{(1)}, S_{1ab}^{(2)}) = f \frac{\phi_{ab}(\bar{n}_1)}{n_0 - 1}.$$

This situation arises, for example, when s_0 is the Sample of Anonymized Records released from the UK Population Census and s_1 is the complete population from which group means are made available for Enumeration Districts (EDs).

For the estimation of the level 1 and marginal parameters the variances are mainly a function of n_0^{-1}. For the estimation of the level 2 parameters both n_0^{-1} and m_1^{-1} are factors. For the estimation of the group-level variance component of variable a the contributions of s_0 and s_1 to the variance are approximately equal when

$$n_0 - 1 = \frac{m_1 - 1}{[1 + (\bar{n}_1 - 1)\delta_{aa}]^2}.$$

Aggregate and unit-level data without group indicatives can also be combined in MLn by treating each unit observation as coming from a group with a sample size of one. This assumes that the unit-level data come from groups different from those for which aggregate data are available. In practice this may not be true, in particular if the aggregate data cover all groups in the population. But the covariances will be small if f is small.

20.4.   SIMULATION STUDIES

To investigate the potential value of combining survey and aggregate data a simulation study was carried out. The values of a two-level bivariate population were generated using the model given by (20.1) and (20.2). We used the District of Reigate as a guide to the group sizes, which had a population of 188 800 people aged 16 and over at the 1991 census and 371 EDs. The total population size was reduced by a factor of 10 to make the simulations manageable, and so the values of $M = 371$ and $N = 18\,880$ were used. Group sizes were also a tenth of the actual EDs, giving an average of $\bar{N} = 50.9$. The variation in group population sizes led to $C^2 = 0.051$ and $C_{-1}^2 = 0.240$.
The values of the parameters ($\times 1000$) used were

$$\Sigma^{(1)} = \begin{pmatrix} \Sigma_{yy}^{(1)} & \Sigma_{yx}^{(1)} \\ \Sigma_{xy}^{(1)} & \Sigma_{xx}^{(1)} \end{pmatrix} = \begin{pmatrix} 78.6 & 22.8 \\ 22.8 & 90.5 \end{pmatrix}$$

$$\Sigma^{(2)} = \begin{pmatrix} \Sigma_{yy}^{(2)} & \Sigma_{yx}^{(2)} \\ \Sigma_{xy}^{(2)} & \Sigma_{xx}^{(2)} \end{pmatrix} = \begin{pmatrix} 0.979 & 1.84 \\ 1.84 & 6.09 \end{pmatrix}.$$

These values were based on values obtained from an analysis of the variables $y = $ limiting long-term illness (LLTI) and $x = $ living in a no-car household (CARO) for Reigate using 1991 census aggregate data and individual-level data obtained from the Sample of Anonymized Records released from the census. The method of combining aggregate and individual-level data described in case (5) in Section 20.3 was used to obtain the variance and covariance components. These parameter values imply the following values for the main parameters of the linear regression relationships relating the response variable y to the explanatory variable x:

$$\Sigma = \begin{pmatrix} \Sigma_{yy} & \Sigma_{yx} \\ \Sigma_{xy} & \Sigma_{xx} \end{pmatrix} = \begin{pmatrix} 79.6 & 24.7 \\ 24.7 & 96.6 \end{pmatrix}$$

$$\beta_{yx} = 0.255, \qquad \beta_{yx}^{(1)} = 0.252, \qquad \beta_{yx}^{(2)} = 0.302,$$
$$\delta_{yy} = 0.0121, \qquad \delta_{xx} = 0.0630, \qquad \delta_{yx} = 0.0208, \qquad a^* = 0.774.$$

The process of generating the finite population values was repeated 100 times. The following cases and designs in terms of numbers of units and groups were considered:

Case (1): unit-level data with group indicators for the population, $n_0 = 18\,880, m_0 = 371$.
Case (1): unit-level data with group indicators for a 20% sample, $n_0 = 3776, m_0 = 371$.
Case (5): aggregate population data $n_1 = 18\,880, m_1 = 371$, sample unit-level data with group indicators; $m_0 = 371$ and $n_0 = 3776, 378, 38$, that is 20%, 2% and 0.2% samples respectively.
Case (3): aggregate population data only, $n_1 = 18\,880, m_1 = 371$.

Estimates of $\Sigma^{(1)}, \Sigma^{(2)}, \Sigma, \beta_{yx}^{(1)}, \beta_{yx}^{(2)}$ and β_{yx} were obtained. Results are given in Table 20.1, which shows the mean and standard deviation of the various estimates of the variance components and the total variance and the associated regression coefficients at each level using the IGLS-based procedures for each case. Moment estimators were also calculated and gave similar results. Table 20.2 summarizes the properties of the direct estimates based on $S^{(1)}$ and $S^{(2)}$.

The results show that for level 2 parameters using aggregate data for the population plus individual-level sample data (case (5)) is practically as good as using all the population data (case (1)) when the individual data are based on a sample much larger than the number of groups. This is because the number of groups is the key factor. As n_0 decreases the variances of the estimates of the level 2 parameters in case (5) increase but not in direct proportion to n_0^{-1}, because of the influence of the term involving $m_1 - 1$. For the level 2 parameters the aggregate data plus 20% individual-level data without group indicators is a lot better than 20% individual-level data with group indicators. This shows the benefit of the group means being based on a larger sample, that is the complete population.

As expected, level 1 parameters are very badly estimated from purely aggregate data (case (3)) and the estimates of the level 2 parameters are poor. For level 1 and level 2 parameters the use of a small amount of individual-level data greatly improves the performance relative to using aggregate data only. For example, case (5) with $n_0 = 38$ gives estimates with a little over half the standard deviation of the estimates of $\Sigma^{(1)}$ and $\Sigma^{(2)}$ compared with using solely aggregate data.

For the level 1 parameters the size of the individual-level sample is the key factor for case (1) and case (5), with the variances varying roughly in proportion to n_0^{-1}. Use of aggregate data for the population plus 20% individual-level sample data without group indicators is as good as using 20% individual-level sample data with group indicators. The estimates of overall individual-level parameters perform in the same way as the estimates of the pure level 1 parameters. We found that the moment and IGLS estimators had similar efficiency.

Table 20.1 Performance of IGLS variance components estimates ($\times 1000$) over 100 replications.

(a) LLTI; $\Sigma_{yy}^{(1)} = 78.6, \Sigma_{yy}^{(2)} = 0.979, \Sigma_{yy} - 79.6$

		$\hat{\Sigma}_{yy}^{(1)}$		$\hat{\Sigma}_{yy}^{(2)}$		$\hat{\Sigma}_{yy}$	
Case	Design	Mean	SD	Mean	SD	Mean	SD
1	$m_0 = 371, n_0 = 18\,880$	78.6	0.820	0.955	0.185	79.6	0.833
1	$m_0 = 371, n_0 = 3776$	78.6	1.856	0.974	0.650	79.6	1.846
5	$n_0 = 3776, m_1 = 371$	78.6	1.851	0.955	0.186	79.6	1.838
5	$n_0 = 378, m_1 = 371$	78.4	5.952	0.959	0.236	79.3	5.804
5	$n_0 = 38, m_1 = 371$	77.3	16.648	0.980	0.374	78.3	16.324
3	$m_1 = 371$	75.6	32.799	0.995	0.690	76.6	32.142

(b) LLTI/CARO; $\Sigma_{yx}^{(1)} = 22.8, \Sigma_{yx}^{(2)} = 1.84, \Sigma_{yx} = 24.7,$

		$\hat{\Sigma}_{yx}^{(1)}$		$\hat{\Sigma}_{yx}^{(2)}$		$\hat{\Sigma}_{yx}$	
Case	Design	Mean	SD	Mean	SD	Mean	SD
1	$m_0 = 371, n_0 = 18\,880$	22.8	0.637	1.81	0.279	24.6	0.648
1	$m_0 = 371, n_0 = 3776$	22.7	1.536	1.72	0.797	24.5	1.494
5	$n_0 = 3776, m_1 = 371$	22.7	1.502	1.81	0.280	24.5	1.480
5	$n_0 = 378, m_1 = 371$	23.0	4.910	1.81	0.307	24.8	4.777
5	$n_0 = 38, m_1 = 371$	21.9	13.132	1.83	0.381	23.7	12.874
3	$m_1 = 371$	29.2	43.361	1.55	0.962	30.8	42.578

(c) CARO; $\Sigma_{xx}^{(1)} = 90.5, \Sigma_{xx}^{(2)} = 6.09, \Sigma_{xx} = 96.6$

		$\hat{\Sigma}_{xx}^{(1)}$		$\hat{\Sigma}_{xx}^{(2)}$		$\hat{\Sigma}_{xx}$	
Case	Design	Mean	SD	Mean	SD	Mean	SD
1	$m_0 = 371, n_0 = 18\,880$	90.6	0.996	6.05	0.642	96.6	1.174
1	$m_0 = 371, n_0 = 3776$	90.5	2.120	5.97	1.285	96.6	2.274
5	$n_0 = 3776, m_1 = 371$	90.5	2.168	6.06	0.639	96.5	2.271
5	$n_0 = 378, m_1 = 371$	90.5	7.713	6.05	0.645	96.6	7.642
5	$n_0 = 38, m_1 = 371$	86.7	21.381	6.13	0.737	92.8	21.020
3	$m_1 = 371$	110.8	99.107	5.47	2.082	116.3	97.329

(d) LLTI vs. CARO; $\beta_{yx}^{(1)} = 252, \beta_{yx}^{(2)} = 302, \beta_{yx} = 255$

		$\hat{\beta}_{yx}^{(1)}$		$\hat{\beta}_{yx}^{(2)}$		$\hat{\beta}_{yx}$	
Case	Design	Mean	SD	Mean	SD	Mean	SD
1	$m_0 = 371, n_0 = 18\,880$	252	6.64	299	34.96	255	6.15
1	$m_0 = 371, n_0 = 3776$	251	16.84	284	124.17	254	14.85
5	$n_0 = 3776, m_1 = 371$	251	15.85	300	34.97	254	14.74
5	$n_0 = 378, m_1 = 371$	254	49.28	298	38.51	257	44.92
5	$n_0 = 38, m_1 = 371$	261	157.32	299	54.30	262	142.15
3	$m_1 = 371$	300	1662.10	294	200.16	254	607.39

Table 20.2 Properties of (a) direct variance estimates, (b) direct covariance estimates, (c) direct variance estimates and (d) direct regression estimates ($\times 1000$) over 100 replications.

(a) LLTI $\sum_{yy}^{(1)} = 78.6$, $\sum_{yy}^{(2)} = 0.979$, $\sum_{yy} = 79.6$

Data	$S_{yy}^{(1)}$		$S_{yy}^{(2)}$	
	Mean	SD	Mean	SD
$m_0 = 371, n_0 = 18\,880$	79.6	0.834	128	9.365
$m_0 = 371, n_0 = 3776$	79.6	1.845	88.5	6.821
$m_0 = 371, n_0 = 378$	79.3	5.780		
$m_0 = 371, n_0 = 38$	77.8	18.673		

(b) LLTI/CARO; $\sum_{yx}^{(1)} = 22.8$, $\sum_{yx}^{(2)} = 1.84$, $\sum_{yx} = 24.7$

Data	$S_{yx}^{(1)}$		$S_{yx}^{(2)}$	
	Mean	SD	Mean	SD
$m_0 = 371, n_0 = 18\,880$	24.7	0.645	115	14.227
$m_0 = 371, n_0 = 3776$	24.5	1.480	41.1	6.626
$m_0 = 371, n_0 = 378$	24.8	4.856		
$m_0 = 371, n_0 = 38$	23.2	14.174		

(c) CARO; $\sum_{xx}^{(1)} = 90.5$, $\sum_{xx}^{(2)} = 6.09$, $\sum_{xx} = 96.6$

Data	$S_{xx}^{(1)}$		$S_{xx}^{(2)}$	
	Mean	SD	Mean	SD
$m_0 = 371, n_0 = 18\,880$	96.6	1.167	399	32.554
$m_0 = 371, n_0 = 3776$	96.5	2.264	152	12.511
$m_0 = 371, n_0 = 378$	96.6	7.546		
$m_0 = 371, n_0 = 38$	92.0	22.430		

(d) LLTI vs. CARO; $\beta_{yx}^{(1)} = 252$, $\beta_{yx}^{(2)} = 302$, $\beta_{yx} = 255$

Data	$B_{yx}^{(1)}$		$B_{yx}^{(2)}$	
	Mean	SD	Mean	SD
$m_0 = 371, n_0 = 18\,880$	255	6.155	288	26.616
$m_0 = 371, n_0 = 3776$	254	14.766	270	36.597
$m_0 = 371, n_0 = 378$	256	45.800		
$m_0 = 371, n_0 = 38$	258	155.517		

20.5. USING AUXILIARY VARIABLES TO REDUCE AGGREGATION EFFECTS

In Section 20.3 it was shown how individual-level data on y and x can be used in combination with aggregate data. In some cases individual-level data may not be available for y and x, but may be available for some auxiliary variables. Steel and Holt (1996a) suggested introducing extra variables to account for the correlations within areas. Suppose that there is a set of auxiliary variables, z, that partially characterize the way in which individuals are clustered within the groups and, conditional on z, the observations for individuals in area g are influenced by random group-level effects. The auxiliary variables in z may only have a small effect on the individual-level relationships and may not be of any direct interest. The auxiliary variables are only included as they may be used in the sampling process or to help account for group effects and we assume that there is no interest in the influence of these variables in their own right. Hence the analysis should focus on relationships averaging out the effects of auxiliary variables. However, because of their strong homogeneity within areas they may affect the ecological analysis greatly. The matrices $z_U = [z_1, \ldots, z_N]'$, $c_U = [c_1, \ldots, c_N]'$ give the values of all units in the population. Both the explanatory and auxiliary variables may contain group-level variables, although there will be identification problems if the mean of an individual-level explanatory variable is used as a group-level variable. This leads to:

Case (6) Data available: d_1 and $\{z_t, t \in s_0\}$, aggregate data and individual-level data for the auxiliary variables. This case could arise, for example, when we have individual-level data on basic demographic variables from a survey and we have information in aggregate form for geographic areas on health or income obtained from the health or tax systems. The survey data may be a sample released from the census, such as the UK Sample of Anonymized Records (SAR).

Steel and Holt (1996a) considered a multi-level model with auxiliary variables and examined its implications for the ecological analysis of covariance matrices and correlation coefficients obtained from them. They also developed a method for adjusting the analysis of aggregate data to provide less biased estimates of covariance matrices and correlation coefficients. Steel, Holt and Tranmer (1996) evaluated this method and were able to reduce the biases by about 70% by using limited amounts of individual-level data for a small set of variables that help characterize the differences between groups. Here we consider the implications of this model for ecological linear regression analysis.

The model given in (20.1) to (20.2) is expanded to include z by assuming the following model conditional on z_U and the groups used:

$$w_t = \mu_{w|z} + \beta'_{wz}z_t + v_g + \epsilon_t, \quad t \in g \tag{20.14}$$

where

$$var(v_g|z_U, c_U) = \Sigma^{(2)}_{|z} \quad \text{and} \quad var(\epsilon_t|z_U, c_U) = \Sigma^{(1)}_{|z}. \tag{20.15}$$

This model implies

$$E(w_t|z_U, c_U) = \mu_{w|z} + \beta'_{wz}z_t,$$ (20.16)

$$var(w_t|z_U, c_U) = \Sigma^{(1)}_{|z} + \Sigma^{(2)}_{|z} = \Sigma_{|z}$$ (20.17)

and

$$Cov(w_t, w_u|z_U, c_U) = \Sigma^{(2)}_{|z} \quad \text{if } c_t = c_u, t \neq u.$$

The random effects in (20.14) are different from those in (20.1) and (20.2) and reflect the within-group correlations after conditioning on the auxiliary variables. The matrix $\Sigma^{(2)}_{|z}$ has components $\Sigma^{(2)}_{xx|z}, \Sigma^{(2)}_{xy|z}$ and $\Sigma^{(2)}_{yy|z}$ and $\beta'_{wz} = (\beta_{xz}, \beta_{yz})'$. Assuming $var(z_t) = \Sigma_{zz}$ then the marginal covariance matrix is

$$\Sigma = \Sigma_{|z} + \beta'_{wz}\Sigma_{zz}\beta_{wz}$$ (20.18)

which has components Σ_{xx}, Σ_{xy} and Σ_{yy}. We assume that the target of inference is $\beta_{yx} = \Sigma^{-1}_{xx}\Sigma_{xy}$, although this approach can be used to estimate regression coefficients at each level.

Under this model, Steel and Holt (1996a) showed

$$E[S^{(2)}_1|z_U, c_U] = \Sigma + \beta'_{wz}(S^{(2)}_{1zz} - \Sigma_{zz})\beta_{wz} + (\bar{n}^*_1 - 1)\Sigma^{(2)}_{|z}.$$ (20.19)

Providing that the variance of $S^{(2)}_1$ is $O(m_1^{-1})$ (see (20.9)) the expectation of the ecological regression coefficients, $B^{(2)}_{1yx} = S^{(2)-1}_{1xx}S^{(2)}_{1xy}$, can be obtained to $O(m_1^{-1})$ by replacing $S^{(2)}_{1xx}$ and $S^{(2)}_{1xy}$ by their expectations.

20.5.1. Adjusted aggregate regression

If individual-level data on the auxiliary variables are available, the aggregation bias due to them may be estimated. Under (20.14), $E[B^{(2)}_{1wz}|z_U, c_U] = \beta_{wz}$ where $B^{(2)}_{1wz} = S^{(2)-1}_{1zz}S^{(2)}_{1zw}$. If an estimate of the individual-level population covariance matrix for z_U was available, possibly from another source such as s_0, Steel and Holt (1996) proposed the following adjusted estimator of Σ:

$$\hat{\Sigma}_6 = S^{(2)}_1 + B^{(2)'}_{1wz}\left(\hat{\Sigma}_{zz} - S^{(2)}_{1zz}\right)B^{(2)}_{1wz}$$

$$= S^{(2)}_{1|z} + B^{(2)'}_{1wz}\hat{\Sigma}_{zz}B^{(2)}_{1wz}$$

where $\hat{\Sigma}_{zz}$ is the estimate of Σ_{zz} calculated from individual-level data. This estimator corresponds to a Pearson-type adjustment, which has been proposed as a means of adjusting for the effect of sampling schemes that depend on a set of design variables (Smith, 1989). This estimator removes the aggregation bias due to the auxiliary variables. This estimator can be used to adjust simultaneously for the effect of aggregation and sample selection involving design variables by including these variables in z_U. For normally distributed data this estimator is MLE.

Adjusted regression coefficients can then be calculated from $\hat{\Sigma}_6$, that is

$$\widehat{\beta}_{6yx} = \widehat{\Sigma}_{6xx}^{-1}\widehat{\Sigma}_{6xy}.$$

The adjusted estimator replaces the components of bias in (20.19) due to $\beta'_{wz}(S^{(2)}_{1zz} \quad \Sigma_{zz})\beta_{wz}$ by $\beta'_{wz}(\widehat{\Sigma}_{zz} - \Sigma_{zz})\beta_{wz}$. If $\widehat{\Sigma}_{zz}$ is an estimate based on an individual-level sample involving m_0 first-stage units then for many sample designs $\widehat{\Sigma}_{zz} - \Sigma_{zz} = O(1/m_0)$, and so $\beta'_{wz}(\widehat{\Sigma}_{zz} - \Sigma_{zz})\beta_{wz}$ is $O(1/m_0)$.

The adjusted estimator can be rewritten as

$$\widehat{\beta}_{6yx} = B^{(2)}_{1yx|z} + \widehat{\beta}_{6zx}B^{(2)}_{1yz|x} \tag{20.20}$$

where $\widehat{\beta}_{6zx} = \widehat{\Sigma}_{6xx}^{-1}B^{(2)'}_{1xz}\widehat{\Sigma}_{zz}$. Corresponding decompositions apply at the group and individual levels:

$$B^{(2)}_{1yx} = B^{(2)}_{1yx|z} + B^{(2)}_{1zx}B^{(2)}_{1yz|x}$$
$$B^{(1)}_{1yx} = B^{(1)}_{1yx|z} + B^{(1)}_{1zx}B^{(1)}_{1yz|x}.$$

The adjustment is trying to correct for the bias in the estimation of β_{zx} by replacing $B^{(2)}_{1zx}$ by $\widehat{\beta}_{6zx}$. The bias due to the conditional variance components $\Sigma^{(2)}_{|z}$ remains.

Steel, Tranmer and Holt (1999) carried out an empirical investigation into the effects of aggregation on multiple regression analysis using data from the Australian 1991 Population Census for the city of Adelaide. Group-level data were available in the form of totals for 1711 census Collection Districts (CDs), which contain an average of about 250 dwellings. The analysis was confined to people aged 15 or over and there was an average of about 450 such people per CD. To enable an evaluation to be carried out data from the households sample file (HSF), which is a 1% sample of households and the people within them, released from the population census were used.

The evaluation considered the dependent variable of personal income. The following explanatory variables were considered: marital status, sex, degree, employed–manual occupation, employed–managerial or professional occupation, employed–other, unemployed, born Australia, born UK and four age categories.

Multiple regression models were estimated using the HSF data and the CD data, weighted by CD population size. The results are summarized in Table 20.3. The R^2 of the CD-level equation, 0.880, is much larger than that of the individual-level equation, 0.496. However, the CD-level R^2 indicates how much of the variation in CD mean income is being explained. The difference between the two estimated models can also be examined by comparing their fit at the individual level. Using the CD-level equation to predict individual-level income gave an R^2 of 0.310. Generally the regression coefficients estimated at the two levels are of the same sign, the exceptions being married, which is non-significant at the individual level, and the coefficient for age 20–29. The values can be very different at the two levels, with the CD-level coefficients being larger than the corresponding individual-level coefficients in some cases and smaller in others. The differences are often considerable: for example, the

Table 20.3 Comparison of individual, CD and adjusted CD-level regression equations.

Variable	Individual level		CD level		Adjusted CD level	
	Coefficient	SE	Coefficient	SE	Coefficient	SE
Intercept	11 876.0	496.0	4 853.6	833.9	1 573.0	1 021.3
Married	−8.5	274.0	4 715.5	430.0	7 770.3	564.4
Female	−6 019.1	245.2	−3 067.3	895.8	2 195.0	915.1
Degree	8 471.5	488.9	21 700.0	1 284.5	23 501.0	1 268.9
Unemp	−962.5	522.1	−390.7	1 287.9	569.5	1 327.2
Manual	9 192.4	460.4	1 457.3	1101.2	2 704.7	1 091.9
ManProf	20 679.0	433.4	23 682.0	1 015.5	23 037.0	1 023.7
EmpOther	11 738.0	347.8	6 383.2	674.9	7 689.9	741.7
Born UK	1 146.3	425.3	2 691.1	507.7	2 274.6	506.0
Born Aust	1 873.8	336.8	2 428.3	464.6	2 898.8	491.8
Age 15–19	−9 860.6	494.7	−481.9	1 161.6	57.8	1 140.6
Age 20–29	−3 529.8	357.6	2 027.0	770.2	1 961.6	758.4
Age 45–59	585.6	360.8	434.3	610.2	1 385.1	1 588.8
Age 60+	255.2	400.1	1 958.0	625.0	2 279.5	1 561.4
R^2	0.496		0.880		0.831	

coefficient for degree increases from 8471 to 21 700. The average absolute difference was 4533. The estimates and associated estimated standard errors obtained at the two levels are different and hence so is the assessment of their statistical significance.

Other variables could be added to the model but the R^2 obtained was considered acceptable and this sort of model is indicative of what researchers might use in practice. The R^2 obtained at the individual level is consistent with those found in other studies of income (e.g. Davies, Joshi and Clarke, 1997). As with all regression models there are likely to be variables with some explanatory power omitted from the model; however, this reflects the world of practical data analysis. This example shows the aggregation effects when a reasonable but not necessarily perfect statistical model is being used. The log transformation was also tried for the income variable but did not result in an appreciably better fit.

Steel, Tranmer and Holt (1999) reported the results of applying the adjusted ecological regression method to the income regression. The auxiliary variables used were: owner occupied, renting from government, housing type, aged 45–59 and aged 60+. These variables were considered because they had relatively high within-CD correlations and hence their variances were subject to strong grouping effects and also it is reasonable to expect that individual-level data might be available for them. Because the adjustment relies on obtaining a good estimate of the unit-level covariance matrix of the adjustment variables, we need to keep the number of variables small. By choosing variables that characterize much of the difference between CDs we hope to have variables that will perform effectively in a range of situations.

These adjustment variables remove between 9 and 75% of the aggregation effect on the variances of the variables in the analysis. For the income variable the reduction was 32% and the average reduction was 52%.

The estimates of the regression equation obtained from $\widehat{\Sigma}_6$, that is β_{6yx}, are given in Table 20.3. In general the adjusted CD regression coefficients are no closer than those for the original CD-level regression equation. The resulting adjustment of R^2 is still considerably higher than that in the individual-level equation indicating that the adjustment is not working well. The measure of fit at the individual level gives an R^2 of 0.284 compared with 0.310 for the unadjusted equation, so according to this measure the adjustment has had a small detrimental effect. The average absolute difference between the CD- and individual-level coefficients has also increased slightly to 4771.

While the adjustment has eliminated about half the aggregation effects in the variables it has not resulted in reducing the difference between the CD- and individual-level regression equations. The adjustment procedure will be effective if $B^{(2)}_{1yx|z} = B^{(1)}_{1yx|z}$, $B^{(2)}_{1yz|x} = B^{(1)}_{1yz|x}$ and $\widehat{\beta}_{6zx}(z) = B^{(1)}_{1zx}$. Steel, Tranmer and Holt (1999) found that the coefficients in $B^{(1)}_{yx|z}$ and $B^{(2)}_{yx|z}$ are generally very different and the average absolute difference is 4919. Inclusion of the auxiliary variables in the regression has had no appreciable effect on the aggregation effect on the regression coefficients and the R^2 is still considerably larger at the CD level than the individual level.

The adjustment procedure replaces $B^{(2)}_{1zx} B^{(2)}_{1yz|x}$ by $\widehat{\beta}_{6zx} B^{(2)}_{1yz|x}$. Analysis of these values showed that the adjusted CD values are considerably closer to the individual-level values than the CD-level values. The adjustment has had some beneficial effect in the estimation of $\beta_{zx}\beta_{yz|x}$ and the bias of the adjusted estimators is mainly due to the difference between the estimates of $\beta_{yx|z}$. The adjustment has altered the component of bias it is designed to reduce. The remaining biases mean that the overall effect is largely unaffected. It appears that conditioning on the auxiliary variables has not sufficiently reduced the biases due to the random effects.

Attempts were made to estimate the remaining variance components from purely aggregate data using MLn but this proved unsuccessful. Plots of the squares of the residuals against the inverse of the population sizes of groups showed that there was not always an increasing trend that would be needed to obtain sensible estimates. Given the results in Section 20.3 concerning the use of purely aggregate data, these results are not surprising.

The multi-level model that incorporates grouping variables and random effects provides a general framework through which the causes of ecological biases can be explained. Using a limited number of auxiliary variables it was possible to explain about half the aggregation effects in income and a number of explanatory variables. Using individual-level data on these adjustment variables the aggregation effects due to these variables can be removed. However, the resulting adjusted regression coefficients are no less biased.

This suggests that we should attempt to find further auxiliary variables that account for a very large proportion of the aggregation effects and for which it would be reasonable to expect that the required individual-level data are available. However, in practice there are always going to be some residual group-level effects and because of the impact of $\bar{n}^*$ in (20.19) there is still the potential for large biases.

20.6. CONCLUSIONS

This chapter has shown the potential for survey and aggregate data to be used together to produce better estimates of parameters at different levels. In particular, survey data may be used to remove biases associated with analysis using group-level aggregate data even if it does not contain indicators for the groups in question. Aggregate data may be used to produce estimates of variance components when the primary data source is a survey that does not contain indicators for the groups. The model and methods described in this chapter are fairly simple. Development of models appropriate to categorical data and more evaluation with real datasets would be worthwhile.

Sampling and nonresponse are two mechanisms that lead to data being missing. The process of aggregation also leads to a loss of information and can be thought of as a problem missing data. The approaches in this chapter could be viewed in this light. Further progress may be possible through use of methods that have been developed to handle incomplete data, such as those discussed by Little in this volume (chapter 18).

ACKNOWLEDGEMENTS

This work was supported by the Economic and Science Research Council (Grant number R 000236135) and the Australian Research Council.

References

Aalen, O. and Husebye, E. (1991) Statistical analysis of repeated events forming renewal processes. *Statistics in Medicine*, **10**, 1227–40.

Abowd, J. M. and Card, D. (1989) On the covariance structure of earnings and hours changes. *Econometrica*, **57**, 411–45.

Achen, C. H. and Shively, W. P. (1995) *Cross Level Inference*. Chicago: The University of Chicago Press.

Agresti, A. (1990) *Categorical Data Analysis*. New York: Wiley.

Allison, P. D. (1982) Discrete-time methods for the anlaysis of event-histories. In *Sociological Methodology 1982* (S. Leinhardt, ed.), pp. 61–98. San Francisco: Jossey-Bass.

Altonji, J. G. and Segal, L. M. (1996) Small-sample bias in GMM estimation of covariance structures. *Journal of Business and Economic Statistics*, **14**, 353–66.

Amemiya, T. (1984) Tobit models: a survey. *Journal of Econometrics*, **24**, 3–61.

Andersen, P. K., Borgan, O., Gill, R. D. and Keiding, N. (1993) *Statistical Models Based on Counting Processes*. New York: Springer-Verlag.

Anderson, T. W. (1957) Maximum likelihood estimates for the multivariate normal distribution when some observations are missing. *Journal of the American Statistical Association*, **52**, 200–3.

Andrews, M. and Bradley, S. (1997) Modelling the transition from school and the demand for training in the United Kingdom. *Economica*, **64**, 387–413.

Assakul, K. and Proctor, C. H. (1967) Testing independence in two- way contingency tables with data subject to misclassification. *Psychometrika*, **32**, 67–76.

Baltagi, B. H. (2001) *Econometric Analysis of Panel Data*. 2nd Edn. Chichester: Wiley.

Basu, D. (1971) An essay on the logical foundations of survey sampling, Part 1. *Foundations of Statistical Inference*, pp. 203–42. Toronto: Holt, Rinehart and Winston.

Bellhouse, D. R. (2000) Density and quantile estimation in large- scale surveys when a covariate is present. Unpublished report.

Bellhouse, D. R. and Stafford, J. E. (1999) Density estimation from complex surveys. *Statistica Sinica*, **9**, 407–24.

Bellhouse, D. R. and Stafford, J. E. (2001) Local polynomial regression techniques in complex surveys. *Survey Methodology*, **27**, 197–203..

Berman, M. and Turner, T. R. (1992) Approximating point process likelihoods with GLIM. *Applied Statistics*, **41**, 31–8.

Berthoud, R. and Gershuny, J. (eds) (2000) *Seven Years in the Lives of British Families*. Bristol: The Policy Press.

Bickel, P. J., Klaassen, C. A. J., Ritov, Y. and Wellner, J. A. (1993) *Efficient and Adaptive Estimation for Semiparametric Models*. Baltimore, Maryland: Johns Hopkins University Press.

Binder, D. A. (1982) Non-parametric Bayesian models for samples from finite populations. *Journal of the Royal Statistical Society, Series B*, **44**, 388–93.

Binder, D. A. (1983) On the variances of asymptotically normal estimators from complex surveys. *International Statistical Review*, **51**, 279–92.

Binder, D. A. (1992) Fitting Cox's proportional hazards models from survey data. *Biometrika*, **79**, 139–47.

Binder, D. A. (1996) Linearization methods for single phase and two-phase samples: a cookbook approach. *Survey Methodology*, **22**, 17–22.

Binder, D. A. (1998) Longitudinal surveys: why are these different from all other surveys? *Survey Methodology*, **24**, 101–8.

Birnbaum, A. (1962) On the foundations of statistical inference (with discussion). *Journal of the American Statistical Association*, **53**, 259–326.

Bishop, Y. M. M., Fienberg, S. E. and Holland, P. W. (1975) *Discrete Multivariate Analysis: Theory and Practice*. Cambrdige, Massachusetts: MIT Press.

Bjørnstad, J. F. (1996) On the generalization of the likelihood function and the likelihood principle. *Journal of the American Statistical Association*, **91**, 791–806.

Blau, D. M. and Robins, P. K. (1987) Training programs and wages – a general equilibrium analysis of the effects of program size. *Journal of Human Resources*, **22**, 113–25.

Blossfeld, H. P., Hamerle, A. and Mayer, K. U. (1989) *Event History Analysis*. Hillsdale, New Jersey: L. Erlbaum Associates.

Boudreau, C. and Lawless, J. F. (2001) Survival analysis based on Cox proportional hazards models and survey data. University of Waterloo, Dept. of Statistics and Actuarial Science, Technical Report.

Box, G. E. P. (1980) Sampling and Bayes' inference in scientific modeling and robustness. *Journal of the Royal Statistical Society, Series A*, **143**, 383–430 (with discussion).

Box, G. E. P. and Cox, D. R. (1964) An analysis of transformations. *Journal of the Royal Statistical Society, Series B*, **26**, 211–52.

Boyd, L. H. and Iversen, G. R. (1979) *Contextual Analysis: Concepts and Statistical Techniques*. Belmont, California: Wadsworth.

Breckling, J. U., Chambers, R. L., Dorfman, A. H., Tam, S. M. and Welsh, A. H. (1994) Maximum likelihood inference from sample survey data. *International Statistical Review*, **62**, 349–63.

Breidt, F. J. and Fuller, W. A. (1993) Regression weighting for multiphase samples. *Sankhya, B*, **55**, 297–309.

Breidt, F. J., McVey, A. and Fuller, W. A. (1996–7) Two-phase estimation by imputation. *Journal of the Indian Society of Agricultural Statistics*, **49**, 79–90.

Breslow, N. E. and Holubkov, R. (1997) Maximum likelihood estimation of logistic regression parameters under two-phase outcome-dependent sampling. *Journal of the Royal Statistical Society, Series B*, **59**, 447–61.

Breunig, R. V. (1999) Nonparametric density estimation for stratified samples. Working Paper, Department of Statistics and Econometrics, The Australian National University.

Breunig, R. V. (2001) Density estimation for clustered data. *Econometric Reviews*, **20**, 353–67.

Brewer, K. R. W. and Mellor, R. W. (1973) The effect of sample structure on analytical surveys. *Australian Journal of Statistics*, **15**, 145–52.

Brick, J. M. and Kalton, G. (1996) Handling missing data in survey research. *Statistical Methods in Medical Research*, **5**, 215–38.

Brier, S. E. (1980) Analysis of contingency tables under cluster sampling. *Biometrika*, **67**, 591–6.

Browne, M. W. (1984) Asymptotically distribution-free methods for the analysis of covariance structures. *British Journal of Mathematical and Statistical Psychology*, **37**, 62–83.

Bryk, A. S. and Raudenbush, S. W. (1992) *Hierarchical Linear Models: Application and Data Analysis Methods*. Newbury Park, California: Sage.

Bull, S. and Pederson, L. L. (1987) Variances for polychotomous logistic regression using complex survey data. *Proceedings of the American Statistical Association, Survey Research Methods Section*, pp. 507–12.

Buskirk, T. (1999) Using nonparametric methods for density estimation with complex survey data. Unpublished Ph. D. thesis, Arizona State University.

Carroll, R. J., Ruppert, D. and Stefanski, L. A. (1995) *Measurement Error in Nonlinear Models*. London: Chapman and Hall.

Cassel, C.-M., Särndal, C.-E. and Wretman, J. H. (1977) *Foundations of Inference in Survey Sampling*. New York: Wiley.

Chamberlain, G. (1982) Multivariate regression models for panel data. *Journal of Econometrics*, **18**, 5–46.

Chambers, R. L. (1996) Robust case-weighting for multipurpose establishment surveys. *Journal of Official Statistics*, **12**, 3–22.

Chambers, R. L. and Dunstan, R. (1986) Estimating distribution functions from survey data. *Biometrika*, **73**, 597–604.

Chambers, R. L. and Steel, D. G. (2001) Simple methods for ecological inference in 2x2 tables. *Journal of the Royal Statistical Society, Series A*, **164**, 175–92.

Chambers, R. L., Dorfman, A. H. and Wang, S. (1998) Limited information likelihood analysis of survey data. *Journal of the Royal Statistical Society, Series B*, **60**, 397–412.

Chambers, R. L., Dorfman, A. H. and Wehrly, T. E. (1993) Bias robust estimation in finite populations using nonparametric calibration. *Journal of the American Statistical Association*, **88**, 268–77.

Chesher, A. (1997) Diet revealed? Semiparametric estimation of nutrient intake-age relationships (with discussion). *Journal of the Royal Statistical Society, Series A*, **160**, 389–428.

Clayton, D. (1978) A model for association in bivariate life tables and its application in epidemiological studies of familial tendency in chronic disease incidence. *Biometrika*, **65**, 14–51.

Cleave, N., Brown, P. J. and Payne, C. D. (1995) Evaluation of methods for ecological inference. *Journal of the Royal Statistical Society, Series A*, **158**, 55–72.

Cochran, W. G. (1977) *Sampling Techniques*. 3rd Edn. New York: Wiley.

Cohen, M. P. (1995) Sample sizes for survey data analyzed with hierarchical linear models. National Center for Educational Statistics, Washington, DC.

Cosslett, S. (1981) Efficient estimation of discrete choice models. In *Structural Analysis of Discrete Data with Econometric Applications* (C. F. Manski and D. McFadden, eds), pp. 191–205. New York: Wiley.

Cox, D. R. (1972) Regression models and life tables (with discussion). *Journal of the Royal Statistical Society, Series B*, **34**, 187–220.

Cox, D. R. and Isham, V. (1980) *Point Processes*. London: Chapman and Hall.

Cox, D. R. and Oakes, D. (1984) *Analysis of Survival Data*. London: Chapman and Hall.

David, M. H., Little, R. J. A., Samuhel, M. E. and Triest, R. K. (1986) Alternative methods for CPS income imputation. *Journal of the American Statistical Association*, **81**, 29–41.

Davies, H., Joshi, H. and Clarke, L. (1997) Is it cash that the deprived are short of? *Journal of the Royal Statistical Society, Series A*, **160**, 107–26.

Deakin, B. M. and Pratten, C. F. (1987) The economic effects of YTS. Employment Gazette, **95**, 491–7.

Decady, Y. J. and Thomas, D. R. (1999) Testing hypotheses on multiple response tables: a Rao-Scott adjusted chi-squared approach. In *Managing on the Digital Frontier* (A. M. Lavack, ed.). *Proceedings of the Administrative Sciences Association of Canada*, **20**, 13–22.

Decady, Y. J. and Thomas, D. R. (2000) A simple test of association for contingency tables with multiple column responses. *Biometrics*, **56**, 893–896.

Deville, J.-C. and Särndal, C.-E. (1992) Calibration estimators in survey sampling. *Journal of the American Statistical Association*, **87**, 376–82.

Diamond, I. D. and McDonald, J. W. (1992) The analysis of current status data. In *Demographic Applications of Event History Analysis* (J. Trussell, R. Hankinson and J. Tilton, eds), pp. 231–52. Oxford: Clarendon Press.

Diggle, P. and Kenward, M. G. (1994) Informative dropout in longitudinal data analysis (with discussion). *Applied Statistics*, **43**, 49–94.

Diggle, P. J., Heagerty, P. J., Liang, K.-Y. and Zeger, S. L. (2002) *Analysis of Longitudinal Data*. 2nd Edn. Oxford: Clarendon Press.

Diggle, P. J., Liang, K.-Y. and Zeger, S. L. (1994) *Analysis of Longitudinal Data*. Oxford: Oxford University Press.

Dolton, P. J. (1993) The economics of youth training in Britain. *Economic Journal*, **103**, 1261–78.

Dolton, P. J., Makepeace, G. H. and Treble, J. G. (1994) The Youth Training Scheme and the school-to-work transition. *Oxford Economic Papers*, **46**, 629–57.

Dumouchel, W. H. and Duncan, G. J. (1983) Using survey weights in multiple regression analysis of stratified samples. *Journal of the American Statistical Association*, **78**, 535–43.

Edwards, A. W. F. (1972) *Likelihood*. London: Cambridge University Press.

Efron, B. and Tibshirani, R. J. (1993) *An Introduction to the Bootstrap*. London: Chapman and Hall.

Elliott, M. R. and Little, R. J. (2000) Model-based alternatives to trimming survey weights. *Journal of Official Statistics*, **16**, 191–209.

Ericson, W. A. (1988) Bayesian inference in finite populations. *Handbook of Statistics*, **6**, 213–46. Amsterdam: North-Holland.

Ericson, W. A. (1969) Subjective Bayesian models in sampling finite populations (with discussion). *Journal of the Royal Statistical Society, Series B*, **31**, 195–233.

Eubank, R. L. (1999) *Nonparametric Regression and Spline Smoothing*. New York: Marcel Dekker.

Ezzati, T. and Khare, M. (1992) Nonresponse adjustments in a National Health Survey. *Proceedings of the American Statistical Association, Survey Research Methods Section*, pp. 00.

Ezzati-Rice, T. M., Khare, M., Rubin, D. B., Little, R. J. A. and Schafer, J. L. (1993) A comparison of imputation techniques in the third national health and nutrition examination survey. *Proceedings of the American Statistical Association, Survey Research Methods Section*, pp. 00.

Ezzati-Rice, T., Johnson, W., Khare, M., Little, R., Rubin, D. and Schafer, J. (1995) A simulation study to evaluate the performance of model-based multiple imputations in NCHS health examination surveys. *Proceedings of the 1995 Annual Research Conference*, US Bureau of the Census, pp. 257–66.

Fahrmeir, L. and Tutz, G. (1994) *Multivariate Statistical Modelling Based on Generalized Linear Models*. New York: Springer-Verlag.

Fan, J. (1992) Design-adaptive nonparametric regression. *Journal of the American Statistical Association*, **87**, 998–1004.

Fan, J. and Gijbels, I. (1996) *Local Polynomial Modelling and its Applications*. London: Chapman and Hall.

Fay, R. E. (1979) On adjusting the Pearson chi-square statistic for clustered sampling. *Proceedings of the American Statistical Association, Social Statistics Section*, pp. 402–6.

Fay, R. E. (1984) Application of linear and log-linear models to data from complex samples. *Survey Methodology*, **10**, 82–96.

Fay, R. E. (1985) A jackknifed chi-square test for complex samples. *Journal of the American Statistical Association*, **80**, 148–57.

Fay, R. E. (1988) CPLX, Contingency table analysis for complex survey designs, unpublished report, U.S. Bureau of the Census.

Fay, R. E. (1996) Alternative paradigms for the analysis of imputed survey data. *Journal of the American Statistical Association*, **91**, 490–8 (with discussion).

Feder, M., Nathan, G. and Pfeffermann, D. (2000) Multilevel modelling of complex survey longitudinal data with time varying random effects. *Survey Methodology*, **26**, 53–65.

Fellegi, I. P. (1980) Approximate test of independence and goodness of fit based on stratified multistage samples. *Journal of the American Statistical Association*, **75**, 261–8.

Fienberg, S. E. and Tanur, J. M. (1986) The design and analysis of longitudinal surveys: controversies and issues of cost and continuity. In *Survey Research Designs: Towards a Better Understanding of the Costs and Benefits* (R. W. Pearson and R. F. Boruch, eds), Lectures Notes in Statistics **38**, 60–93. New York: Springer-Verlag.

Fisher, R. (1994) Logistic regression analysis of CPS overlap survey split panel data. *Proceedings of the American Statistical Association, Survey Research Methods Section*, pp. 620–5.

Folsom, R., LaVange, L. and Williams, R. L. (1989) A probability sampling perspective on panel data analysis. In *Panel Surveys* (D. Kasprzyk G. Duncan, G. Kalton and M. P. Singh, eds), pp. 108–38. New York: Wiley.

Fuller, W. A. (1984) Least-squares and related analyses for complex survey designs. *Survey Methodology*, **10**, 97–118.

Fuller, W. A. (1987) *Measurement Error Models*. New York: Wiley.

Fuller, W. A. (1990) Analysis of repeated surveys. *Survey Methodology*, **16**, 167–80.

Fuller, W. A. (1998) Replicating variance estimation for two-phase sample. *Statistica Sinica*, **8**, 1153–64.

Fuller, W. A. (1999) Environmental surveys over time. *Journal of Agricultural, Biological and Environmental Statistics*, **4**, 331–45.

Gelman, A., Carlin, J. B., Stern, H. S. and Rubin, D. B. (1995) *Bayesian Data Analysis*. London: Chapman and Hall.

Ghosh, M. and Meeden, G. (1986) Empirical Bayes estimation of means from stratified samples. *Journal of the American Statistical Association*, **81**, 1058–62.

Ghosh, M. and Meeden, G. (1997) *Bayesian Methods for Finite Population Sampling*. London: Chapman and Hall.

Godambe, V. P. (ed.) (1991) *Estimating Functions*. Oxford: Oxford University Press.

Godambe, V. P. and Thompson, M. E. (1986) Parameters of super populations and survey population: their relationship and estimation. *International Statistical Review*, **54**, 37–59.

Goldstein, H. (1987) *Multilevel Models in Educational and Social Research*. London: Griffin.

Goldstein, H. (1995) *Multilevel Statistical Models*. 2nd Edn. London: Edward Arnold.

Goldstein, H., Healy, M. J. R. and Rasbash, J. (1994) Multilevel time series models with applications to repeated measures data. *Statistics in Medicine*, **13**, 1643–55.

Goodman, L. (1961) Statistical methods for the mover-stayer model. *Journal of the American Statistical Association*, **56**, 841–68.

Gourieroux, C. and Monfort, A. (1996) *Simulation-Based Econometric Methods*. Oxford: Oxford University Press.

Graubard, B. I., Fears, T. I. and Gail, M. H. (1989) Effects of cluster sampling on epidemiologic analysis in population-based case-control studies. *Biometrics*, **45**, 1053–71.

Greenland, S., Robins, J. M. and Pearl, J. (1999) Confounding and collapsibility in causal inference. *Statistical Science*, **14**, 29–46.

Greenlees, W. S., Reece, J. S. and Zieschang, K. D. (1982) Imputation of missing values when the probability of nonresponse depends on the variable being imputed. *Journal of the American Statistical Association*, **77**, 251–61.

Gritz, M. (1993) The impact of training on the frequency and duration of employment. *Journal of Econometrics*, **57**, 21–51.

Groves, R. M. (1989) *Survey Errors and Survey Costs*. New York: Wiley.

Guo, G. (1993) Event-history analysis for left-truncated data. *Sociological Methodology*, **23**, 217–43.

Hacking, I. (1965) *Logic of Statistical Inference*. New York: Cambridge University Press.

Hansen, M. H., Madow, W. G. and Tepping, B. J. (1983) An evaluation of model-dependent and probability-sampling inferences in sample surveys. *Journal of the American Statistical Association*, **78**, 776–93.

Hansen, M. H., Hurwitz, W. N. and Madow, W. G. (1953) *Sample Survey Methods and Theory*. New York: Wiley.

Härdle, W. (1990) *Applied Nonparametric Regression Analysis*. Cambridge: Cambridge University Press.

Hartley, H. O. and Rao, J. N. K. (1968) A new estimation theory for sample surveys, II. In *New Developments in Survey Sampling* (N. L. Johnson and H. Smith, eds), pp. 147–69. New York: Wiley Interscience.

Hartley, H. O. and Sielken, R. L., Jr (1975) A 'super-population viewpoint' for finite population sampling. *Biometrics*, **31**, 411–22.

Heckman, J. (1976) The common structure of statistical models of truncation, sample selection and limited dependent variables, and a simple estimator for such models. *Annals of Economic and Social Measurement*, **5**, 475–92.

Heckman, J. J. and Singer, B. (1984) A method for minimising the impact of distributional assumptions in econometric models for duration data. *Econometrica*, **52**, 271–320.

Heeringa, S. G., Little, R. J. A. and Raghunathan, T. E. (2002) Multivariate imputation of coarsened survey data on household wealth. In *Survey Nonresponse* (R. M. Groves, D. A. Dillman, J. L. Eltinge and R. J. A. Little, eds), pp. 357–371. New York: Wiley.

Heitjan, D. F. (1994) Ignorability in general complete-data models. *Biometrika*, **81**, 701–8.

Heitjan, D. F. and Rubin, D. B. (1991) Ignorability and coarse data. *Annals of Statistics*, **19**, 2244–53.

Hidiroglou, M. A. (2001) Double sampling. *Survey Methodology*, **27**, 143–54.

Hinkins, S., Oh, F. L. and Scheuren, F. (1997) Inverse sampling design algorithms, *Survey Methodology*, **23**, 11–21.

Hoem, B. and Hoem, J. (1992) The disruption of marital and non- marital unions in contemporary Sweden. In *Demographic Applications of Event History Analysis* (J. Trussell, R. Hankinson and J. Tilton, eds), pp. 61–93. Oxford: Clarendon Press.

Hoem, J. (1989) The issue of weights in panel surveys of individual behavior. In *Panel Surveys* (D. Kasprzyk *et al.*, eds), pp. 539–65. New York: Wiley.

Hoem, J. M. (1985) Weighting, misclassification, and other issues in the analysis of survey samples of life histories. In *Longitudinal Analysis of Labor Market Data* (J. J. Heckman and B. Singer, eds), Ch. 5. Cambridge: Cambridge University Press.

Holland, P. (1986) Statistics and causal inference. *Journal of the American Statistical Association*, **81**, 945–61.

Holt, D. (1989) Aggregation versus disaggregation. In *Analysis of Complex Surveys* (C. Skinner, D. Holt and T. M. F. Smith, eds), Ch. 10. 1. Chichester: Wiley.

Holt, D. and Smith, T. M. F. (1979) Poststratification. *Journal of the Royal Statistical Society, Series A*, **142**, 33–46.

Holt, D., McDonald, J. W. and Skinner, C. J. (1991) The effect of measurement error on event history analysis. In *Measurement Errors in Surveys* (P. P. Biemer *et al.*, eds), pp. 665–85. New York: Wiley.

Holt, D., Scott, A. J. and Ewings, P. D. (1980) Chi-squared test with survey data. *Journal of the Royal Statistical Society, Series A*, **143**, 303–20.

Holt, D., Smith, T. M. F. and Winter, P. D. (1980) Regression analysis of data from complex surveys. *Journal of the Royal Statistical Society, Series A*, **143**, 474–87.

Holt, D., Steel, D. and Tranmer, M. (1996) Area homogeneity and the modifiable areal unit problem. *Geographical Systems*, **3**, 181–200.

Horvitz, D. G. and Thompson, D. J. (1952) A generalization of sampling without replacement from a finite universe. *Journal of the American Statistical Association*, **47**, 663–85.

Hougaard, P. (2000) *Analysis of Multivariate Survival Data*. New York: Springer-Verlag.

Hsiao, C. (1986) *Analysis of Panel Data*. Cambridge: Cambridge University Press.

Huster, W. J., Brookmeyer, R. L. and Self, S. G. (1989) Modelling paired survival data with covariates. *Biometrics*, **45**, 145–56.

Jeffreys, H. (1961) *Theory of Probability*. 3rd Edn. Oxford: Oxford University Press.

Joe, H. (1997) *Multivariate Models and Dependence Concepts*. London: Chapman and Hall.

Johnson, G. E. and Layard, R. (1986) The natural rate of unemployment: explanation and policy. In *Handbook of Labour Economics* (O. Ashenfelter and R. Layard, eds). Amsterdam: North-Holland.

Jones, I. (1988) An evaluation of YTS. *Oxford Review of Economic Policy*, **4**, 54–71.

Jones, M. C. (1989) Discretized and interpolated kernel density estimates. *Journal of the American Statistical Association*, **84**, 733–41.

Jones, M. C. (1991) Kernel density estimation for length biased data. *Biometrika*, **78**, 511–19.

Kalbfleisch, J. D. and Lawless, J. F. (1988) Likelihood analysis of multi-state models for disease incidence and mortality. *Statistics in Medicine*, **7**, 149–60.

Kalbfleisch, J. D. and Lawless, J. F. (1989) Some statistical methods for panel life history data. *Proceedings of the Statistics Canada Symposium on the Analysis of Data in Time*, pp. 185–92. Ottawa: Statistics Canada.

Kalbfleisch, J. D. and Prentice, R. L. (2002) *The Statistical Analysis of Failure Time Data*. 2nd Edn. New York: Wiley.

Kalbfleisch, J. D. and Sprott, D. A. (1970) Application of likelihood methods to problems involving large numbers of parameters (with discussion). *Journal of Royal Statistical Society, Series B*, **32**, 175–208.

Kalton, G. and Citro, C. (1993) Panel surveys: adding the fourth dimension. *Survey Methodology*, **19**, 205–15.

Kalton, G. and Kasprzyk, D. (1986) The treatment of missing survey data. *Survey Methodology*, **12**, 1–16.

Kalton, G. and Kish, L. (1984) Some efficient random imputation methods. *Communications in Statistics Theory and Methods*, **13**, 1919–39.

Kasprzyk, D., Duncan, G. J., Kalton, G. and Singh, M. P. (1989) *Panel Surveys*. New York: Wiley.

Kass, R. E. and Raftery, A. E. (1995) Bayes factors. *Journal of the American Statistical Association*, **90**, 773–95.

Khare, M., Little, R. J. A., Rubin, D. B. and Schafer, J. L. (1993) Multiple imputation of NHANES III. *Proceedings of the American Statistical Association, Survey Research Methods Section*, pp. 00.

Kim, J. K. and Fuller, W. A. (1999) Jackknife variance estimation after hot deck imputation. *Proceedings of the American Statistical Association, Survey Research Methods Section*, pp. 825–30.

King, G. (1997) *A Solution to the Ecological Inference Problem: Reconstructing Individual Behavior from Aggregate Data*. Princeton, New Jersey: Princeton University Press.

Kish, L. and Frankel, M. R. (1974) Inference from complex samples (with discussion). *Journal of the Royal Statistical Society, Series B*, **36**, 1–37.

Klein, J. P. and Moeschberger, M. L. (1997) *Survival Analysis*. New York: Springer-Verlag.

Koehler, K. J. and Wilson, J. R. (1986) Chi-square tests for comparing vectors of proportions for several cluster samples. *Communications in Statistics, Part A – Theory and Methods*, **15**, 2977–90.

Konijn, H. S. (1962) Regression analysis in sample surveys. *Journal of the American Statistical Association*, **57**, 590–606.

Korn, E. L. and Graubard, B. I. (1990) Simultaneous testing of regression coefficients with complex survey data: use of Bonferroni *t*-statistics. *American Statistician*, **44**, 270–6.

Korn, E. L. and Graubard, B. I. (1998a) Variance estimation for superpopulation parameters. *Statistica Sinica*, **8**, 1131–51.

Korn, E. L. and Graubard, B. I. (1998b) Scatterplots with survey data. *American Statistician*, **52**, 58–69.

Korn, E. L. and Graubard, B. I. (1999) *Analysis of Health Surveys*. New York: Wiley.

Korn, E. L. Graubard, B. I. and Midthune, D. (1997) Time-to-event analysis of longitudinal follow-up of a survey: choice of the time-scale. *American Journal of Epidemiology*, **145**, 72–80.

Kott, P. S. (1990) Variance estimation when a first phase area sample is restratified. *Survey Methodology*, **16**, 99–103.

Kott, P. S. (1995) Can the jackknife be used with a two-phase sample? *Proceedings of the Survey Research Methods Section, Statistical Society of Canada*, pp. 107–10.

Kreft, I. and De Leeuw, J. (1998) *Introducing Multilevel Modeling*. London: Sage.

Krieger, A. M. and Pfeffermann, D. (1992) Maximum likelihood from complex sample surveys. *Survey Methodology*, **18**, 225–39.

Krieger, A. M. and Pfeffermann, D. (1997) Testing of distribution functions from complex sample surveys. *Journal of Official Statistics*, **13**, 123–42.

Lancaster, T. (1990) *The Econometric Analysis of Transition Data*. Cambridge: Cambridge University Press.

Lawless, J. F. (1982) *Statistical Models and Methods for Lifetime Data*. New York: Wiley.

Lawless, J. F. (1987) Regression methods for Poisson process data. *Journal of the American Statistical Association*, **82**, 808–15.

Lawless, J. F. (1995) The analysis of recurrent events for multiple subjects. *Applied Statistics*, **44**, 487–98.

Lawless, J. F. (2002) *Statistical Models and Methods for Lifetime Data*. 2nd Edn. New York: Wiley.

Lawless, J. F. and Nadeau, C. (1995) Some simple robust methods for the analysis of recurrent events. *Technometrics*, **37**, 158–68.

Lawless, J. F., Kalbfleisch, J. D. and Wild, C. J. (1999) Semiparametric methods for response-selective and missing data problems in regression. *Journal of the Royal Statistical Society, Series B*, **61**, 413–38.

Lazarsfeld, P. F. and Menzel, H. (1961) On the relation between individual and collective properties. In *Complex Organizations: A Sociological Reader* (A. Etzioni, ed.). New York: Holt, Rinehart and Winston.

Lazzeroni, L. C. and Little, R. J. A. (1998) Random-effects models for smoothing post-stratification weights. *Journal of Official Statistics*, **14**, 61–78.

Lee, E. W., Wei, L. J. and Amato, D. A. (1992) Cox-type regression analysis for large numbers of small groups of correlated failure time observations. In *Survival Analysis: State of the Art* (J. P. Klein and P. K. Goel, eds), pp. 237–47. Dordrecht: Kluwer.

Lehtonen, R. and Pahkinen, E. J. (1995) *Practical Methods for Design and Analysis of Complex Surveys*. Chichester: Wiley.

Lepkowski, J. M. (1989) Treatment of wave nonresponse in panel surveys. In *Panel Surveys* (D. Kasprzyk et al., eds), pp. 348–74. New York: Wiley.

Lessler, J. T. and Kalsbeek, W. D. (1992) *Nonsampling Errors in Surveys*. New York: Wiley.

Liang, K.-Y. and Zeger, S. L. (1986) Longitudinal data analysis using generalized linear models. *Biometrika*, **73**, 13–22.

Lillard, L., Smith, J. P. and Welch, F. (1982) What do we really know about wages? The importance of nonreporting and census imputation. The Rand Corporation, Santa Monica, California.

Lillard, L., Smith, J. P. and Welch, F. (1986) What do we really know about wages? The importance of nonreporting and census imputation. *Journal of Political Economy*, **94**, 489–506.

Lillard, L. A. and Willis, R. (1978) Dynamic aspects of earnings mobility. *Econometrica*, **46**, 985–1012.

Lin, D. Y. (1994) Cox regression analysis of multivariate failure time data: the marginal approach. *Statistics in Medicine*, **13**, 2233–47.

Lin, D. Y. (2000) On fitting Cox's proportional hazards models to survey data. *Biometrika*, **87**, 37–47.

Lindeboom, M. and van den Berg, G. (1994) Heterogeneity in models for bivariate survival: the importance of the mixing distribution. *Journal of the Royal Statistical Society, Series B*, **56**, 49–60.

Lindsey, J. K. (1993) *Models for Repeated Measurements*. Oxford: Clarendon Press.

Lipsitz, S. R. and Ibrahim, J. (1996) Using the E-M algorithm for survival data with incomplete categorical covariates. *Lifetime Data Analysis*, **2**, 5–14.

Little, R. J. A. (1982) Models for nonresponse in sample surveys. *Journal of the American Statistical Association*, **77**, 237–50.

Little, R. J. A. (1983a) Comment on 'An evaluation of model dependent and probability sampling inferences in sample surveys' by M. H. Hansen, W. G. Madow and B. J. Tepping. *Journal of the American Statistical Association*, **78**, 797–9.

Little, R. J. A. (1983b) Estimating a finite population mean from unequal probability samples. *Journal of the American Statistical Association*, **78**, 596–604.

Little, R. J. A. (1985) A note about models for selectivity bias. *Econometrica*, **53**, 1469–74.

Little, R. J. A. (1988) Missing data in large surveys. *Journal of Business and Economic Statistics*, **6**, 287–301 (with discussion).

Little, R. J. A. (1989) On testing the equality of two independent binomial proportions. *The American Statistician*, **43**, 283–8.

Little, R. J. A. (1991) Inference with survey weights. *Journal of Official Statistics*, **7**, 405–24.

Little, R. J. A. (1992) Incomplete data in event history analysis. In *Demographic Applications of Event History Analysis* (J. Trussell, R. Hankinson and J. Tilton, eds), Ch. 8. Oxford: Clarendon Press.

Little, R. J. A. (1993a) Poststratification: a modeler's perspective. *Journal of the American Statistical Association*, **88**, 1001–12.

Little, R. J. A. (1993b) Pattern-mixture models for multivariate incomplete data. *Journal of the American Statistical Association*, **88**, 125–34.

Little, R. J. A. (1994) A class of pattern-mixture models for normal missing data. *Biometrika*, **81**, 471–83.

Little, R. J. A. (1995) Modeling the drop-out mechanism in repeated-measures studies. *Journal of the American Statistical Association*, **90**, 1112–21.

Little, R. J. A. and Rubin, D. B. (1983) Discussion of 'Six approaches to enumerative sampling' by K. R. W. Brewer and C. E. Sarndal. In *Incomplete Data in Sample Surveys, Vol. 3: Proceedings of the Symposium* (W. G. Madow and I. Olkin, eds). New York: Academic Press.

Little, R. J. A. and Rubin, D. B. (1987) *Statistical Analysis with Missing Data*. New York: Wiley.

Little, R. J. A. and Rubin, D. B. (2002) *Statistical Analysis with Missing Data*, 2nd. Ed. New York: Wiley

Little, R. J. A. and Wang, Y.-X. (1996) Pattern-mixture models for multivariate incomplete data with covariates. *Biometrics*, **52**, 98–111.

Lohr, S. L. (1999) *Sampling: Design and Analysis*. Pacific Grove, California: Duxbury.

Longford, N. (1987) A fast scoring algorithm for maximum likelihood estimation in unbalanced mixed models with nested random effects. *Biometrika*, **74**, 817–27.

Longford, N. (1993) *Random Coefficient Models*. Oxford: Oxford University Press.

Loughin, T. M. and Scherer, P. N. (1998) Testing association in contingency tables with multiple column responses. *Biometrics*, **54**, 630–7.

Main, B. G. M. and Shelly, M. A. (1988) The effectiveness of YTS as a manpower policy. *Economica*, **57**, 495–514.

McCullagh, P. and Nelder, J. A. (1983) *Generalized Linear Models*. London: Chapman and Hall.

McCullagh, P. and Nelder, J. A. (1989) *Generalized Linear Models*. 2nd Edn. London: Chapman and Hall.

Mealli, F. and Pudney, S. E. (1996) Occupational pensions and job mobility in Britain: estimation of a random-effects competing risks model. *Journal of Applied Econometrics*, **11**, 293–320.

Mealli, F. and Pudney, S. E. (1999) Specification tests for random-effects transition models: an application to a model of the British Youth Training Scheme. *Lifetime Data Analysis*, **5**, 213–37.

Mealli, F., Pudney, S. E. and Thomas, J. M. (1996) Training duration and post-training outcomes: a duration limited competing risks model. *Economic Journal*, **106**, 422–33.

Meyer, B. (1990) Unemployment insurance and unemployment spells. *Econometrica*, **58**, 757–82.

Molina, E. A., Smith, T. M. F and Sugden, R. A. (2001) Modelling overdispersion for complex survey data. *International Statistical Review*, **69**, 373–84.

Morel, J. G. (1989) Logistic regression under complex survey designs. *Survey Methodology*, **15**, 203–23.

Mote, V. L. and Anderson, R. L. (1965) An investigation of the effect of misclassification on the properties of chi-square tests in the analysis of categorical data. *Biometrika*, **52**, 95–109.

Narendranathan, W. and Stewart, M. B. (1993) Modelling the probability of leaving unemployment: competing risks models with flexible baseline hazards. *Journal of the Royal Statistical Society, Series C*, **42**, 63–83.

Nascimento Silva, P. L. D. and Skinner, C. J. (1995) Estimating distribution functions with auxiliary information using poststratification. *Journal of Official Statistics*, **11**, 277–94.

Neuhaus, J. M., Kalbfleisch, J. D. and Hauck, W. W. (1991) A comparison of cluster-specific and population-averaged approaches for analyzing correlated binary data. *International Statistical Review*, **59**, 25–36.

Neyman, J. (1938) Contribution to the theory of sampling human populations. *Journal of the American Statistical Association*, **33**, 101–16.

Ng, E. and Cook, R. J. (1999) Robust inference for bivariate point processes. *Canadian Journal of Statistics*, **27**, 509–24.

Nguyen, H. H. and Alexander, C. (1989) On χ^2 test for contingency tables from complex sample surveys with fixed cell and marginal design effects. *Proceedings of the American Statistical Association, Survey Research Methods Section*, pp. 753–6.

Nusser, S. M. and Goebel, J. J. (1977) The National Resource Inventory: a long-term multi-resource monitoring programme. *Environmental and Ecological Statistics*, **4**, 181–204.

Obuchowski, N. A. (1998) On the comparison of correlated proportions for clustered data. *Statistics in Medicine*, **17**, 1495–1507.

Oh, H. L. and Scheuren, F. S. (1983) Weighting adjustments for unit nonresponse. In *Incomplete Data in Sample Surveys, Vol. II: Theory and Annotated Bibliography* (W. G. Madow, I. Olkin and D. B. Rubin, eds). New York: Academic Press.

Ontario Ministry of Health (1992) *Ontario Health Survey: User's Guide, Volumes I and II*. Queen's Printer for Ontario.

Østbye, T., Pomerleau, P., Speechley, M., Pederson, L. L. and Speechley, K. N. (1995) Correlates of body mass index in the 1990 Ontario Health Survey. *Canadian Medical Association Journal*, **152**, 1811–17.

Pfeffermann, D. (1993) The role of sampling weights when modeling survey data. *International Statistical Review*, **61**, 317–37.

Pfeffermann, D. (1996) The use of sampling weights for survey data analysis. *Statistical Methods in Medical Research*, **5**, 239–61.

Pfeffermann, D. and Holmes, D. J. (1985) Robustness considerations in the choice of method of inference for regression analysis of survey data. *Journal of the Royal Statistical Society, Series A*, **148**, 268–78.

Pfeffermann, D. and Sverchkov, M. (1999) Parametric and semi- parametric estimation of regression models fitted to survey data. *Sankhya, B*, **61**, 166–86.

Pfeffermann, D., Krieger, A. M. and Rinott, Y. (1998) Parametric distributions of complex survey data under informative probability sampling. *Statistica Sinica*, **8**, 1087– 1114.

Pfeffermann, D., Skinner, C. J., Holmes, D. J., Goldstein, H. and Rasbash, J. (1998) Weighting for unequal selection probabilities in multilevel models. *Journal of the Royal Statistical Society, Series B*, **60**, 23–40.

Platek, R. and Gray, G. B. (1983) Imputation methodology: total survey error. In *Incomplete Data in Sample Surveys* (W. G. Madow, I. Olkin and D. B. Rubin, eds), Vol. 2, pp. 249–333. New York: Academic Press.

Potter, F. (1990) A study of procedures to identify and trim extreme sample weights. *Proceedings of the American Statistical Association, Survey Research Methods Section*, pp. 225–30.

Prentice, R. L. and Pyke, R. (1979) Logistic disease incidence models and case-control studies. *Biometrika*, **66**, 403–11.

Pudney, S. E. (1981) An empirical method of approximating the separable structure of consumer preferences. *Review of Economic Studies*, **48**, 561–77.

Pudney, S. E. (1989) *Modelling Individual Choice: The Econometrics of Corners, Kinks and Holes*. Oxford: Basil Blackwell.

Ramos, X. (1999) The covariance structure of earnings in Great Britain. British Household Panel Survey Working Paper 99–5, University of Essex.

Rao, J. N. K. (1973) On double sampling for stratification and analytic surveys. *Biometrika*, **60**, 125–33.

Rao, J. N. K. (1996) On variance estimation with imputed survey data. *Journal of the American Statistical Association*, **91**, 499–506 (with discussion).

Rao, J. N. K. (1999) Some current trends in sample survey theory and methods. *Sankhya, B*, **61**, 1–57.

Rao, J. N. K. and Scott, A. J. (1981) The analysis of categorical data from complex sample surveys: chi-squared tests for goodness of fit and independence in two-way tables. *Journal of the American Statistical Association*, **76**, 221–30.

Rao, J. N. K. and Scott, A. J. (1984) On chi-squared tests for multi-way tables with cell proportions estimated from survey data. *Annals of Statistics*, **12**, 46–60.

Rao, J. N. K. and Scott, A. J. (1992) A simple method for the analysis of clustered data. *Biometrics*, **48**, 577–85

Rao, J. N. K. and Scott, A. J. (1999a) A simple method for analyzing overdispersion in clustered Poisson data. *Statistics in Medicine*, **18**, 1373–85.

Rao, J. N. K. and Scott, A. J. (1999b) Analyzing data from complex sample surveys using repeated subsampling. Unpublished Technical Report.

Rao, J. N. K. and Shao, J. (1992) Jackknife variance estimation with survey data under hot deck imputation. *Biometrika*, **79**, 811–22.

Rao, J. N. K. and Sitter, R. R. (1995) Variance estimation under two-phase sampling with application to imputation for missing data, *Biometrika*, **82**, 452–60.

Rao, J. N. K. and Thomas, D. R. (1988) The analysis of cross- classified categorical data from complex sample surveys. *Sociological Methodology*, **18**, 213–69.

Rao, J. N. K. and Thomas, D. R. (1989) Chi-squared tests for contingency tables. In *Analysis of Complex Surveys* (C. J. Skinner, D. Holt and T. M. F. Smith, eds), pp. 89–114. Chichester: Wiley.

Rao, J. N. K. and Thomas, D. R. (1991) Chi-squared tests with complex survey data subject to misclassification error. In *Measurement Errors in Surveys* (P. P. Biemer *et al.*, eds), pp. 637–63. New York: Wiley.

Rao, J. N. K., Hartley, H. O. and Cochran, W. G. (1962) On a simple procedure of unequal probability sampling without replacement. *Journal of the Royal Statistical Society, Series B*, **24**, 482–91.

Rao, J. N. K., Kumar, S. and Roberts, G. (1989) Analysis of sample survey data involving categorical response variables: methods and software. *Survey Methodology*, **15**, 161–85.

Rao, J. N. K., Scott, A. J. and Skinner, C. J. (1998) Quasi-score tests with survey data. *Statistica Sinica*, **8**, 1059–70.

Rice, J. A. (1995) *Mathematical Statistics and Data Analysis*. 2nd Edn. Belmont, California: Duxbury.

Ridder, G. (1987) The sensitivity of duration models to misspecified unobserved heterogeneity and duration dependence. *Mimeo*, University of Amsterdam.

Riley, M. W. (1964) Sources and types of sociological data. In *Handbook of Modern Sociology* (R. Farris, ed.). Chicago: Rand McNally.

Roberts, G., Rao, J. N. K. and Kumar, S. (1987) Logistic regression analysis of sample survey data. *Biometrika*, **74**, 1–12.

Robins, J. M., Greenland, S. and Hu, F. C. (1999) Estimation of the causal effect of a time-varying exposure on the marginal mean of a repeated binary outcome. *Journal of the American Statistical Association*, **94**, 447, 687–700.

Ross, S. M. (1983) *Stochastic Processes*. New York: Wiley.

Rotnitzky, A. and Robins, J. M. (1997) Analysis of semi-parametric regression models with nonignorable nonresponse. *Statistics in Medicine*, **16**, 81–102.

Royall, R. M. (1976) Likelihood functions in finite population sampling theory. *Biometrika*, **63**, 606–14.

Royall, R. M. (1986) Model robust confidence intervals using maximum likelihood estimators. *International Statistical Review*, **54**, 221–6.

Royall, R. M. (1997) *Statistical Evidence: A Likelihood Paradigm*. London: Chapman and Hall.

Royall, R. M. (2000) On the probability of observing misleading statistical evidence. *Journal of the American Statistical Association*, **95**, 760–8.

Royall, R. M. and Cumberland, W. G. (1981) An empirical study of the ratio estimator and estimators of its variance (with discussion). *Journal of the American Statistical Association*, **76**, 66–88.

Rubin, D. B. (1976) Inference and missing data. *Biometrika*, **63**, 581–92.

Rubin, D. B. (1977) Formalizing subjective notions about the effect of nonrespondents in sample surveys. *Journal of the American Statistical Association*, **72**, 538–43.

Rubin, D. B. (1983) Comment on 'An evaluation of model dependent and probability sampling inferences in sample surveys' by M. H. Hansen, W. G. Madow and B. J. Tepping. *Journal of the American Statistical Association*, **78**, 803–5.

Rubin, D. B. (1984) Bayesianly justifiable and relevant frequency calculations for the applied statistician. *Annals of Statistics*, **12**, 1151–72.

Rubin, D. B. (1985) The use of propensity scores in applied Bayesian inference. In *Bayesian Statistics 2* (J. M. Bernado *et al.*, eds). Amsterdam: Elsevier Science.

Rubin, D. B. (1987) *Multiple Imputation for Nonresponse in Surveys*. New York: Wiley.

Rubin, D. B. (1996) Multiple imputation after 18+ years. *Journal of the American Statistical Association*, **91**, 473–89 (with discussion).

Rubin, D. B. and Schenker, N. (1986) Multiple imputation for interval estimation from simple random samples with ignorable nonresponse. *Journal of the American Statistical Association*, **81**, 366–74.

Särndal, C. E. and Swensson, B. (1987) A general view of estimation for two-phases of selection with application to two- phase sampling and nonresponse. *International Statistical Review*, **55**, 279–94.

Särndal, C.-E., Swensson, B. and Wretman, J. (1992) *Model Assisted Survey Sampling*. New York: Springer-Verlag.

SAS (1992) The mixed procedure. In *SAS/STAT Software: Changes and Enhancements, Release 6.07*. Technical Report P-229, SAS Institute, Inc., Cary, North Carolina.

SAS (2001) SAS/STAT Release 8.2. SAS Institute, Inc., Cary, North Carolina.

Schafer, J. L. (1996) *Analysis of Incomplete Multivariate Data*. London: Chapman and Hall.

Scharfstein, D. O., Robins, J. M. and Rotnitsky, A. (1999) Adjusting for nonignorable drop-out using semiparametric models (with discussion). *Journal of the American Statistical Association*, **94**, 1096–1146.

Scott, A. J. (1977a) Some comments on the problem of randomisation in survey sampling. *Sankhya, C*, **39**, 1–9.

Scott, A. J. (1977b) Large-sample posterior distributions for finite populations. *Annals of Mathematical Statistics*, **42**, 1113–17.

Scott, A. J. and Smith, T. M. F. (1969) Estimation in multistage samples. *Journal of the American Statistical Association*, **64**, 830–40.

Scott, A. J. and Wild, C. J. (1986) Fitting logistic models under case-control or choice based sampling. *Journal of the Royal Statistical Society, Series B*, **48**, 170–82.

Scott, A. J. and Wild, C. J. (1989) Selection based on the response in logistic regression. In *Analysis of Complex Surveys* (C. J. Skinner, D. Holt and T. M. F. Smith, eds), pp. 191–205. New York: Wiley.

Scott, A. J. and Wild, C. J. (1997) Fitting regression models to case-control data by maximum likelihood. *Biometrika*, **84**, 57–71.

Scott, A. J. and Wild, C. J. (2001) Maximum likelihood for generalised case-control studies. *Journal of Statistical Planning and Inference*, **96**, 3–27.

Scott, A. J., Rao, J. N. K. and Thomas, D. R. (1990) Weighted least squares and quasilikelihood estimation for categorical data under singular models. *Linear Algebra and its Applications*, **127**, 427–47.

Scott, D. W. (1992) *Multivariate Density Estimation: Theory, Practice and Visualization*. New York: Wiley.

Sen, P. K. (1988) Asymptotics in finite populations. *Handbook of Statistics, Vol. 6, Sampling* (P. R. Krishnaiah and C. R. Rao, eds). Amsterdam: North-Holland.

Servy, E., Hachuel, L. and Wojdyla, D. (1998) Analisis de tablas de contingencia para muestras de diseno complejo. Facultad De Ciencias Economicas Y Estadistica De La Universidad Nacional De Rosario, Argentina.

Shin, H. (1994) An approximate Rao-Scott modification factor in two-way tables with only known marginal deffs. *Proceedings of the American Statistical Association, Survey Research Methods Section*, pp. 600–1.

Silverman, B. W. (1986) *Density Estimation for Statistics and Data Analysis*. London: Chapman and Hall.

Simonoff, J. S. (1996) *Smoothing Methods in Statistics*. New York: Springer-Verlag.

Singh, A. C. (1985) On optimal asymptotic tests for analysis of categorical data from sample surveys. Methodology Branch Working Paper SSMD 86-002, Statistics Canada.

Singh, A. C., Stukel, D. M. and Pfeffermann, D. (1998) Bayesian versus frequentist measures of error in small area estimation. *Journal of the Royal Statistical Society, Series B*, **60**, 377–96.

Sitter, R. V. (1992) Bootstrap methods for survey data. *Canadian Journal of Statistics*, **20**, 135–54.

Skinner, C. (1989) Introduction to Part A. In *Analysis of Complex Surveys* (C. Skinner, D. Holt and T. M. F. Smith, eds), Ch. 2. Chichester: Wiley.

Skinner, C. J. (1994) Sample models and weights. *Proceedings of the American Statistical Association, Survey Research Methods Section*, pp. 133–42.

Skinner, C. J. (1998) Logistic modelling of longitudinal survey data with measurement error. *Statistica Sinica*, **8**, 1045–58.

Skinner, C. J. (2000) Dealing with measurement error in panel analysis. In *Researching Social and Economic Change: the Use of Household Panel Studies* (D. Rose, ed.), pp. 113–25. London: Routledge.

Skinner, C. J. and Humphreys, K. (1999) Weibull regression for lifetimes measured with error. *Lifetime Data Analysis*, **5**, 23–37.

Skinner, C. J., Holt, D. and Smith, T. M. F. (eds) (1989) *Analysis of Complex Surveys.* Chichester: Wiley.

Smith, T. M.F (1976) The foundations of survey sampling: a review (with discussion), *Journal of the Royal Statistical Society, Series A*, **139**, 183–204.

Smith, T. M. F. (1984) Sample surveys. *Journal of the Royal Statistical Society*, **147**, 208–21.

Smith, T. M. F. (1988) To weight or not to weight, that is the question. In *Bayesian Statistics 3* (J. M. Bernado, M. H. DeGroot and D. V. Lindley, eds), pp. 437–51. Oxford: Oxford University Press.

Smith, T. M. F. (1989) Introduction to Part B: aggregated analysis. In *Analysis of Complex Surveys* (C. Skinner, D. Holt and T. M. F. Smith, eds). Chichester: Wiley.

Smith, T. M. F. (1994) Sample surveys 1975–1990; an age of reconciliation? *International Statistical Review*, **62**, 5–34 (with discussion).

Smith, T. M. F. and Holt, D. (1989) Some inferential problems in the analysis of surveys over time. *Proceedings of the 47th Session of the ISI*, Vol. 1, pp. 405–23.

Solon, G. (1989) The value of panel data in economic research. In *Panel Surveys* (D. Kasprzyk *et al.*, eds) , pp. 486–96. New York: Wiley.

Spiekerman, C. F. and Lin, D. Y. (1998) Marginal regression models for multivariate failure time data. *Journal of the American Statistical Association*, **93**, 1164–75.

Stata Corporation (2001) *Stata Reference Manual Set.* College Station, Texas: Stata Press.

Statistical Sciences (1990) *S-PLUS Reference Manual.* Seattle: Statistical Sciences.

Steel, D. (1999) Variances of estimates of covariance components from aggregate and unit level data. School of Mathematics and Applied Statistics, University of Wollongong, Preprint.

Steel, D. and Holt, D. (1996a) Analysing and adjusting aggregation effects: the ecological fallacy revisited. *International Statistical Review*, **64**, 39–60.

Steel, D. and Holt, D. (1996b) Rules for random aggregation. *Environment and Planning, A*, **28**, 957–78.

Steel, D., Holt, D. and Tranmer, M. (1996) Making unit-level inference from aggregate data. *Survey Methodology*, **22**, 3–15.

Steel, D., Tranmer, M. and Holt, D. (1999) Unravelling ecological analysis. School of Mathematics and Applied Statistics, University of Wollongong, Preprint.

Steel, D. G. (1985) Statistical analysis of populations with group structure. *Ph. D. thesis*, Department of Social Statistics, University of Southampton.

Stolzenberg, R. M. and Relles, D. A. (1990) Theory testing in a world of constrained research design – the significance of Heckman's censored sampling bias correction for nonexperimental research. *Sociological Methods and Research*, **18**, 395–415.

Sugden, R. A. and Smith, T. M. F. (1984) Ignorable and informative designs in survey sampling inference. *Biometrika*, **71**, 495–506.

Sverchkov, M. and Pfeffermann, D. (2000) Prediction of finite population totals under informative sampling utilizing the sample distribution. *Proceedings of the American Statistical Association, Survey Research Methods Section*, pp. 41–6.

Tanner, M. A. (1996) *Tools for Statistical Inference: Methods for the Exploration of Posterior Distributions and Likelihood Functions.* 3rd Edn. New York: Springer-Verlag.

Tanner, M. A. and Wong, W. H. (1987) The calculation of posterior distributions by data augmentation. *Journal of the American Statistical Association*, **82**, 528–50.

Therneau, T. M. and Grambsch, P. M. (2000) *Modeling Survival Data: Extending the Cox Model.* New York: Springer-Verlag.

Therneau, T. M. and Hamilton, S. A. (1997) rhDNase as an example of recurrent event analysis. *Statistics in Medicine*, **16**, 2029–47.

Thomas, D. R. (1989) Simultaneous confidence intervals for proportions under cluster sampling. *Survey Methodology*, **15**, 187–201.

Thomas, D. R. and Rao, J. N. K. (1987) Small-sample comparisons of level and power for simple goodness-of-fit statistics under cluster sampling. *Journal of the American Statistical Association*, **82**, 630–6.

Thomas, D. R., Singh, A. C. and Roberts, G. R. (1995) Tests of independence on two-way tables under cluster sampling: an evaluation. Working Paper WPS 95-04, School of Business, Carleton University, Ottawa.

Thomas, D. R., Singh, A. C. and Roberts, G. R. (1996) Tests of independence on two-way tables under cluster sampling: an evaluation. *International Statistical Review*, **64**, 295–311.

Thompson, M. E. (1997) *Theory of Sample Surveys*. London: Chapman and Hall.

Trivellato, U. and Torelli, N. (1989) Analysis of labour force dynamics from rotating panel survey data. *Proceedings of the 47th Session of the ISI*, Vol. 1, pp. 425–44.

Trussell, J., Rodriguez, G. and Vaughan, B. (1992) Union dissolution in Sweden. In *Demographic Applications of Event History Analysis* (J. Trussell, R. Hankinson and J. Tilton, eds), pp. 38–60. Oxford: Clarendon Press.

Tukey, J. W. (1977) *Exploratory Data Analysis*. Reading, Massachusetts: Addison-Wesley.

US Department of Health and Human Services (1990) *The Health Benefits of Smoking Cessation*. Public Health Service, Centers for Disease Control, Center for Chronic Disease Prevention and Health Promotion, Office on Smoking and Health. DHSS Publication No. (CDC) 90–8416.

Valliant, R., Dorfman, A. H. and Royall, R. M. (2000) *Finite Population Sampling and Inference: a Prediction Approach*. New York: Wiley.

van den Berg, G. J. (1997) Association measures for durations in bivariate hazard models. *Journal of Econometrics (Annals of Econometrics)*, **79**, 221–45.

Wand, M. P. and Jones, M. C. (1995) *Kernel Smoothing*. London: Chapman and Hall.

Weeks, M. (1995) Circumventing the curse of dimensionality in applied work using computer intensive methods. *Economic Journal*, **105**, 520–30.

Weisberg, S. (1985) *Applied Linear Regression*. 2nd Edn. New York: Wiley.

White, H. S. (1994) *Estimation, Inference and Specification Analysis*. Cambridge: Cambridge University Press.

Williams, R. L. (1995) Product-limit survival functions with correlated survival times. *Lifetime Data Analysis*, **1**, 17–86.

Wolter, K. M. (1985) *Introduction to Variance Estimation*. New York: Springer-Verlag.

Woodruff, R. S. (1952) Confidence intervals for medians and other position measures. *Journal of the American Statistical Association*, **47**, 635–46.

Xue, X. and Brookmeyer, R. (1996) Bivariate frailty model for the analysis of multivariate failure time. *Lifetime Data Analysis*, **2**, 277–89.

Yamaguchi, K. (1991) *Event History Analysis*. Newbury Park, California: Sage.

Zeger, S. L. and Liang, K. (1986) Longitudinal data analysis for discrete and continuous outcomes. *Biometrics*, **42**, 121–30.

T. M. F. Smith: Publications up to 2002

JOURNAL PAPERS

1. R. P. Brooker and T. M. F. Smith. (1965), Business failures – the English insolvency statistics, *Abacus*, **1**, 131–49.
2. T. M. F. Smith. (1966), Ratios of ratios and their applications, *Journal of the Royal Statistical Society, Series A*, **129**, 531–3.
3. T. M. F. Smith. (1966), Some comments on the index of retail prices, *Applied Statistics*, **5**, 128–35.
4. R. P. Brooker and T. M. F. Smith. (1967), Share transfer audits and the use of statistics, *Secretaries Chronicle*, **43**, 144–7.
5. T. M. F. Smith. (1967), A comparison of some models for predicting time series subject to seasonal variation, *The Statistician*, **17**, 301–5.
6. F. G. Foster and T. M. F. Smith. (1969), The computer as an aid in teaching statistics, *Applied Statistics*, **18**, 264–70.
7. A. J. Scott and T. M. F. Smith. (1969), A note on estimating secondary characteristics in multivariate surveys, *Sankhya, Series A*, **31**, 497–8.
8. A. J. Scott and T. M. F. Smith. (1969), Estimation in multi- stage surveys, *Journal of the American Statistical Association*, **64**, 830–40.
9. T. M. F. Smith. (1969), A note on ratio estimates in multi- stage sampling, *Journal of the Royal Statistical Society, Series A*, **132**, 426–30.
10. A. J. Scott and T. M. F. Smith. (1970), A note on Moran's approximation to Student's t, *Biometrika*, **57**, 681–2.
11. A. J. Scott and T. M. F. Smith. (1971), Domains of study in stratified sampling, *Journal of the American Statistical Association*, **66**, 834–6.
12. A. J. Scott and T. M. F. Smith. (1971), Interval estimates for linear combinations of means, *Applied Statistics*, **20**, 276–85.
13. J. A. John and T. M. F. Smith. (1972), Two factor experiments in non-orthogonal designs, *Journal of the Royal Statistical Society, Series B*, **34**, 401–9.
14. S. C. Cotter, J. A. John and T. M. F. Smith. (1973), Multi-factor experiments in non-orthogonal designs, *Journal of the Royal Statistical Society, Series B*, **35**, 361–7.
15. A. J. Scott and T. M. F. Smith. (1973), Survey design, symmetry and posterior distributions, *Journal of the Royal Statistical Society, Series B*, **35**, 57–60.
16. J. A. John and T. M. F. Smith. (1974), Sum of squares in non-full rank general linear hypotheses, *Journal of the Royal Statistical Society, Series B*, **36**, 107–8.
17. A. J. Scott and T. M. F. Smith. (1974), Analysis of repeated surveys using time series models, *Journal of the American Statistical Association*, **69**, 674–8.
18. A. J. Scott and T. M. F. Smith. (1974), Linear superpopulation models in survey sampling, *Sankhya, Series C*, **36**, 143–6.
19. A. J. Scott and T. M. F. Smith. (1975), Minimax designs for sample surveys, *Biometrika*, **62**, 353–8.

20. A. J. Scott and T. M. F. Smith. (1975), The efficient use of supplementary information in standard sampling procedures, *Journal of the Royal Statistical Society, Series B*, **37**, 146–8.

21. D. Holt and T. M. F. Smith. (1976), The design of survey for planning purposes, *The Australian Journal of Statistics*, **18**, 37 44.

22. T. M. F. Smith. (1976), The foundations of survey sampling: a review, *Journal of the Royal Statistical Society, Series A*, **139**, 183–204 (with discussion) [Read before the Society, January 1976].

23. A. J. Scott, T. M. F. Smith and R. Jones. (1977), The application of time series methods to the analysis of repeated surveys, *International Statistical Review*, **45**, 13–28.

24. T. M. F. Smith. (1978), Some statistical problems in accountancy, *Bulletin of the Institute of Mathematics and its Applications*, **14**, 215–19.

25. T. M. F. Smith. (1978), Statistics: the art of conjecture, *The Statistician*, **27**, 65–86.

26. D. Holt and T. M. F. Smith. (1979), Poststratification, *Journal of the Royal Statistical Society, Series A*, **142**, 33–46.

27. D. Holt, T. M. F. Smith and T. J. Tomberlin. (1979), A model-based approach to estimation for small subgroups of a population, *Journal of the American Statistical Association*, **74**, 405–10.

28. T. M. F. Smith. (1979), Statistical sampling in auditing: a statistician's viewpoint, *The Statistician*, **28**, 267–80.

29. D. Holt, T. M. F. Smith and P. D. Winter. (1980), Regression analysis of data from complex surveys, *Journal of the Royal Statistical Society, Series A*, **143**, 474–87.

30. B. Gomes da Costa, T. M. F. Smith and D. Whitley. (1981), German language proficiency levels attained by language majors: a comparison of U. S. A. and England and Wales results, *The Incorporated Linguist*, **20**, 65–7.

31. I. Diamond and T. M. F. Smith. (1982), Whither mathematics? Comments on the report by Professor D. S. Jones. *Bulletin of the Institute of Mathematics and its Applications*, **18**, 189–92.

32. G. Hoinville and T. M. F. Smith. (1982), The Rayner Review of Government Statistical Service, *Journal of the Royal Statistical Society, Series A*, **145**, 195–207.

33. T. M. F. Smith. (1983), On the validity of inferences from non-random samples, *Journal of the Royal Statistical Society, Series A*, **146**, 394–403.

34. T. M. F. Smith and R. W. Andrews. (1983), Pseudo-Bayesian and Bayesian approach to auditing, *The Statistician*, **32**, 124–6.

35. I. Diamond and T. M. F. Smith. (1984), Demand for higher education: comments on the paper by Professor P. G. Moore. *Bulletin of the Institute of Mathematics and its Applications*, **20**, 124–5.

36. T. M. F. Smith. (1984), Sample surveys: present position and potential developments: some personal views. *Journal of the Royal Statistical Society, Series A*, **147**, 208–21.

37. R. A. Sugden and T. M. F. Smith. (1984), Ignorable and informative designs in survey sampling inference, *Biometrika*, **71**, 495–506.

38. T. J. Murrells, T. M. F. Smith, J. C. Catford and D. Machin. (1985), The use of logit models to investigate social and biological factors in infant mortality I: methodology. *Statistics in Medicine*, **4**, 175–87.

39. T. J. Murrells, T. M. F. Smith, J. C. Catford and D. Machin. (1985), The use of logit models to investigate social and biological factors in infant mortality II: stillbirths, *Statistics in Medicine*, **4**, 189–200.

40. D. Pfeffermann and T. M. F. Smith. (1985), Regression models for grouped populations in cross-section surveys, *International Statistical Review*, **53**, 37–59.

41. T. M. F. Smith. (1985), Projections of student numbers in higher education, *Journal of the Royal Statistical Society, Series A*, **148**, 175–88.

42. E. A. Molina C. and T. M. F. Smith. (1986), The effect of sample design on the comparison of associations, *Biometrika*, **73**, 23–33.

43. T. Murrells, T. M. F. Smith, D. Machin and J. Catford. (1986), The use of logit models to investigate social and biological factors in infant mortality III: neonatal mortality, *Statistics in Medicine*, **5**, 139–53.

44. C. J. Skinner, T. M. F. Smith and D. J. Holmes. (1986), The effect of sample design on principal component analysis. *Journal of the American Statistical Association*, **81**, 789–98.

45. T. M. F. Smith. (1987), Influential observations in survey sampling, *Journal of Applied Statistics*, **14**, 143–52.

46. E. A. Molina Cuevas and T. M. F. Smith. (1988), The effect of sampling on operative measures of association with a ratio structure, *International Statistical Review*, **56**, 235–42.

47. T. Murrells, T. M. F. Smith, D. Machin and J. Catford. (1988), The use of logit models to investigate social and biological factors in infant mortality IV: post-neonatal mortality, *Statistics in Medicine*, **7**, 155–69.

48. T. M. F. Smith and R. A. Sugden. (1988), Sampling and assignment mechanisms in experiments, surveys and observational studies, *International Statistical Review*, **56**, 165–80.

49. T. J. Murrells, T. M. F. Smith and D. Machin. (1990), The use of logit models to investigate social and biological factors in infant mortality V: a multilogit analysis of stillbirths, neonatal and post-neonatal mortality, *Statistics in Medicine*, **9**, 981–98.

50. T. M. F. Smith. (1991), Post-stratification, *The Statistician*, **40**, 315–23.

51. T. M. F. Smith and E. Njenga. (1992), Robust model-based methods for analytic surveys, *Survey Methodology*, **18**, 187–208.

52. T. M. F. Smith. (1993), Populations and selection: limitations of statistics, *Journal of the Royal Statistical Society, Series A*, **156**, 145–66 [Presidential address to the Royal Statistical Society].

53. T. M. F. Smith and P. G. Moore. (1993), The Royal Statistical Society: current issues, future prospects, *Journal of Official Statistics*, **9**, 245–53.

54. T. M. F. Smith. (1994), Sample surveys 1975–90; an age of reconciliation?, *International Statistical Review*, **62**, 5–34 [First Morris Hansen Lecture with discussion].

55. T. M. F. Smith. (1994), Taguchi methods and sample surveys, *Total Quality Management*, **5**, No. 5.

56. D. Bartholomew, P. Moore and T. M. F. Smith. (1995), The measurement of unemployment in the UK, *Journal of the Royal Statistical Society, Series A*, **158**, 363–17.

57. Sujuan Gao and T. M. F. Smith. (1995), On the nonexistence of a global non-negative minimum bias invariant quadratic estimator of variance components, *Statistics and Probability Letters*, **25**, 117–120.

58. T. M. F. Smith. (1995), The statistical profession and the Chartered Statistician (CStat), *Journal of Official Statistics*, **11**, 117–20.

59. J. E. Andrew, P. Prescott and T. M. F. Smith. (1996), Testing for adverse reactions using prescription event monitoring, *Statistics in Medicine*, **15**, 987–1002.

60. T. M. F. Smith. (1996), Public opinion polls: the UK general election, 1992, *Journal of the Royal Statistical Society, Series A*, **159**, 535–45.

61. R. A. Sugden, T. M. F. Smith and R. Brown. (1996), Chao's list sequential scheme for unequal probability sampling, *Journal of Applied Statistics*, **23**, 413–21.

62. T. M. F. Smith. (1997), Social surveys and social science, *The Canadian Journal of Statistics*, **25**, 23–44.

63. R. A. Sugden and T. M. F. Smith. (1997), Edgeworth approximations to the distribution of the sample mean under simple random sampling, *Statistics and Probability Letters*, **34**, 293–9.

64. T. M. F. Smith and T. M. Brunsdon. (1998), Analysis of compositional time series, *Journal of Official Statistics*, **14**, 237–54.

65. Sujuan Gao and T. M. F. Smith. (1998), A constrained MINQU estimator of correlated response variance from unbalanced data in complex surveys, *Statistica Sinica*, **8**, 1175–88.

66. R. A. Sugden, T. M. F. Smith and R. P. Jones. (2000), Cochran's rule for simple random sampling, *Journal of the Royal Statistical Society, Series B*, **62**, 787–94.

67. V. Barnett, J. Haworth and T. M. F. Smith. (2001), A two-phase sampling scheme with applications to auditing or *sed quis custodiet ipsos custodes?*, *Journal of the Royal Statistical Society, Series A*, **164**, 407–22.

68. E. A. Molina, T. M. F. Smith and R. A. Sugden. (2001), Modelling overdispersion for complex survey data. *International Statistical Review*, **69**, 373–84.

69. D. B. N. Silva and T. M. F. Smith. (2001), Modelling compositional time series from repeated surveys, *Survey Methodology*, **27**, 205–15.

70. T. M. F. Smith. (2001), *Biometrika* centenary: sample surveys, *Biometrika*, **88**, 167–94.

71. R. A. Sugden and T. M. F. Smith. (2002), Exact linear unbiased estimation in survey sampling, *Journal of Statistical Planning and Inference*, **102**, 25–38 (with discussion).

BOOKS

1. B. Gomes da Costa, T. M. F. Smith and D. Whiteley. (1975), *German Language Attainment: a Sample Survey of Universities and Colleges in the U.K.*, Heidelberg: Julius Groos Verlag.

2. T. M. F. Smith. (1976), *Statistical Sampling for Accountants*, London: Haymarket Press, 255pp.

3. C. J. Skinner, D. Holt and T. M. F. Smith (eds). (1989), *Analysis of Complex Surveys*, Chichester: Wiley, 309pp.

BOOK CONTRIBUTIONS

1. T. M. F. Smith. (1971), Appendix 0, Second Survey of Aircraft Noise Annoyance around London (Heathrow) Airport, London: HMSO.

2. A. Bebbington and T. M. F. Smith. (1977), The effect of survey design on multivariate analysis. In O'Muircheartaigh, C. A. and Payne, C. (eds) *The Analysis of Survey Data, Vol. 2: Model Fitting*, New York: Wiley, pp. 175–92.

3. T. M. F. Smith. (1978), Principles and problems in the analysis of repeated surveys. In N. K. Namboodiri (ed.) *Survey Sampling and Measurement*, New York: Academic Press, Ch. 13, pp. 201–16.

4. T. M. F. Smith. (1981), Regression analysis for complex surveys. In D. Krewski, J. N. K. Rao and R. Platek (eds) *Current Topics in Survey Sampling*, New York: Academic Press, pp. 267–92.

5. T. M. F. Smith. (1987), Survey sampling. Unit 12, Course M345, Milton Keyres: Open University Press, 45pp.

6. T. M. F. Smith. (1988), To weight or not to weight, that is the question. In J. M. Bernardo, M. H. DeGroot, D. V. Lindley and A. F. M. Smith (eds) *Bayesian Statistics 3*, Oxford: Oxford University Press, pp. 437–51.

7. G. Nathan and T. M. F. Smith. (1989), The effect of selection on regression analysis. In C. J. Skinner, D. Holt and T. M. F. Smith (eds) *Analysis of Complex Surveys*, Chichester, Wiley, Ch. 7, pp. 149–64.

8. T. M. F. Smith. (1989), Introduction to Part B: aggregated analysis. In C. J. Skinner, D. Holt and T. M. F. Smith (eds) *Analysis of Complex Surveys*, Chichester: Wiley, Ch. 6 pp. 135–148.

9. T. M. F. Smith and D. J. Holmes. (1989), Multivariate analysis. In C. J. Skinner, D. Holt and T. M. F. Smith (eds) *Analysis of Complex Surveys*, Chichester: Wiley, Ch. 8, pp. 165–90.

PUBLISHED DISCUSSION AND COMMENTS

1. T. M. F. Smith. (1969), Discussion of 'A theory of consumer behaviour derived from repeat paired preference testing' by G. Horsnell, *Journal of the Royal Statistical Society, Series A*, **132**, 186–7.
2. T. M. F. Smith. (1976), Comments on 'Some results on generalized difference estimation and generalized regression estimation for finite populations' by C. M. Cassel, C.-E. Särndal and J. H. Wretman, *Biometrika*, **63**, 620.
3. T. M. F. Smith. (1979), Discussion of 'Public Opinion Polls' by A. Stuart, N. L. Webb and D. Butler, *Journal of the Royal Statistical Society, Series A*, **142**, 460–1.
4. T. M. F. Smith. (1981), Comment on 'An empirical study of the ratio estimator and estimators of its variance' by R. M. Royall and W. G. Cumberland, *Journal of the American Statistical Association*, **76**, 83.
5. T. M. F. Smith. (1983), Comment on 'An Evaluation of Model-Dependent and Probability-Sampling Inference in Sample Surveys' by M. H. Hansen, W. G. Madow, and B. J. Tepping, *Journal of the American Statistical Association*, **78**, 801–2.
6. T. M. F. Smith. (1983), Invited discussion on 'Six approaches to enumerative survey sampling' by K. R. W. Brewer and C.-E. Särndal, in W. G. Madow and I. Olkin (eds) *Incomplete Data in Sample Surveys*, Vol. 3, New York: Academic Press, pp. 385–7.
7. T. M. F. Smith. (1990), Comment on 'History and development of the theoretical foundations of surveys' by J. N. K. Rao and D. R. Bellhouse, *Survey Methodology*, **16**, 26–9.
8. T. M. F. Smith. (1990), Discussion of 'Public confidence in the integrity and validity of official statistics' by J. Hibbert, *Journal of the Royal Statistical Society, Series A*, **153**, 137.
9. T. M. F. Smith. (1991), Discussion of A Hasted et al 'Statistical analysis of public lending right loans', *Journal of the Royal Statistical Society, Series A*, **154**, 217–19.
10. T. M. F. Smith. (1992), Discussion of 'A National Statistical Commission' by P. G. Moore, *Journal of the Royal Statistical Society, Series A*, **155**, 24.
11. T. M. F. Smith. (1992), Discussion of 'Monitoring the health of unborn populations' by C. Thunhurst and A. MacFarlane. *Journal of the Royal Statistical Society, Series A*, **155**, 347.
12. T. M. F. Smith. (1992), The Central Statistical Office and agency status, *Journal of the Royal Statistical Society, Series A*, **155**, 181–4.

CONFERENCE PROCEEDINGS

1. T. M. F. Smith and M. H. Schueth. (1983), Validation of survey results in market research, *Proceedings of Business and Economics Section of the American Statistical Association*, 757–63.
2. T. M. F. Smith and R. A. Sugden. (1985), Inference and ignorability of selection for experiments and surveys, *Bulletin of the International Statistical Institute, 45th Session*, Vol. LI, Book 2, 10.2, 1–12.
3. T. M. F. Smith and T. M. Brunsdon. (1987), The time series analysis of compositional data, *Bulletin of the International Statistical Institute, 46th Session*, Vol. LII, 417–18.

4. T. M. F. Smith. (1989), Invited discussion of 'History, development, and emerging methods in survey sampling', *Proceedings of the American Statistical Association Sesquicentennial invited paper sessions*, 429–32.

5. T. M. F. Smith and R. W. Andrews. (1989), A Bayesian analysis of the audit process. *Bulletin of the International Statistical Insitute, 47th Session*, Vol. LIII, Book 1, 66–7.

6. T. M. F. Smith and R. W. Andrews. (1989), Statistical analysis of auditing, *Proceedings of the American Statistical Association*.

7. T. M. F. Smith and T. M. Brunsdon. (1989), Analysis of compositional time series, *Proceedings of the American Statistical Association*.

8. T. M. F. Smith and D. Holt (1989), Some inferential problems in the analysis of surveys over time, *Bulletin of the International Statistical Institute, 47th Session*, Vol. LIII, Book 2, 405–24.

9. T. M. F. Smith and E. Njenga. (1991), Robust model-based methods for sample surveys. *Proceedings of a Symposium in honour of V. P. Godambe*, University of Waterloo.

10. T. M. F. Smith. (1993), The Chartered Statistician (CStat), *Bulletin of the International Statistical Institute, 49th Session*, Vol. LV, Book 1, 67–78.

11. T. M. F. Smith. (1995), Problem of resource allocation, *Proceedings of Statistics Canada Symposium 95. From Data to Information – Methods and Systems*.

12. T. M. F. Smith. (1995), Public opinion polls: The UK General Election 1992, *Bulletin of the International Statistical Institute, 50th Session*, Vol. LVI, Book 2, 112–13.

13. T. M. F. Smith. (1995), Social surveys and social science, *Proceedings of the Survey Methods Section, Statistical Society of Canada Annual Meeting, July 1995*.

14. T. M. F. Smith. (1996), Disaggregation and inference from sample surveys, in *100 Anni di Indagini Campionare*, Centro d'Informazione e Stampa Universitaria, Rome, 29–34.

15. N. Davies and T. M. F. Smith. (1999), A strategy for continuing professional education in statistics, *Proceedings of the 5th International Conference on Teaching of Statistics*, Vol. 1, 379–84.

16. T. M. F. Smith. (1999), Defining parameters of interest in longitudinal studies and some implications for design, *Bulletin of the International Statistical Institute, 52nd Session*, Vol. LVIII, Book 2, 307–10.

17. T. M. F. Smith. (1999), Recent developments in sample survey theory and their impact on official statistics, *Bulletin of the International Statistical Institute, 52nd Session*, Vol. LVIII, Book 1, 7–10 [President's invited lecture].

OTHER PUBLICATIONS

1. T. M. F. Smith. (1966), The variances of the 1966 sample census, Report to the G. R. O., October 1966, 30pp.

2. T. M. F. Smith. (1977), Statistics: a universal discipline, Inaugural Lecture, University of Southampton.

3. T. M. F. Smith. (1991), The measurement of value added in higher education, Report for the Committee of Vice-Chancellors and Principals.

4. T. M. F. Smith. (1995), Chartered status – a sister society's experience. How the Royal Statistical Society fared, *OR Newsletter*, February, 11–16.

5. N. Davies and T. M. F. Smith. (1999), Continuing professional development: view from the statistics profession. *Innovation*, 87–90.

Author Index

Subject Index

WILEY SERIES IN SURVEY METHODOLOGY
Established in Part by WALTER A. SHEWHART AND SAMUEL S. WILKS

Editors: *Robert M. Groves, Graham Kalton, J. N. K. Rao, Norbert Schwarz, Christopher Skinner*

Wiley Series in Survey Methodology covers topics of current research and practical interests in survey methodology and sampling. While the emphasis is on application, theoretical discussion is encouraged when it supports a broader understanding of the subject matter.

The authors are leading academics and researchers in survey methodology and sampling. The readership includes professionals in, and students of, the fields of applied statistics, biostatistics, public policy, and government and corporate enterprises.

BIEMER, GROVES, LYBERG, MATHIOWETZ, and SUDMAN   Measurement Errors in Surveys
COCHRAN   Sampling Techniques, *Third Edition*
COUPER, BAKER, BETHLEHEM, CLARK, MARTIN, NICHOLLS, and O'REILLY (editors)   Computer Assisted Survey Information Collection
COX, BINDER, CHINNAPPA, CHRISTIANSON, COLLEDGE, and KOTT (editors)   Business Survey Methods
*DEMING   Sample Design in Business Research
DILLMAN   Mail and Telephone Surveys: The Total Design Method, *Second Edition*
DILLMAN   Mail and Internet Surveys: The Tailored Design Method
GROVES and COUPER   Nonresponse in Household Interview Surveys
GROVES   Survey Errors and Survey Costs
GROVES, DILLMAN, ELTINGE, and LITTLE   Survey Nonresponse
GROVES, BIEMER, LYBERG, MASSEY, NICHOLLS, and WAKSBERG   Telephone Survey Methodology
*HANSEN, HURWITZ, and MADOW   Sample Survey Methods and Theory, Volume 1: Methods and Applications
*HANSEN, HURWITZ, and MADOW   Sample Survey Methods and Theory, Volume II: Theory
KISH   Statistical Design for Research
*KISH   Survey Sampling
KORN and GRAUBARD   Analysis of Health Surveys
LESSLER and KALSBEEK   Nonsampling Error in Surveys
LEVY and LEMESHOW   Sampling of Populations: Methods and Applications, *Third Edition*
LYBERG, BIEMER, COLLINS, de LEEUW, DIPPO, SCHWARZ, TREWIN (editors)   Survey Measurement and Process Quality
MAYNARD, HOUTKOOP-STEENSTRA, SCHAEFFER, VAN DER ZOUWEN   Standardization and Tacit Knowledge: Interaction and Practice in the Survey Interview
SIRKEN, HERRMANN, SCHECHTER, SCHWARZ, TANUR, and TOURANGEAU (editors)   Cognition and Survey Research
VALLIANT, DORFMAN, and ROYALL   Finite Population Sampling and Inference: A Prediction Approach
CHAMBERS and SKINNER (editors).   Analysis of Survey Data

*Now available in a lower priced paperback edition in the Wiley Classics Library.